T0180391

Lecture Notes in Computer Science 12132

More information about this series at http://www.springer.com/series/7412

Aurélio Campilho · Fakhri Karray ·
Zhou Wang (Eds.)

Image Analysis and Recognition

17th International Conference, ICIAR 2020
Póvoa de Varzim, Portugal, June 24–26, 2020
Proceedings, Part II

 Springer

Editors
Aurélio Campilho (iD)
University of Porto
Porto, Portugal

Fakhri Karray (iD)
University of Waterloo
Waterloo, ON, Canada

Zhou Wang (iD)
University of Waterloo
Waterloo, ON, Canada

ISSN 0302-9743   ISSN 1611-3349   (electronic)
Lecture Notes in Computer Science
ISBN 978-3-030-50515-8   ISBN 978-3-030-50516-5   (eBook)
https://doi.org/10.1007/978-3-030-50516-5

LNCS Sublibrary: SL6 – Image Processing, Computer Vision, Pattern Recognition, and Graphics

This Springer imprint is published by the registered company Springer Nature Switzerland AG
The registered company address is: Gewerbestrasse 11, 6330 Cham, Switzerland

Preface

ICIAR 2020 was the 17th edition of the series of annual conferences on Image Analysis and Recognition, organized, this year, as a virtual conference due to the pandemic outbreak of Covid-19 affecting all the world, with an intensity never felt by the humanity in the last hundred years. These are difficult and challenging times, nevertheless the situation provides new opportunities for disseminating science and technology to an even wider audience, through powerful online mediums. Although organized as a virtual conference, ICIAR 2020 kept a forum for the participants to interact and present their latest research contributions in theory, methodology, and applications of image analysis and recognition. ICIAR 2020, the International Conference on Image Analysis and Recognition took place during June 24–26, 2020. ICIAR is organized by the Association for Image and Machine Intelligence (AIMI), a not-for-profit organization registered in Ontario, Canada.

We received a total of 123 papers from 31 countries. The review process was carried out by members of the Program Committee and other external reviewers. Each paper was reviewed by at least two reviewers, and checked by the conference co-chairs. A total of 73 papers were accepted and appear in these proceedings. We would like to sincerely thank the authors for their excellent research work and for responding to our call, and to thank the reviewers for dedicating time to the review process and for the careful evaluation and the feedback provided to the authors. It is this collective effort that resulted in a strong conference program and a high-quality proceedings.

We were very pleased to include three outstanding keynote talks: "Deep Learning and The Future of Radiology" by Daniel Rueckert (Imperial College London, UK); "Towards Human-Friendly Explainable Artificial Intelligence" by Hani Hagras (University of Essex, UK); and "Embedded Computer Vision and Machine Learning for Drone Imaging" by Ioannis Pitas (Aristotle University of Thessaloniki, Greece). We would like to express our gratitude to the keynote speakers for accepting our invitation to share their vision and recent advances made in their areas of expertise.

This virtual conference was organized in two parallel tracks, corresponding to nine sessions, each one corresponding to the following chapters in this proceedings with two volumes:

1. Image Processing and Analysis
2. Video Analysis
3. Computer Vision
4. 3D Computer Vision
5. Machine Learning
6. Medical Image Analysis
7. Analysis of Histopathology Images
8. Diagnosis and Screening of Ophthalmic Diseases
9. Grand Challenge on Automatic Lung Cancer Patient Management

Chapter 8 and 9 correspond to two successful parallel events: Special Session on "Novel Imaging Methods for Diagnosis and Screening of Ophthalmic Diseases" co-chaired by Ana Mendonça (University of Porto, Portugal) and Koen Vermeer (Roterdam Eye Hospital, The Netherlands); and "Grand Challenge on Automatic Lung Cancer Patient Management" organized by João Pedrosa, Carlos Ferreira, and Guilherme Aresta from Institute for Systems and Computer Engineering, Technology and Science (INESC TEC), Portugal.

We would like to thank the program area chairs: Armando Pinho (University of Aveiro, Portugal) and Ed Vrscay (University of Waterloo, Canada), chairs for the area on Image Processing and Analysis; José Santos Victor (Instituto Superior Técnico, University of Lisbon, Portugal) and Petia Radeva (University of Barcelona, Spain), chairs for the area on Computer Vision; Jaime Cardoso (University of Porto, Portugal) and J. Salvador Sanchez Garreta (University of Jaume I, Spain), chairs for the area on Machine Learning; and Ana Mendonça (University of Porto, Portugal) and Roberto Hornero (University of Valladolid, Spain), chairs for the area on Medical Image Analysis; who have secured a high-quality program. We also would like to thank the members of the Organizing Committee from INESC TEC, for helping with the local logistics, and the publications and web chairs, Carlos Ferreira and Khaled Hammouda, for maintaining the website, interacting with the authors, and preparing the proceedings. We are also grateful to Springer's editorial staff, for supporting this publication in the LNCS series. As well, we would like to thank the precious sponsorship and support of the INESC TEC, the Faculty of Engineering at the University of Porto, Portugal, the Waterloo Artificial Intelligence Institute, the Faculty of Engineering of the University of Waterloo, and the Center for Pattern Analysis and Machine Intelligence at the University of Waterloo. We also appreciate the valuable co-sponsorship of the IEEE EMB Portugal Chapter, the IEEE Computational Intelligence Society, Kitchener-Waterloo Chapter, and the Portuguese Association for Pattern Recognition. We also would like to acknowledge Lurdes Catalino from Abreu Events for managing the registrations.

We were very pleased to welcome all the participants to ICIAR 2020, a virtual conference edition. For those who were not able to attend, we hope this publication provides a good overview into the research presented at the conference.

June 2020

Aurélio Campilho
Fakhri Karray
Zhou Wang

Organization

General Chairs

Aurélio Campilho   University of Porto, Portugal
Fakhri Karray   University of Waterloo, Canada
Zhou Wang   University of Waterloo, Canada

Local Organizing Committee

Catarina Carvalho   INESC TEC, Portugal
João Pedrosa   INESC TEC, Portugal
Luís Teixeira   University of Porto, Portugal

Program Chairs

Image Processing and Analysis

Armando Pinho   University of Aveiro, Portugal
Ed Vrscay   University of Waterloo, Canada

Computer Vision

J. Santos Victor   Instituto Superior Técnico, Portugal
Petia Radeva   University of Barcelona, Spain

Machine Learning

Jaime Cardoso   University of Porto, Portugal
J. Salvador Garreta   University of Jaume I, Spain

Medical Image Analysis

Ana Mendonça   University of Porto, Portugal
Roberto Hornero   University of Valladolid, Spain

Grand Challenge on Automatic Lung Cancer Patient Management

João Pedrosa   INESC TEC, Portugal
Carlos Ferreira   INESC TEC, Portugal
Guilherme Aresta   INESC TEC, Portugal

Novel Imaging Methods for Diagnosis and Screening of Ophthalmic Diseases

Ana Mendonça   University of Porto, Portugal
Koen Vermeer   Rotterdam Eye Hospital, The Netherlands

Publication and Web Chairs

Carlos Ferreira   INESC TEC, Portugal
Khaled Hammouda   Shopify, Canada

Supported and Co-sponsored by

AIMI – Association for Image and Machine Intelligence

Center for Biomedical Engineering Research
INESC TEC – Institute for Systems and Computer Engineering,
Technology and Science
Portugal

Department of Electrical and Computer Engineering
Faculty of Engineering
University of Porto
Portugal

Faculty of Engineering
University of Waterloo
Canada

CPAMI – Centre for Pattern Analysis and Machine Intelligence
University of Waterloo
Canada

Waterloo AI Institute
Canada

IEEE Engineering in Medicine and Biology Society
Portugal

IEEE Computational Intelligence Society
Kitchener-Waterloo Chapter

APRP - Portuguese Association for Pattern Recognition
Portugal

Program Committee

Alaa El Khatib	University of Waterloo, Canada
Alberto Taboada-Crispi	Universidad Central Marta Abreu de Las Villas, Cuba
Alexander Wong	University of Waterloo, Canada
Ambra Demontis	Università di Cagliari, Italy
Ana Filipa Sequeira	INESC TEC, Portugal
Ana Maria Mendonça	University of Porto, Portugal
Andreas Uhl	University of Salzburg, Austria
Angel Sappa	ESPOL Polytechnic University, Ecuador, and Computer Vision Center, Spain
António Cunha	University of Trás-os-Montes and Alto Douro, Portugal
Arjan Kujiper	TU Darmstadt, Fraunhofer IGD, Germany
Armando Pinho	University of Aveiro, Portugal
Aurélio Campilho	University of Porto, Portugal
Beatriz Remeseiro	Universidad de Oviedo, Spain
Bob Zhang	University of Macau, Macau
Carlos Thomaz	FEI, Brazil
Catarina Carvalho	INESC TEC, Portugal
Chaojie Ou	University of Waterloo, Canada
Dariusz Frejlichowski	West Pomeranian University of Technology, Poland
Dipti Sarmah	Sarmah's Algorithmic Intelligence Research Lab, The Netherlands
Dominique Brunet	Environment and Climate Change Canada (Toronto Area), Canada
Edward Vrscay	University of Waterloo, Canada
Fabian Falck	Imperial College London, UK
Fakhri Karray	University of Waterloo, Canada
Farzad Khalvati	University of Toronto, Canada
Francesco Camastra	University of Naples Parthenope, Italy
Francesco Renna	University of Porto, Portugal
Francesco Tortorella	Universita' degli Studi di Salerno, Italy
Gerald Schaefer	Loughborough University, UK
Giang Tran	University of Waterloo, Canada
Gilson Giraldi	LNCC, Brazil
Giuliano Grossi	University of Milan, Italy
Guillaume Noyel	International Prevention Research Institute, France
Hasan Ogul	Baskent University, Turkey
Hassan Rivaz	Concordia University, Canada
Hélder Oliveira	INESC TEC, Portugal
Hicham Sekkati	National Research Council of Canada, Canada
Howard Li	University of New Brunswick, Canada
Huiyu Zhou	Queen's University Belfast, UK
Jaime Cardoso	University of Porto, Portugal
Jinghao Xue	University College London, UK
João Pedrosa	INESC TEC, Portugal

João Rodrigues	University of the Algarve, Portugal
Johan Debayle	École nationale supérieure des Mines de Saint-Étienne, France
Jonathan Boisvert	CNRC, Canada
Jorge Batista	University of Coimbra, Portugal
Jorge Marques	University of Lisbon, Portugal
Jorge Silva	University of Porto, Portugal
José Alba Castro	University of Vigo, Spain
José Garreta	University of Jaume I, Spain
José Rouco	University of Coruña, Spain
José Santos-Victor	University of Lisbon, Portugal
Jose-Jesus Fernandez	CNB-CSIC, Spain
Juan José Rodríguez	Universidad de Burgos, Spain
Juan Lorenzo Ginori	Universidad Central Marta Abreu de Las Villas, Cuba
Kaushik Roy	North Carolina A&T State University, USA
Kelwin Fernandes	NILG.AI, Portugal
Koen Vermeer	Rotterdam Eye Hospital, The Netherlands
Linlin Xu	University of Waterloo, Canada
Luc Duong	École de technologie supérieure, Canada
Luís Alexandre	University of Beira Interior, Portugal
Luís Teixeira	University of Porto, Portugal
Mahmoud El-Sakka	University of Western Ontario, Canada
Mahmoud Hassaballah	South Valley University, Egypt
Mahmoud Melkemi	Univeristé de Haute-Alsace, France
Manuel Penedo	University of Coruña, Spain
María García	University of Valladolid, Spain
Marie Muller	North Carolina State University, USA
Mariella Dimiccoli	Institut de Robòtica i Informàtica Industrial, Spain
Mario Vento	Università di Salerno, Italy
Markus Koskela	CSC - IT Center for Science, Finland
Mehran Ebrahimi	University of Ontario, Canada
Mohammad Shafiee	University of Waterloo, Canada
Nicola Strisciuglio	University of Groningen, The Netherlands
Oliver Montesdeoca	Universitat de Barcelona, Spain
Parthipan Siva	Sportlogiq, Canada
Pascal Fallavollita	University of Ottawa, Canada
Pavel Zemčík	Brno University of Technology, Czech Republic
Pedro Carvalho	INESC TEC, Portugal
Pedro Pina	University of Lisbon, Portugal
Petia Radeva	University of Barcelona, Spain
Philip Morrow	Ulster University, UK
Radim Kolář	Brno University of Technology, Czech Republic
Reyer Zwiggelaar	Aberystwyth University, UK
Robert Fisher	University of Edinburgh, UK
Robert Sablatnig	TU Wien, Austria
Roberto Hornero	University of Valladolid, Spain

Rosa María Valdovinos	Universidad Autónoma del Estado de México, Mexico
Rui Bernardes	University of Coimbra, Portugal
Sajad Saeedi	Imperial College London, UK
Sébai Dorsaf	National School of Computer Science, Tunisia
Shamik Sural	Indian Institute of Technology, India
Vicente García-Jiménez	Universidad Autónoma de Ciudad Juérez, Mexico
Víctor González-Castro	Universidad de Leon, Spain
Xosé Pardo	CiTIUS, Universidade de Santiago de Compostela, Spain
Yasuyo Kita	National Institute AIST, Japan
Yun-Qian Miao	General Motors, Canada
Zhou Wang	University of Waterloo, Canada

Additional Reviewers

Américo Pereira	INESC TEC, Portugal
Audrey Chung	University of Waterloo, Canada
Devinder Kumar	Stanford University, USA
Dongdong Ma	Tsinghua University, China
Guilherme Aresta	INESC TEC, Portugal
Honglei Su	Qingdao University, China
Isabel Rio-Torto	University of Porto, Portugal
Juncheng Zhang	Tsinghua University, China
Khashayar Namdar	University of Toronto, Canada
Lu Zhang	INSA Rennes, France
Mafalda Falcão	INESC TEC, Portugal
Pedro Costa	INESC TEC, Portugal
Saman Motamed	University of Toronto, Canada
Tânia Pereira	INESC TEC, Portugal
Tom Vicar	Brno University of Technology, Czech Republic
Youcheng Zhang	Tsinghua University, China

Contents – Part II

Medical Image Analysis

Analysis of Histopathology Images

Diagnosis and Screening of Ophthalmic Diseases

Grand Challenge on Automatic Lung Cancer Patient Management

Contents – Part I

3D Computer Vision

Machine Learning

Weighted Fisher Discriminant Analysis in the Input and Feature Spaces

Benyamin Ghojogh[1]([✉])[iD], Milad Sikaroudi[2][iD], H. R. Tizhoosh[2][iD], Fakhri Karray[1][iD], and Mark Crowley[1][iD]

[1] Department of Electrical and Computer Engineering,
University of Waterloo, Waterloo, ON, Canada
{bghojogh,karray,mcrowley}@uwaterloo.ca
[2] KIMIA Lab, University of Waterloo, Waterloo, ON, Canada
{msikaroudi,tizhoosh}@uwaterloo.ca

Abstract. Fisher Discriminant Analysis (FDA) is a subspace learning method which minimizes and maximizes the intra- and inter-class scatters of data, respectively. Although, in FDA, all the pairs of classes are treated the same way, some classes are closer than the others. Weighted FDA assigns weights to the pairs of classes to address this shortcoming of FDA. In this paper, we propose a cosine-weighted FDA as well as an automatically weighted FDA in which weights are found automatically. We also propose a weighted FDA in the feature space to establish a weighted kernel FDA for both existing and newly proposed weights. Our experiments on the ORL face recognition dataset show the effectiveness of the proposed weighting schemes.

Keywords: Fisher Discriminant Analysis (FDA) · Kernel FDA · Cosine-weighted FDA · Automatically weighted FDA · Manually weighted FDA

1 Introduction

Fisher Discriminant Analysis (FDA) [1], first proposed in [2], is a powerful subspace learning method which tries to minimize the intra-class scatter and maximize the inter-class scatter of data for better separation of classes. FDA treats all pairs of the classes the same way; however, some classes might be much further from one another compared to other classes. In other words, the distances of classes are different. Treating closer classes need more attention because classifiers may more easily confuse them whereas classes far from each other are generally easier to separate. The same problem exists in Kernel FDA (KFDA) [3] and in most of subspace learning methods that are based on generalized eigenvalue problem such as FDA and KFDA [4]; hence, a weighting procedure might be more appropriate.

In this paper, we propose several weighting procedures for FDA and KFDA. The contributions of this paper are three-fold: (1) proposing Cosine-Weighted

© Springer Nature Switzerland AG 2020
A. Campilho et al. (Eds.): ICIAR 2020, LNCS 12132, pp. 3–15, 2020.
https://doi.org/10.1007/978-3-030-50516-5_1

FDA (CW-FDA) as a new modification of FDA, (2) proposing Automatically Weighted FDA (AW-FDA) as a new version of FDA in which the weights are set automatically, and (3) proposing Weighted KFDA (W-KFDA) to have weighting procedures in the feature space, where both the existing and the newly proposed weighting methods can be used in the feature space.

The paper is organized as follows: In Sect. 2, we briefly review the theory of FDA and KFDA. In Sect. 3, we formulate the weighted FDA, review the existing weighting methods, and then propose CW-FDA and AW-FDA. Section 4 proposes weighted KFDA in the feature space. In addition to using the existing methods for weighted KFDA, two versions of CW-KFDA and also AW-KFDA are proposed. Section 5 reports the experiments. Finally, Sect. 6 concludes the paper.

2 Fisher and Kernel Discriminant Analysis

2.1 Fisher Discriminant Analysis

Let $\{x_i^{(r)} \in \mathbb{R}^d\}_{i=1}^{n_r}$ denote the samples of the r-th class where n_r is the class's sample size. Suppose $\boldsymbol{\mu}^{(r)} \in \mathbb{R}^d$, c, n, and $U \in \mathbb{R}^{d \times d}$ denote the mean of r-th class, the number of classes, the total sample size, and the projection matrix in FDA, respectively. Although some methods solve FDA using least squares problem [5,6], the regular FDA [2] maximizes the Fisher criterion [7]:

$$\underset{U}{\text{maximize}} \quad \frac{\text{tr}(U^\top S_B U)}{\text{tr}(U^\top S_W U)}, \tag{1}$$

where $\text{tr}(\cdot)$ is the trace of matrix. The Fisher criterion is a generalized Rayleigh-Ritz Quotient [8]. We may recast the problem to [9]:

$$\begin{aligned} \underset{U}{\text{maximize}} \quad & \text{tr}(U^\top S_B U), \\ \text{subject to} \quad & U^\top S_W U = I, \end{aligned} \tag{2}$$

where the $S_W \in \mathbb{R}^{d \times d}$ and $S_B \in \mathbb{R}^{d \times d}$ are the intra- (within) and inter-class (between) scatters, respectively [9]:

$$S_W := \sum_{r=1}^{c} \sum_{i=1}^{n_r} n_r (x_i^{(r)} - \boldsymbol{\mu}^{(r)})(x_i^{(r)} - \boldsymbol{\mu}^{(r)})^\top = \sum_{r=1}^{c} n_r \, \breve{X}_r \, \breve{X}_r^\top, \tag{3}$$

$$S_B := \sum_{r=1}^{c} \sum_{\ell=1}^{c} n_r \, n_\ell (\boldsymbol{\mu}^{(r)} - \boldsymbol{\mu}^{(\ell)})(\boldsymbol{\mu}^{(r)} - \boldsymbol{\mu}^{(\ell)})^\top = \sum_{r=1}^{c} n_k \, M_r \, N \, M_r^\top, \tag{4}$$

where $\mathbb{R}^{d \times n_r} \ni \breve{X}_r := [x_1^{(r)} - \boldsymbol{\mu}^{(r)}, \dots, x_{n_r}^{(r)} - \boldsymbol{\mu}^{(r)}]$, $\mathbb{R}^{d \times c} \ni M_r := [\boldsymbol{\mu}^{(r)} - \boldsymbol{\mu}^{(1)}, \dots, \boldsymbol{\mu}^{(r)} - \boldsymbol{\mu}^{(c)}]$, and $\mathbb{R}^{c \times c} \ni N := \text{diag}([n_1, \dots, n_c]^\top)$. The mean of the r-th class is $\mathbb{R}^d \ni \boldsymbol{\mu}^{(r)} := (1/n_r) \sum_{i=1}^{n_r} x_i^{(r)}$. The Lagrange relaxation [10] of the optimization problem is: $\mathcal{L} = \text{tr}(U^\top S_B U) - \text{tr}\big(\Lambda^\top (U^\top S_W U - I)\big)$, where Λ is a diagonal matrix which includes the Lagrange multipliers. Setting the derivative of Lagrangian to zero gives:

$$\frac{\partial \mathcal{L}}{\partial U} = 2S_B U - 2S_W U \Lambda \overset{\text{set}}{=} 0 \implies S_B U = S_W U \Lambda, \tag{5}$$

which is the generalized eigenvalue problem (S_B, S_W) where the columns of U and the diagonal of Λ are the eigenvectors and eigenvalues, respectively [11]. The p leading columns of U (so to have $U \in \mathbb{R}^{d \times p}$) are the FDA projection directions where p is the dimensionality of the subspace. Note that $p \le \min(d, n-1, c-1)$ because of the ranks of the inter- and intra-class scatter matrices [9].

2.2 Kernel Fisher Discriminant Analysis

Let the scalar and matrix kernels be denoted by $k(x_i, x_j) := \phi(x_i)^\top \phi(x_j)$ and $K(X_1, X_2) := \Phi(X_1)^\top \Phi(X_2)$, respectively, where $\phi(.)$ and $\Phi(.)$ are the pulling functions. According to the representation theory [12], any solution must lie in the span of all the training vectors, hence, $\Phi(U) = \Phi(X) Y$ where $Y \in \mathbb{R}^{n \times d}$ contains the coefficients. The optimization of kernel FDA is [3,9]:

$$\begin{aligned} \underset{Y}{\text{maximize}} \quad & \text{tr}(Y^\top \Delta_B Y), \\ \text{subject to} \quad & Y^\top \Delta_W Y = I, \end{aligned} \tag{6}$$

where $\Delta_W \in \mathbb{R}^{n \times n}$ and $\Delta_B \in \mathbb{R}^{n \times n}$ are the intra- and inter-class scatters in the feature space, respectively [3,9]:

$$\Delta_W := \sum_{r=1}^{c} n_r K_r H_r K_r^\top, \tag{7}$$

$$\Delta_B := \sum_{r=1}^{c} \sum_{\ell=1}^{c} n_r n_\ell (\xi^{(r)} - \xi^{(\ell)})(\xi^{(r)} - \xi^{(\ell)})^\top = \sum_{r=1}^{c} n_r \Xi_r N \Xi_r^\top, \tag{8}$$

where $\mathbb{R}^{n_r \times n_r} \ni H_r := I - (1/n_r)\mathbf{1}\mathbf{1}^\top$ is the centering matrix, the (i,j)-th entry of $K_r \in \mathbb{R}^{n \times n_r}$ is $K_r(i,j) := k(x_i, x_j^{(r)})$, the i-th entry of $\xi^{(r)} \in \mathbb{R}^n$ is $\xi^{(r)}(i) := (1/n_r) \sum_{j=1}^{n_r} k(x_i, x_j^{(r)})$, and $\mathbb{R}^{n \times c} \ni \Xi_r := [\xi^{(r)} - \xi^{(1)}, \dots, \xi^{(r)} - \xi^{(c)}]$.

The p leading columns of Y (so to have $Y \in \mathbb{R}^{n \times p}$) are the KFDA projection directions which span the subspace. Note that $p \le \min(n, c-1)$ because of the ranks of the inter- and intra-class scatter matrices in the feature space [9].

3 Weighted Fisher Discriminant Analysis

The optimization of Weighted FDA (W-FDA) is as follows:

$$\begin{aligned} \underset{U}{\text{maximize}} \quad & \text{tr}(U^\top \widehat{S}_B U), \\ \text{subject to} \quad & U^\top S_W U = I, \end{aligned} \tag{9}$$

where the weighted inter-class scatter, $\widehat{S}_B \in \mathbb{R}^{d \times d}$, is defined as:

$$\widehat{S}_B := \sum_{r=1}^{c} \sum_{\ell=1}^{c} \alpha_{r\ell}\, n_r\, n_\ell (\boldsymbol{\mu}^{(r)} - \boldsymbol{\mu}^{(\ell)})(\boldsymbol{\mu}^{(r)} - \boldsymbol{\mu}^{(\ell)})^\top = \sum_{r=1}^{c} n_r\, \boldsymbol{M}_r\, \boldsymbol{A}_r\, \boldsymbol{N}\, \boldsymbol{M}_r^\top, \tag{10}$$

where $\mathbb{R} \ni \alpha_{r\ell} \geq 0$ is the weight for the pair of the r-th and ℓ-th classes, $\mathbb{R}^{c \times c} \ni \boldsymbol{A}_r := \mathbf{diag}([\alpha_{r1}, \dots, \alpha_{rc}])$. In FDA, we have $\alpha_{r\ell} = 1$, $\forall r, \ell \in \{1, \dots, c\}$. However, it is better for the weights to be decreasing with the distances of classes to concentrate more on the nearby classes. We denote the distances of the r-th and ℓ-th classes by $d_{r\ell} := ||\boldsymbol{\mu}^{(r)} - \boldsymbol{\mu}^{(\ell)}||_2$. The solution to Eq. (9) is the generalized eigenvalue problem $(\widehat{S}_B, S_W)$ and the p leading columns of U span the subspace.

3.1   Existing Manual Methods

In the following, we review some of the existing weights for W-FDA.

Approximate Pairwise Accuracy Criterion: The Approximate Pairwise Accuracy Criterion (APAC) method [13] has the weight function:

$$\alpha_{r\ell} := \frac{1}{2\, d_{r\ell}^2} \mathrm{erf}\left(\frac{d_{r\ell}}{2\sqrt{2}}\right), \tag{11}$$

where $\mathrm{erf}(x)$ is the error function:

$$[-1, 1] \ni \mathrm{erf}(x) := \frac{2}{\sqrt{\pi}} \int_0^x e^{-t^2}\, dt. \tag{12}$$

This method approximates the Bayes error for class pairs.

Powered Distance Weighting: The powered distance (POW) method [14] uses the following weight function:

$$\alpha_{r\ell} := \frac{1}{d_{r\ell}^m}, \tag{13}$$

where $m > 0$ is an integer. As $\alpha_{r\ell}$ is supposed to drop faster than the increase of $d_{k\ell}$, we should have $m \geq 3$ (we use $m = 3$ in the experiments).

Confused Distance Maximization: The Confused Distance Maximization (CDM) [15] method uses the confusion probability among the classes as the weight function:

$$\alpha_{r\ell} := \begin{cases} \frac{n_{\ell|r}}{n_r} & \text{if } k \neq \ell, \\ 0 & \text{if } r = \ell, \end{cases} \tag{14}$$

where $n_{\ell|r}$ is the number of points of class r classified as class ℓ by a classifier such as quadratic discriminant analysis [15,16]. One problem of the CDM method is that if the classes are classified perfectly, all weights become zero. Conditioning the performance of a classifier is also another flaw of this method.

k-**Nearest Neighbors Weighting:** The *k*-Nearest Neighbor (*k*NN) method [17] tries to put every class away from its *k*-nearest neighbor classes by defining the weight function as

$$\alpha_{r\ell} := \begin{cases} 1 & \text{if } \boldsymbol{\mu}^{(\ell)} \in k\text{NN}(\boldsymbol{\mu}^{(r)}), \\ 0 & \text{otherwise.} \end{cases} \tag{15}$$

The *k*NN and CDM methods are sparse to make use of the betting on sparsity principle [1,18]. However, these methods have some shortcomings. For example, if two classes are far from one another in the input space, they are not considered in *k*NN or CDM, but in the obtained subspace, they may fall close to each other, which is not desirable. Another flaw of *k*NN method is the assignment of 1 to all *k*NN pairs, but in the *k*NN, some pairs might be comparably closer.

3.2   Cosine Weighted Fisher Discriminant Analysis

Literature has shown that cosine similarity works very well with the FDA, especially for face recognition [19,20]. Moreover, according to the opposition-based learning [21], capturing similarity and dissimilarity of data points can improve the performance of learning. A promising operator for capturing similarity and dissimilarity (opposition) is cosine. Hence, we propose CW-FDA, as a manually weighted method, with cosine to be the weight defined as

$$\alpha_{r\ell} := 0.5 \times \left[1 + \cos\left(\measuredangle(\boldsymbol{\mu}^{(r)}, \boldsymbol{\mu}^{(\ell)})\right)\right] = 0.5 \times \left[1 + \frac{\boldsymbol{\mu}^{(r)\top}\boldsymbol{\mu}^{(\ell)}}{||\boldsymbol{\mu}^{(r)}||_2||\boldsymbol{\mu}^{(\ell)}||_2}\right], \tag{16}$$

to have $\alpha_{r\ell} \in [0,1]$. Hence, the *r*-th weight matrix is $\boldsymbol{A}_r := \mathbf{diag}(\alpha_{r\ell}, \forall \ell)$, which is used in Eq. (10). Note that as we do not care about $\alpha_{r,r}$, because inter-class scatter for $r = \ell$ is zero, we can set $\alpha_{rr} = 0$.

3.3   Automatically Weighted Fisher Discriminant Analysis

In AW-FDA, there are $c + 1$ matrix optimization variables which are $\boldsymbol{V}$ and $\boldsymbol{A}_k \in \mathbb{R}^{c \times c}, \forall k \in \{1, \ldots, c\}$ because at the same time where we want to maximize the Fisher criterion, the optimal weights are found. Moreover, to use the betting on sparsity principle [1,18], we can make the weight matrix sparse, so we use "ℓ_0" norm for the weights to be sparse. The optimization problem is as follows

$$\begin{aligned} \underset{\boldsymbol{U}, \boldsymbol{A}_r}{\text{maximize}} \quad & \mathbf{tr}(\boldsymbol{U}^\top \widehat{\boldsymbol{S}}_B \boldsymbol{U}), \\ \text{subject to} \quad & \boldsymbol{U}^\top \boldsymbol{S}_W \boldsymbol{U} = \boldsymbol{I}, \\ & ||\boldsymbol{A}_r||_0 \leq k, \quad \forall r \in \{1, \ldots, c\}. \end{aligned} \tag{17}$$

We use alternating optimization [22] to solve this problem:

$$U^{(\tau+1)} := \arg\max_{U} \left(\mathbf{tr}(U^\top \widehat{S}_B^{(\tau)} U) \,|\, U^\top S_W U = I \right), \tag{18}$$

$$A_r^{(\tau+1)} := \arg\min_{A_r} \left(-\mathbf{tr}(U^{(\tau+1)\top} \widehat{S}_B U^{(\tau+1)}) \,|\, ||A_r||_0 \le k \right), \forall r, \tag{19}$$

where τ denotes the iteration.

Since we use an iterative solution for the optimization, it is better to normalize the weights in the weighted inter-class scatter; otherwise, the weights gradually explode to maximize the objective function. We use ℓ_2 (or Frobenius) norm for normalization for ease of taking derivatives. Hence, for OW-FDA, we slightly modify the weighted inter-class scatter as

$$\widehat{S}_B := \sum_{r=1}^{c} \sum_{\ell=1}^{c} \frac{\alpha_{r\ell}}{\sum_{\ell'=1}^{c} \alpha_{r\ell'}^2} n_r n_\ell (\mu^{(r)} - \mu^{(\ell)})(\mu^{(r)} - \mu^{(\ell)})^\top \tag{20}$$

$$= \sum_{r=1}^{c} n_r M_r \breve{A}_r N M_r^\top, \tag{21}$$

where $\breve{A}_r := A_r / ||A_r||_F^2$ because A_k is diagonal, and $||.||_F$ is Frobenius norm.

As discussed before, the solution to Eq. (18) is the generalized eigenvalue problem $(\widehat{S}_B^{(\tau)}, S_W)$. We use a step of gradient descent [23] to solve Eq. (19) followed by satisfying the "ℓ_0" norm constraint [22]. The gradient is calculated as follows. Let $\mathbb{R} \ni f(U, A_k) := -\mathbf{tr}(U^\top \widehat{S}_B U)$. Using the chain rule, we have:

$$\mathbb{R}^{c \times c} \ni \frac{\partial f}{\partial A_r} = \mathbf{vec}_{c \times c}^{-1} \left[(\frac{\partial \breve{A}_r}{\partial A_r})^\top (\frac{\partial \widehat{S}_B}{\partial \breve{A}_r})^\top \mathbf{vec}(\frac{\partial f}{\partial \widehat{S}_B}) \right], \tag{22}$$

where we use the Magnus-Neudecker convention in which matrices are vectorized, $\mathbf{vec}(.)$ vectorizes the matrix, and $\mathbf{vec}_{c \times c}^{-1}$ is de-vectorization to $c \times c$ matrix. We have $\mathbb{R}^{d \times d} \ni \partial f / \partial \widehat{S}_B = -UU^\top$ whose vectorization has dimensionality d^2. For the second derivative, we have:

$$\mathbb{R}^{d^2 \times c^2} \ni \frac{\partial \widehat{S}_B}{\partial \breve{A}_r} = n_r (M_r N^\top) \otimes M_r, \tag{23}$$

where $\otimes$ denotes the Kronecker product. The third derivative is:

$$\mathbb{R}^{c^2 \times c^2} \ni \frac{\partial \breve{A}_r}{\partial A_r} = \frac{1}{||A_r||_F^2} \left(\frac{-2}{||A_r||_F^2} (A_r \otimes A_r) + I_{c^2} \right). \tag{24}$$

The learning rate of gradient descent is calculated using line search [23].

After the gradient descent step, to satisfy the condition $||A_r||_0 \le k$, the solution is projected onto the set of this condition. Because $-f$ should be maximized, this projection is to set the $(c-k)$ smallest diagonal entries of A_r to zero [22]. In case $k = c$, the projection of the solution is itself, and all the weights are kept.

After solving the optimization, the p leading columns of U are the OW-FDA projection directions that span the subspace.

4 Weighted Kernel Fisher Discriminant Analysis

We define the optimization for Weighted Kernel FDA (W-KFDA) as:

$$\underset{Y}{\text{maximize}} \quad \mathbf{tr}(Y^\top \widehat{\Delta}_B\, Y),$$

$$\text{subject to} \quad Y^\top \Delta_W\, Y = I, \tag{25}$$

where the weighted inter-class scatter in the feature space, $\widehat{\Delta}_B \in \mathbb{R}^{n \times n}$, is defined as:

$$\widehat{\Delta}_B := \sum_{r=1}^{c}\sum_{\ell=1}^{c} \alpha_{r\ell}\, n_r\, n_\ell (\xi^{(r)} - \xi^{(\ell)})(\xi^{(r)} - \xi^{(\ell)})^\top = \sum_{r=1}^{c} n_r\, \Xi_r\, A_r\, N\, \Xi_r^\top. \tag{26}$$

The solution to Eq. (25) is the generalized eigenvalue problem $(\widehat{\Delta}_B, \Delta_W)$ and the p leading columns of Y span the subspace.

4.1 Manually Weighted Methods in the Feature Space

All the existing weighting methods in the literature for W-FDA can be used as weights in W-KFDA to have W-FDA in the feature space. Therefore, Eqs. (11), (13), (14), and (15) can be used as weights in Eq. (26) to have W-KFDA with APAC, POW, CDM, and kNN weights, respectively. To the best of our knowledge, W-KFDA is novel and has not appeared in the literature. Note that there is a weighted KFDA in the literature [24], but that is for data integration, which is for another purpose and has an entirely different approach.

The CW-FDA can be used in the feature space to have CW-KFDA. For this, we propose two versions of CW-KFDA: (I) In the first version, we use Eq. (16) or $A_r := \mathbf{diag}(\alpha_{r\ell}, \forall \ell)$ in the Eq. (26). (II) In the second version, we notice that cosine is based on inner product so the normalized kernel matrix between the means of classes can be used instead to use the similarity/dissimilarity in the feature space rather than in the input space. Let $\mathbb{R}^{d \times c} \ni M := [\mu_1, \ldots, \mu_c]$. Let $\widehat{K}_{i,j} := K_{i,j}/\sqrt{K_{i,i}K_{j,j}}$ be the normalized kernel matrix [25] where $K_{i,j}$ denotes the (i,j)-th element of the kernel matrix $\mathbb{R}^{c \times c} \ni K(M, M) = \Phi(M)^\top \Phi(M)$. The weights are $[0, 1] \ni \alpha_{r\ell} := \widehat{K}_{r,\ell}$ or $A_r := \mathbf{diag}(\widehat{K}_{r,\ell}, \forall \ell)$. We set $\alpha_{r,r} = 0$.

4.2 Automatically Weighted Kernel Fisher Discriminant Analysis

Similar to before, the optimization in AW-KFDA is:

$$\underset{Y, A_r}{\text{maximize}} \quad \mathbf{tr}(Y^\top \widehat{\Delta}_B\, Y),$$

$$\text{subject to} \quad Y^\top \Delta_W\, Y = I,$$

$$\|A_r\|_0 \le k, \quad \forall r \in \{1, \ldots, c\}, \tag{27}$$

where $\widehat{\boldsymbol{\Delta}}_B := \sum_{r=1}^c n_r \, \boldsymbol{\Xi}_r \, \breve{\boldsymbol{A}}_r \, \boldsymbol{N} \, \boldsymbol{\Xi}_r^\top$. This optimization is solved similar to how Eq. (17) was solved where we have $\boldsymbol{Y} \in \mathbb{R}^{n \times d}$ rather than $\boldsymbol{U} \in \mathbb{R}^{d \times d}$. Here, the solution to Eq. (18) is the generalized eigenvalue problem $(\widehat{\boldsymbol{\Delta}}_B^{(\tau)}, \boldsymbol{\Delta}_W)$. Let $f(\boldsymbol{Y}, \boldsymbol{A}_k) := -\mathbf{tr}(\boldsymbol{Y}^\top \widehat{\boldsymbol{\Delta}}_B \, \boldsymbol{Y})$. The Eq. (19) is solved similarly but we use $\mathbb{R}^{n \times n} \ni \partial f / \partial \widehat{\boldsymbol{\Delta}}_B = -\boldsymbol{Y}\boldsymbol{Y}^\top$ and

$$\mathbb{R}^{c \times c} \ni \frac{\partial f}{\partial \boldsymbol{A}_r} = \mathbf{vec}_{c \times c}^{-1}\left[\left(\frac{\partial \breve{\boldsymbol{A}}_r}{\partial \boldsymbol{A}_r}\right)^\top \left(\frac{\partial \widehat{\boldsymbol{\Delta}}_B}{\partial \breve{\boldsymbol{A}}_r}\right)^\top \mathbf{vec}\left(\frac{\partial f}{\partial \widehat{\boldsymbol{\Delta}}_B}\right)\right], \tag{28}$$

$$\mathbb{R}^{d^2 \times c^2} \ni \frac{\partial \widehat{\boldsymbol{\Delta}}_B}{\partial \breve{\boldsymbol{A}}_r} = n_r \, (\boldsymbol{\Xi}_r \, \boldsymbol{N}^\top) \otimes \boldsymbol{\Xi}_r. \tag{29}$$

After solving the optimization, the p leading columns of $\boldsymbol{Y}$ span the OW-KFDA subspace. Recall $\boldsymbol{\Phi}(\boldsymbol{U}) = \boldsymbol{\Phi}(\boldsymbol{X})\,\boldsymbol{Y}$. The projection of some data $\boldsymbol{X}_t \in \mathbb{R}^{d \times n_t}$ is $\mathbb{R}^{p \times n_t} \ni \widetilde{\boldsymbol{X}}_t = \boldsymbol{\Phi}(\boldsymbol{U})^\top \boldsymbol{\Phi}(\boldsymbol{X}_t) = \boldsymbol{Y}^\top \boldsymbol{\Phi}(\boldsymbol{X})^\top \boldsymbol{\Phi}(\boldsymbol{X}_t) = \boldsymbol{Y}^\top \boldsymbol{K}(\boldsymbol{X}, \boldsymbol{X}_t)$.

5   Experiments

5.1   Dataset

For experiments, we used the public ORL face recognition dataset [26] because face recognition has been a challenging task and FDA has numerously been used for face recognition (e.g., see [19,20,27]). This dataset includes 40 classes, each having ten different poses of the facial picture of a subject, resulting in 400 total images. For computational reasons, we selected the first 20 classes and resampled the images to 44×36 pixels. Please note that massive datasets are not feasible for the KFDA/FDA because of having a generalized eigenvalue problem in it. Some samples of this dataset are shown in Fig. 1. The data were split into training and test sets with 66%/33% portions and were standardized to have mean zero and variance one.

Fig. 1. Sample images of the classes in the ORL face dataset. Numbers are the class indices.

5.2   Evaluation of the Embedding Subspaces

For the evaluation of the embedded subspaces, we used the 1-Nearest Neighbor (1NN) classifier because it is useful to evaluate the subspace by the closeness of the projected data samples. The training and out-of-sample (test) accuracy of classifications are reported in Table 1. In the input space, kNN with $k = 1, 3$ have the best results but in $k = c - 1$, AW-FDA outperforms it in generalization (test) result. The performances of CW-FDA and AW-FDA with $k = 1, 3$ are promising, although not the best. For instance, AW-FDA with $k = 1$ outperforms weighted FDA with APAC, POW, and CDM methods in the training embedding, while has the same performance as kNN. In most cases, AW-FDA with all k values has better performance than the FDA, which shows the effectiveness of the obtained weights compared to equal weights in FDA. Also, the sparse k in AWF-FDA outperforming FDA (with dense weights equal to one) validates the betting on sparsity.

Table 1. Accuracy of 1 NN classification for different obtained subspaces. In each cell of input or feature spaces, the first and second rows correspond to the classification accuracy of training and test data, respectively.

	FDA	APAC	POW	CDM	kNN $(k = 1)$	kNN $(k = 3)$	kNN $(k = c - 1)$	CW-FDA version 1	CW-FDA version 2	AW-FDA $(k = 1)$	AW-FDA $(k = 3)$	AW-FDA $(k = c - 1)$
Input space	97.01%	97.01%	97.01%	74.62%	97.76%	97.76%	97.01%	97.01%	–	97.76%	97.01%	96.26%
	92.42%	93.93%	96.96%	45.45%	96.96%	98.48%	92.42%	92.42%	–	87.87%	93.93%	93.93%
Feature space	97.01%	97.01%	97.01%	91.79%	95.52%	97.76%	97.01%	97.01%	97.01%	100%	100%	100%
	83.33%	86.36%	89.39%	77.27%	80.30%	83.33%	83.33%	84.84%	87.87%	100%	100%	100%

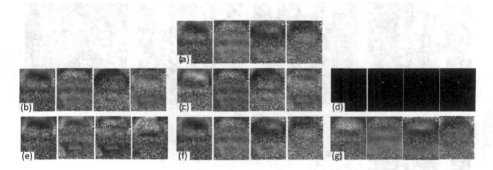

Fig. 2. The leading Fisherfaces in (a) FDA, (b) APAC, (c) POW, (d) CDM, (e) kNN, (f) CW-FDA, and (g) AW-FDA.

In the feature space, where we used the radial basis kernel, AW-KFDA has the best performance with entirely accurate recognition. Both versions of CW-KFDA outperform regular KFDA and KFDA with CDM, and kNN (with $k = 1, c - 1$) weighting. They also have better generalization than APAC, kNN with all k values. Overall, the results show the effectiveness of the proposed weights in the input and feature spaces. Moreover, the existing weighting methods, which were for the input space, have outstanding performance when used in our proposed weighted KFDA (in feature space). This shows the validness of the proposed weighted KFDA even for the existing weighting methods.

5.3   Comparison of Fisherfaces

Figure 2 depicts the four leading eigenvectors obtained from the different methods, including the FDA itself. These ghost faces, or so-called Fisherfaces [27],

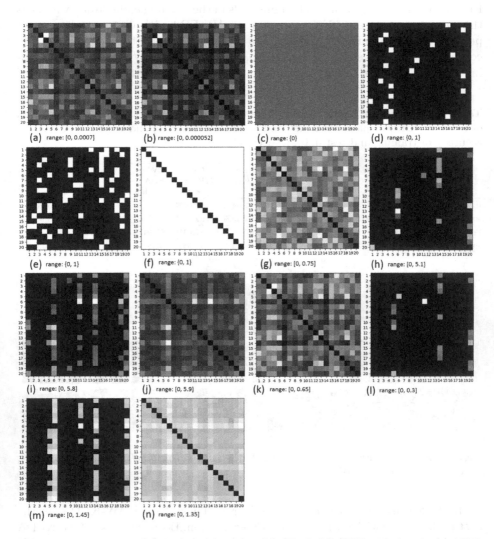

Fig. 3. The weights in (a) APAC, (b) POW, (c) CDM, (d) kNN with $k = 1$, (e) kNN with $k = 3$, (f) kNN with $k = c - 1$, (g) CW-FDA, (h) AW-FDA with $k = 1$, (i) AW-FDA with $k = 3$, (j) AW-FDA with $k = c - 1$, (k) CW-KFDA, (l) AW-KFDA with $k = 1$, (m) AW-KFDA with $k = 3$, (n) AW-KFDA with $k = c - 1$. The rows and columns index the classes.

capture the critical discriminating facial features to discriminant the classes in subspace. Note that Fisherfaces cannot be shown in kernel FDA as its projection directions are n dimensional. CDM has captured some pixels as features because its all weights have become zero for its explained flaw (see Sect. 3.1 and Fig. 3). The Fisherfaces, in most of the methods including CW-FDA, capture information of facial organs such as hair, forehead, eyes, chin, and mouth.

The features of AW-FDA are more akin to the Haar wavelet features, which are useful for facial feature detection [28].

5.4   Comparison of the Weights

We show the obtained weights in different methods in Fig. 3. The weights of APAC and POW are too small, while the range of weights in the other methods is more reasonable. The weights of CDM have become all zero because the samples were purely classified (recall the flaw of CDM). The weights of kNN method are only zero and one, which is a flaw of this method because, amongst the neighbors, some classes are closer. This issue does not exist in AW-FDA with different k values. Moreover, although not all the obtained weights are visually interpretable, some non-zero weights in AW-FDA or AW-KFDA, with e.g. $k = 1$, show the meaningfulness of the obtained weights (noticing Fig. 1). For example, the non-zero pairs $(2, 20)$, $(4, 14)$, $(13, 6)$, $(19, 20)$, $(17, 6)$ in AW-FDA and the pairs $(2, 20)$, $(4, 14)$, $(19, 20)$, $(17, 14)$ in AW-KFDA make sense visually because of having glasses so their classes are close to one another.

6   Conclusion

In this paper, we discussed that FDA and KFDA have a fundamental flaw, and that is treating all pairs of classes in the same way while some classes are closer to each other and should be processed with more care for a better discrimination. We proposed CW-FDA with cosine weights and also AW-FDA in which the weights are found automatically. We also proposed a weighted KFDA to weight FDA in the feature space. We proposed AW-KFDA and two versions of CW-KFDA as well as utilizing the existing weighting methods for weighted KFDA. The experiments in which we evaluated the embedding subspaces, the Fisherfaces, and the weights, showed the effectiveness of the proposed methods. The proposed weighted FDA methods outperformed regular FDA and many of the existing weighting methods for FDA. For example, AW-FDA with $k = 1$ outperformed weighted FDA with APAC, POW, and CDM methods in the training embedding. In feature space, AW-KFDA obtained perfect discrimination.

References

1. Friedman, J., Hastie, T., Tibshirani, R.: The Elements of Statistical Learning. Springer Series in Statistics, vol. 1. Springer, New York (2001). https://doi.org/10.1007/978-0-387-84858-7

2. Fisher, R.A.: The use of multiple measurements in taxonomic problems. Ann. Eugen. **7**(2), 179–188 (1936)
3. Mika, S., Ratsch, G., Weston, J., Scholkopf, B., Mullers, K.R.: Fisher discriminant analysis with kernels. In: Neural Networks for Signal Processing IX: Proceedings of the 1999 IEEE Signal Processing Society Workshop, pp. 41–48. IEEE (1999)
4. Ghojogh, B., Karray, F., Crowley, M.: Roweis discriminant analysis: a generalized subspace learning method. arXiv preprint arXiv:1910.05437 (2019)
5. Zhang, Z., Dai, G., Xu, C., Jordan, M.I.: Regularized discriminant analysis, ridge regression and beyond. J. Mach. Learn. Res. **11**(Aug), 2199–2228 (2010)
6. Díaz-Vico, D., Dorronsoro, J.R.: Deep least squares Fisher discriminant analysis. IEEE Trans. Neural Netw. Learn. Syst. (2019)
7. Xu, Y., Lu, G.: Analysis on Fisher discriminant criterion and linear separability of feature space. In: 2006 International Conference on Computational Intelligence and Security, vol. 2, pp. 1671–1676. IEEE (2006)
8. Parlett, B.N.: The Symmetric Eigenvalue Problem, vol. 20. Society for Industrial and Applied Mathematics (SIAM), Philadelphia (1998)
9. Ghojogh, B., Karray, F., Crowley, M.: Fisher and kernel Fisher discriminant analysis: tutorial. arXiv preprint arXiv:1906.09436 (2019)
10. Boyd, S., Vandenberghe, L.: Convex Optimization. Cambridge University Press, Cambridge (2004)
11. Ghojogh, B., Karray, F., Crowley, M.: Eigenvalue and generalized eigenvalue problems: tutorial. arXiv preprint arXiv:1903.11240 (2019)
12. Alperin, J.L.: Local Representation Theory: Modular Representations as an Introduction to the Local Representation Theory of Finite Groups, vol. 11. Cambridge University Press, Cambridge (1993)
13. Loog, M., Duin, R.P., Haeb-Umbach, R.: Multiclass linear dimension reduction by weighted pairwise Fisher criteria. IEEE Trans. Pattern Anal. Mach. Intell. **23**(7), 762–766 (2001)
14. Lotlikar, R., Kothari, R.: Fractional-step dimensionality reduction. IEEE Trans. Pattern Anal. Mach. Intell. **22**(6), 623–627 (2000)
15. Zhang, X.Y., Liu, C.L.: Confused distance maximization for large category dimensionality reduction. In: 2012 International Conference on Frontiers in Handwriting Recognition, pp. 213–218. IEEE (2012)
16. Ghojogh, B., Crowley, M.: Linear and quadratic discriminant analysis: tutorial. arXiv preprint arXiv:1906.02590 (2019)
17. Zhang, X.Y., Liu, C.L.: Evaluation of weighted Fisher criteria for large category dimensionality reduction in application to Chinese handwriting recognition. Pattern Recogn. **46**(9), 2599–2611 (2013)
18. Hastie, T., Tibshirani, R., Wainwright, M.: Statistical Learning with Sparsity: The Lasso and Generalizations. Chapman and Hall/CRC, London (2015)
19. Perlibakas, V.: Distance measures for PCA-based face recognition. Pattern Recogn. Lett. **25**(6), 711–724 (2004)
20. Mohammadzade, H., Hatzinakos, D.: Projection into expression subspaces for face recognition from single sample per person. IEEE Trans. Affect. Comput. **4**(1), 69–82 (2012)
21. Tizhoosh, H.R.: Opposition-based learning: a new scheme for machine intelligence. In: International Conference on Computational Intelligence for Modelling, Control and Automation, vol. 1, pp. 695–701. IEEE (2005)
22. Jain, P., Kar, P.: Non-convex optimization for machine learning. Found. Trends® Mach. Learn. **10**(3–4), 142–336 (2017)

23. Nocedal, J., Wright, S.: Numerical Optimization. Springer, Berlin (2006). https://doi.org/10.1007/978-0-387-40065-5

24. Hamid, J.S., Greenwood, C.M., Beyene, J.: Weighted kernel Fisher discriminant analysis for integrating heterogeneous data. Comput. Stat. Data Anal. **56**(6), 2031–2040 (2012)

25. Ah-Pine, J.: Normalized kernels as similarity indices. In: Zaki, M.J., Yu, J.X., Ravindran, B., Pudi, V. (eds.) PAKDD 2010. LNCS (LNAI), vol. 6119, pp. 362–373. Springer, Heidelberg (2010). https://doi.org/10.1007/978-3-642-13672-6_36

26. AT&T Laboratories Cambridge: ORL Face Dataset. http://cam-orl.co.uk/facedatabase.html. Accessed 2019

27. Belhumeur, P.N., Hespanha, J.P., Kriegman, D.J.: Eigenfaces vs. Fisherfaces: recognition using class specific linear projection. IEEE Trans. Pattern Anal. Mach. Intell. **19**(7), 711–720 (1997)

28. Wang, Y.Q.: An analysis of the Viola-Jones face detection algorithm. Image Process. On Line **4**, 128–148 (2014)

Backprojection for Training Feedforward Neural Networks in the Input and Feature Spaces

Benyamin Ghojogh$^{(\boxtimes)}$(ID), Fakhri Karray(ID), and Mark Crowley(ID)

Department of Electrical and Computer Engineering, University of Waterloo,
Waterloo, ON, Canada
{bghojogh,karray,mcrowley}@uwaterloo.ca

Abstract. After the tremendous development of neural networks trained by backpropagation, it is a good time to develop other algorithms for training neural networks to gain more insights into networks. In this paper, we propose a new algorithm for training feedforward neural networks which is fairly faster than backpropagation. This method is based on projection and reconstruction where, at every layer, the projected data and reconstructed labels are forced to be similar and the weights are tuned accordingly layer by layer. The proposed algorithm can be used for both input and feature spaces, named as backprojection and kernel backprojection, respectively. This algorithm gives an insight to networks with a projection-based perspective. The experiments on synthetic datasets show the effectiveness of the proposed method.

Keywords: Neural network · Backprojection · Kernel backprojection · Projection · Training

1 Introduction

In one of his recent seminars, Geoffrey Hinton mentioned that after all of the developments of neural networks [1] and deep learning [2], perhaps it is time to move on from backpropagation [3] to newer algorithms for training neural networks. Especially, now that we know why shallow [4] and deep [5] networks work very well and why local optima are fairly good in networks [6], other training algorithms can help improve the insights into neural nets. Different training methods have been proposed for neural networks, some of which are backpropagation [3], genetic algorithms [7,8], and belief propagation as in restricted Boltzmann machines [9].

A neural network can be viewed from a manifold learning perspective [10]. Most of the spectral manifold learning methods can be reduced to kernel principal component analysis [11] which is a projection-based method [12]. Moreover, at its initialization, every layer of a network can be seen as a random projection [13]. Hence, a promising direction could be a projection view of training neural

© Springer Nature Switzerland AG 2020
A. Campilho et al. (Eds.): ICIAR 2020, LNCS 12132, pp. 16–24, 2020.
https://doi.org/10.1007/978-3-030-50516-5_2

networks. In this paper, we propose a new training algorithm for feedforward neural networks based on projection and *backprojection* (or so-called reconstruction). In the backprojection algorithm, we update the weights layer by layer. For updating a layer m, we project the data from the input, until the layer m. We also backproject the labels of data from the last layer to the layer m. The projected data and backprojected labels at layer m should be equal because in a perfectly trained network, projection of data by the entire layers should result in the corresponding labels. Thus, minimizing a loss function over the projected data and backprojected labels would correctly tune the layer's weights. This algorithm is proposed for both the input and feature spaces where in the latter, the kernel of data is fed to the network.

2 Backprojection Algorithm

2.1 Projection and Backprojection in Network

In a neural network, every layer without its activation function acts as a linear projection. Without the nonlinear activation functions, a network/autoencoder is reduced to a linear projection/principal component analysis [12]. If U denotes the projection matrix (i.e., the weight matrix of a layer), $U^\top x$ projects x onto the column space of U. The reverse operation of projection is called reconstruction or backprojection and is formulated as $UU^\top x$ which shows the projected data in the input space dimensionality (note that it is $Uf^{-1}(f(U^\top x))$ if we have a nonlinear function $f(.)$ after the linear projection). At the initialization, a layer acts as a random projection [13] which is a promising feature extractor according to the Johnson-Lindenstrauss lemma [14]. Fine tuning the weights using labels makes the features more useful for discrimination of classes.

2.2 Definitions

Let us have a training set $\mathcal{X} := \{x_i \in \mathbb{R}^d\}_{i=1}^n$ and their one-hot encoded labels $\mathcal{Y} := \{y_i \in \mathbb{R}^p\}_{i=1}^n$ where n, d, and p are the sample size, dimensionality of data, and dimensionality of labels, respectively. We denote the dimensionality or the number of neurons in layer m by d_m. By convention, we have $d_0 := d$ and $d_{n_\ell} = p$ where n_ℓ is the number of layers and p is the dimensionality of the output layer. Let the data after the activation function of the m-th layer be denoted by $x^{(m)} \in \mathbb{R}^{d_m}$. Let the projected data in the m-th layer be $\mathbb{R}^{d_m} \ni z^{(m)} := U_m^\top x^{(m-1)}$ where $U_m \in \mathbb{R}^{d_{m-1} \times d_m}$ is the weight matrix of the m-th layer. Note that $x^{(m)} = f_m(z^{(m)})$ where $f_m(.)$ is the activation function in the m-th layer. By convention, $x^{(0)} := x$. The data are projected and passed through the activation functions layer by layer; hence, $x^{(m)}$ is calculated as:

$$\mathbb{R}^{d_m} \ni x^{(m)} := f_m(U_m^\top f_{m-1}(U_{m-1}^\top \cdots f_1(U_1^\top x))) = f_m(U_m^\top x^{(m-1)}). \quad (1)$$

In a mini-batch gradient descent set-up, let $\{x_i\}_{i=1}^b$ be a batch of size b. For a batch, we denote the outputs of activation functions at the m-th layer by $\mathbb{R}^{d_m \times b} \ni X^{(m)} := [x_1^{(m)}, \ldots, x_b^{(m)}]$.

Now, consider the one-hot encoded labels of batch, denoted by $\boldsymbol{y} \in \mathbb{R}^p$. We take the inverse activation function of the labels and then reconstruct or *backproject* them to the previous layer to obtain $\boldsymbol{y}^{(n_\ell - 1)}$. We do similarly until the layer m. Let $\boldsymbol{y}^{(m)} \in \mathbb{R}^{d_m}$ denote the backprojected data at the m-th layer, calculated as:

$$\boldsymbol{y}^{(m)} := \boldsymbol{U}_{m+1} \boldsymbol{f}_{m+1}^{-1}(\boldsymbol{U}_{m+2} \boldsymbol{f}_{m+2}^{-1}(\cdots \boldsymbol{U}_{n_\ell} \boldsymbol{f}_{n_\ell}^{-1}(\boldsymbol{y}))) = \boldsymbol{U}_{m+1} \boldsymbol{f}_{m+1}^{-1}(\boldsymbol{y}^{(m+1)}). \quad (2)$$

By convention, $\boldsymbol{y}^{(n_\ell)} := \boldsymbol{y}$. The backprojected batch at the m-th layer is $\mathbb{R}^{d_m \times b} \ni \boldsymbol{Y}^{(m)} := [\boldsymbol{y}_1^{(m)}, \ldots, \boldsymbol{y}_b^{(m)}]$. We use $\boldsymbol{X} \in \mathbb{R}^{d \times b}$ and $\boldsymbol{Y} \in \mathbb{R}^{p \times b}$ to denote the column-wise batch matrix and its one-hot encoded labels.

2.3   Optimization

In the backprojection algorithm, we optimize the layers' weights one by one. Consider the m-th layer whose loss we denote by $\mathcal{L}_m$:

$$\underset{\boldsymbol{U}_m}{\text{minimize}} \quad \mathcal{L}_m := \sum_{i=1}^{b} \ell(\boldsymbol{x}_i^{(m)} - \boldsymbol{y}_i^{(m)}) = \sum_{i=1}^{b} \ell(\boldsymbol{f}_m(\boldsymbol{U}_m^\top \boldsymbol{x}_i^{(m-1)}) - \boldsymbol{y}_i^{(m)}), \quad (3)$$

where $\ell(.)$ is a loss function such as the squared ℓ_2 norm (or Mean Squared Error (MSE)), cross-entropy, etc. The loss $\mathcal{L}_m$ tries to make the projected data $\boldsymbol{x}_i^{(m)}$ as similar as possible to the backprojected data $\boldsymbol{y}_i^{(m)}$ by tuning the weights $\boldsymbol{U}_m$. This is because the output of the network is supposed to be equal to the labels, i.e., $\boldsymbol{x}^{(n_\ell)} \approx \boldsymbol{y}$. In order to tune the weights for Eq. (3), we use a step of gradient descent. Using chain rule, the gradient is:

$$\mathbb{R}^{d_{m-1} \times d_m} \ni \frac{\partial \mathcal{L}_m}{\partial \boldsymbol{U}_m} = \sum_{i=1}^{b} \mathbf{vec}_{d_{m-1} \times d_m}^{-1} \left[\left(\frac{\partial \boldsymbol{z}_i^{(m)}}{\partial \boldsymbol{U}_m} \right)^\top \left(\frac{\partial \boldsymbol{f}_m(\boldsymbol{z}_i^{(m)})}{\partial \boldsymbol{z}_i^{(m)}} \right)^\top \frac{\partial \ell(\boldsymbol{f}_m(\boldsymbol{z}_i^{(m)}))}{\partial \boldsymbol{f}_m(\boldsymbol{z}_i^{(m)})} \right],$$
$$(4)$$

where we use the Magnus-Neudecker convention in which matrices are vectorized and $\mathbf{vec}_{d_{m-1} \times d_m}^{-1}$ is de-vectorization to $d_{m-1} \times d_m$ matrix. If the loss function is MSE or cross-entropy for example, the derivatives of the loss function w.r.t. the activation function, respectively, are:

$$\mathbb{R}^{d_m} \ni \frac{\partial \ell(\boldsymbol{f}_m(\boldsymbol{z}_i^{(m)}))}{\partial \boldsymbol{f}_m(\boldsymbol{z}_i^{(m)})} = 2(\boldsymbol{f}_m(\boldsymbol{z}_i^{(m)}) - \boldsymbol{y}_i^{(m)}), \text{ and} \quad (5)$$

$$\mathbb{R}^{d_m} \ni \frac{\partial \ell(\boldsymbol{f}_m(\boldsymbol{z}_i^{(m)}))}{\partial \boldsymbol{f}_m(\boldsymbol{z}_i^{(m)})} = -\left[\frac{\boldsymbol{y}_{i,j}^{(m)}}{\boldsymbol{f}_m(\boldsymbol{z}_{i,j}^{(m)})}, \forall j \in \{1, \ldots, d_m\} \right]^\top, \quad (6)$$

where $\boldsymbol{y}_{i,j}^{(m)}$ and $\boldsymbol{z}_{i,j}^{(m)}$ are the j-th dimension of $\boldsymbol{y}_i^{(m)}$ and $\boldsymbol{z}_i^{(m)} = \boldsymbol{U}_m^\top \boldsymbol{x}_i^{(m-1)}$, respectively.

---

1 **Procedure:** UpdateLayerWeights($\mathcal{U}$, $\boldsymbol{X}$, $\boldsymbol{Y}$, m)
2 **Input:** weights: $\mathcal{U} := \{\boldsymbol{U}_r\}_{r=1}^{n_\ell}$, batch data: $\boldsymbol{X} \in \mathbb{R}^{d \times b}$, batch labels:
$\boldsymbol{Y} \in \mathbb{R}^{p \times b}$, layer: $m \in [1, n_\ell]$
3 $\boldsymbol{X}^{(0)} := \boldsymbol{X}$
4 **for** *layer r from* 1 *to* $(m-1)$ **do**
5 $\quad$ | $\quad \boldsymbol{Z}^{(r)} := \boldsymbol{U}_r^\top \boldsymbol{X}^{(r-1)}$
6 $\quad$ | $\quad \boldsymbol{X}^{(r)} := \boldsymbol{f}_r(\boldsymbol{Z}^{(r)})$
7 $\boldsymbol{Y}^{(n_\ell)} := \boldsymbol{Y}$
8 **for** *layer r from* $(n_\ell - 1)$ *to* m **do**
9 $\quad$ | $\quad \boldsymbol{Y}^{(r+1)} := \varPi(\boldsymbol{Y}^{(r+1)})$
10 $\quad$ | $\quad \boldsymbol{Y}^{(r)} := \boldsymbol{U}_{r+1} \boldsymbol{f}_{r+1}^{-1}(\boldsymbol{Y}^{(r+1)})$
11 $\boldsymbol{U}_m := \boldsymbol{U}_m - \eta\,(\partial \mathcal{L}_m / \partial \boldsymbol{U}_m)$
12 **Return** $\boldsymbol{U}_m$

Algorithm 1: Updating the weights of a layer in backprojection

For the activation functions in which the nodes are independent, such as linear, sigmoid, and hyperbolic tangent, the derivative of the activation function w.r.t. its input is a diagonal matrix:

$$\mathbb{R}^{d_m \times d_m} \ni \frac{\partial \boldsymbol{f}_m(\boldsymbol{z}_i^{(m)})}{\partial \boldsymbol{z}_i^{(m)}} = \mathbf{diag}\left(\frac{\partial \boldsymbol{f}_m(z_{i,j}^{(m)})}{\partial z_{i,j}^{(m)}}, \forall j \in \{1, \ldots, d_m\}\right), \qquad (7)$$

where $\mathbf{diag}(.)$ makes a matrix with its input as diagonal.

The derivative of the projected data before the activation function (i.e., the input of the activation function) w.r.t. the weights of the layer is:

$$\mathbb{R}^{d_m \times (d_m d_{m-1})} \ni \frac{\partial \boldsymbol{z}_i^{(m)}}{\partial \boldsymbol{U}_m} = \frac{\partial \boldsymbol{U}_m^\top \boldsymbol{x}_i^{(m-1)}}{\partial \boldsymbol{U}_m} = \boldsymbol{I}_{d_m} \otimes \boldsymbol{x}_i^{(m-1)\top}, \qquad (8)$$

where $\otimes$ denotes the Kronecker product and $\boldsymbol{I}_{d_m}$ is the $d_m \times d_m$ identity matrix.

The procedure for updating weights in the m-the layer is shown in Algorithm 1. Until the layer m, data is projected and passed through activation functions layer by layer. Also, the label is backprojected and passed through inverse activation functions until the layer m. A step of gradient descent is used to update the layer's weights where $\eta > 0$ is the learning rate. Note that the backprojected label at a layer may not be in the feasible domain of its inverse activation function. Hence, at every layer, we should project the backprojected label onto the feasible domain [15]. We denote projection onto the feasible set by $\varPi(.)$.

2.4   Different Procedures

So far, we explained how to update the weights of a layer. Here, we detail updating the entire network layers. In terms of the order of updating layers, we can have three different procedures for a backprojection algorithm. One possible procedure is to update the first layer first and move to next layers one by one until

---

1 **Procedure:** Backprojection($\mathcal{X}$, $\mathcal{Y}$, b, e)
2 **Input:** training data: $\mathcal{X}$, training labels: $\mathcal{Y}$, batch size: b, number of epochs: e
3 Initialize $\mathcal{U} = \{U_r\}_{r=1}^{n_\ell}$
4 **for** *epoch from* 1 *to* e **do**
5   **for** *batch from* 1 *to* $\lceil n/b \rceil$ **do**
6         $X, Y \leftarrow$ take batch from $\mathcal{X}$ and $\mathcal{Y}$
7   **if** *procedure is forward* **then**
8   **for** *layer m from* 1 *to* n_ℓ **do**
9                 $U_m \leftarrow$ UpdateLayerWeights($\{U_r\}_{r=1}^{n_\ell}$, X, Y, m)
10   **else if** *procedure is backward* **then**
11   **for** *layer m from* n_ℓ *to* 1 **do**
12                 $U_m \leftarrow$ UpdateLayerWeights($\{U_r\}_{r=1}^{n_\ell}$, X, Y, m)
13   **else if** *procedure is forward-backward* **then**
14   **if** *batch index is odd* **then**
15   **for** *layer m from* 1 *to* n_ℓ **do**
16                     $U_m \leftarrow$ UpdateLayerWeights($\{U_r\}_{r=1}^{n_\ell}$, X, Y, m)
17   **else**
18   **for** *layer m from* n_ℓ *to* 1 **do**
19                     $U_m \leftarrow$ UpdateLayerWeights($\{U_r\}_{r=1}^{n_\ell}$, X, Y, m)

Algorithm 2: Backprojection

we reach the last layer. Repeating this procedure for the batches results in the *forward procedure*. In an opposite direction, we can have the *backward procedure* where, for each batch, we update the layers from the last layer to the first layer one by one. If we have both directions of updating, i.e., forward update for a batch and backward update for the next batch, we call it the *forward-backward procedure*. Algorithm 2 shows how to update the layers in different procedures of the backprojection algorithm. Note that in this algorithm, an updated layer impacts the update of next/previous layer. One alternative approach is to make updating of layers dependent only on the weights tuned by previous mini-batch. In that approach, the training of layers can be parallelized within mini-batch.

3   Kernel Backprojection Algorithm

Suppose $\phi : \mathcal{X} \rightarrow \mathcal{H}$ is the pulling function to the feature space. Let t denote the dimensionality of the feature space, i.e., $\phi(x) \in \mathbb{R}^t$. Let the matrix-form of $\mathcal{X}$ and $\mathcal{Y}$ be denoted by $\mathbb{R}^{d \times n} \ni \breve{X} := [x_1, \ldots, x_n]$ and $\mathbb{R}^{p \times n} \ni \breve{Y} := [y_1, \ldots, y_n]$. The kernel matrix [16] for the training data $\breve{X}$ is defined as $\mathbb{R}^{n \times n} \ni \breve{K} := \Phi(\breve{X})^\top \Phi(\breve{X})$ where $\mathbb{R}^{t \times n} \ni \Phi(\breve{X}) := [\phi(x_1), \ldots, \phi(x_n)]$. We normalize the kernel matrix [17] as $\breve{K}(i, j) := \breve{K}(i, j) / [\breve{K}(i, i) \breve{K}(j, j)]^{1/2}$ where $\breve{K}(i, j)$ denotes the (i, j)-th element of the kernel matrix.

According to representation theory [18], the projection matrix $U_1 \in \mathbb{R}^{d \times d_1}$ can be expressed as a linear combination of the projected training data. Hence, we have $\mathbb{R}^{t \times d_1} \ni \Phi(U_1) = \Phi(\check{X}) \Theta$ where every column of $\Theta := [\theta_1, \ldots, \theta_{d_1}] \in \mathbb{R}^{n \times d_1}$ is the vector of coefficients for expressing a projection direction as a linear combination of projected training data. The projection of the pulled data is $\mathbb{R}^{d_1 \times n} \ni \Phi(U_1)^\top \Phi(\check{X}) = \Theta^\top \Phi(\check{X})^\top \Phi(\check{X}) = \Theta^\top \check{K}$.

In the kernel backprojection algorithm, in the first network layer, we project the pulled data from the feature space with dimensionality t to another feature space with dimensionality d_1. The projections of the next layers are the same as in backprojection. In other words, *kernel backprojection applies backprojection in the feature space rather than the input space*. In a mini-batch set-up, we use the columns of the normalized kernel corresponding to the batch samples, denoted by $\{k_i \in \mathbb{R}^n\}_{i=1}^b$. Therefore, the projection of the i-th data point in the batch is $\mathbb{R}^{d_1} \ni \Theta^\top k_i$. In kernel backprojection, the dimensionality of the input is n and the kernel vector k_i is fed to the network as input. If we replace the x_i by k_i, Algorithms 1 and 2 are applicable for kernel backprojection.

In the test phase, we normalize the kernel over the matrix $[\check{X}, x_t]$ where $x_t \in \mathbb{R}^d$ is the test data point. Then, we take the portion of normalized kernel which correspond to the kernel over the training versus test data, denoted by $\mathbb{R}^n \ni k_t := \Phi(\check{X})^\top \Phi(x_t)$. The projection at the first layer is then $\mathbb{R}^{d_1} \ni \Theta^\top k_t$.

4 Experiments

Datasets: For experiments, we created two synthetic datasets with 300 data points each, one for binary-class and one for three-class classification (see Figs. 1 and 2). For more difficulty, we set different variances for the classes. The data were standardized as a preprocessing. For this conference short-paper, we limit ourselves to introduction of this new approach and small synthetic experiments. Validation on larger real-world datasets is ongoing for future publication.

Neural Network Settings: We implemented a neural network with three layers whose number of neurons are $\{15, 20, p\}$ where $p = 1$ and $p = 3$ for the binary and ternary classification, respectively. In different experiments, we used MSE loss for the middle layers and MSE or cross-entropy losses for the last layer. Moreover, we used Exponential Linear Unit (ELU) [19] or linear functions for activation functions of the middle layers while sigmoid or hyperbolic tangent (tanh) were used for the last layer. The derivative and inverse of these activation functions are as the following:

ELU: $f(z) = \begin{cases} e^z - 1, z \leq 0 \\ z, z > 0 \end{cases}, f'(z) = \begin{cases} e^z, z \leq 0 \\ 1, z > 0 \end{cases}, f^{-1}(y) = \begin{cases} \ln(y+1), y \leq 0 \\ y, y > 0 \end{cases},$

Linear: $f(z) = z, \quad f'(z) = 1, \quad f^{-1}(y) = y,$

Sigmoid: $f(z) = \dfrac{1}{1 + e^{-z}}, \quad f'(z) = f(1-f), \quad f^{-1}(y) = \ln(\dfrac{y}{1-y}),$

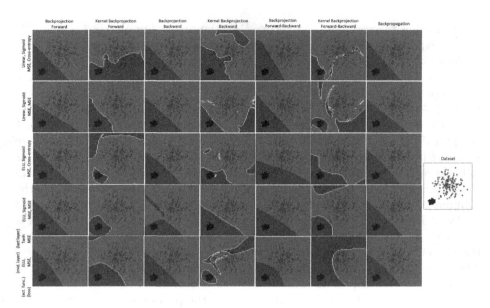

Fig. 1. Discrimination of two classes by different training algorithms with various activation functions and loss functions. The label for each row indicates the activation functions and the loss functions for the middle then the last layers. (Color figure online)

$$\text{Tanh: } f(z) = \frac{e^z - e^{-z}}{e^z + e^{-z}}, \quad f'(z) = 1 - f^2, \quad f^{-1}(y) = 0.5\ln(\frac{1+y}{1-y}),$$

where in the inverse functions, we bound the output values for computational reasons in computer. Mostly, a learning rate of $\eta = 10^{-4}$ was used for backprojection and backpropagation and $\eta = 10^{-5}$ was used for kernel backprojection.

Comparison of Procedures: The performance of different forward, backward, and forward-backward procedures in backprojection and kernel backprojection are illustrated in Fig. 1. In these experiments, the Radial Basis Function (RBF) kernel was used in kernel backprojection. Although the performance of these procedures are not identical but all of them are promising discrimination of classes. This shows that all three proposed procedures work well for backprojection in the input and feature spaces. In other words, the algorithm is fairly robust to the order of updating layers.

Comparison to Backpropagation: The performances of backprojection, kernel backprojection, and backpropagation are compared in the binary and ternary classification, shown in Figs. 1 and 2, respectively. In Fig. 2, the linear kernel was used. In Fig. 1, most often, kernel backprojection considers a spherical class around the blue (or even red) class which is because of the choice of RBF kernel. Comparison to backpropagation in the two figures shows that backprojection's performance nearly matches that of backpropagation.

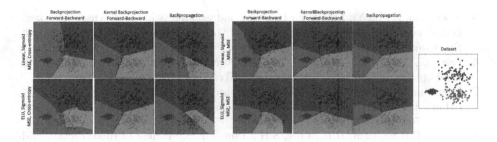

Fig. 2. Discrimination of three classes by different training algorithms with various activation functions and loss functions.

In the different experiments, the mean time of every epoch was often 0.08, 0.11, and 0.2 s for backprojection, kernel backprojection, and backpropagation, respectively, where the number of epochs were fairly similar in the experiments. This shows that backprojection is *faster* than backpropagation. This is because backpropagation updates the weights one by one while backprojection updates layer by layer.

5   Conclusion and Future Direction

In this paper, we proposed a new training algorithm for feedforward neural network named backprojection. The proposed algorithm, which can be used for both the input and feature spaces, tries to force the projected data to be similar to the backprojected labels by tuning the weights layer by layer. This training algorithm, which is moderately faster than backpropagation in our initial experiments, can be used with either forward, backward, or forward-backward procedures. It is noteworthy that adding a penalty term for weight decay [20] to Eq. (3) can regularize the weights in backprojection [21]. Moreover, batch normalization can be used in backprojection by standardizing the batch at the layers [22]. This paper concentrated on feedforward neural networks. As a future direction, we can develop backprojection for other network structures such as convolutional networks [23] and carry more expensive validation experiments on real-world data.

References

1. Fausett, L.: Fundamentals of Neural Networks: Architectures, Algorithms, and Applications. Prentice-Hall Inc, Upper Saddle River (1994)
2. Goodfellow, I., Bengio, Y., Courville, A.: Deep Learning. MIT Press, Cambridge (2016)
3. Rumelhart, D.E., Hinton, G.E., Williams, R.J.: Learning representations by back-propagating errors. Nature **323**(6088), 533–536 (1986)
4. Soltanolkotabi, M., Javanmard, A., Lee, J.D.: Theoretical insights into the optimization landscape of over-parameterized shallow neural networks. IEEE Trans. Inf. Theory **65**(2), 742–769 (2018)

5. Allen-Zhu, Z., Li, Y., Liang, Y.: Learning and generalization in overparameterized neural networks, going beyond two layers. In: Advances in Neural Information Processing Systems, pp. 6155–6166 (2019)
6. Feizi, S., Javadi, H., Zhang, J., Tse, D.: Porcupine neural networks: (almost) all local optima are global. arXiv preprint arXiv:1710.02196 (2017)
7. Montana, D.J., Davis, L.: Training feedforward neural networks using genetic algorithms. In: International Joint Conference on Artificial Intelligence (IJCAI), vol. 89, pp. 762–767 (1989)
8. Leung, F.H.F., Lam, H.K., Ling, S.H., Tam, P.K.S.: Tuning of the structure and parameters of a neural network using an improved genetic algorithm. IEEE Trans. Neural Networks **14**(1), 79–88 (2003)
9. Hinton, G.E., Salakhutdinov, R.R.: Reducing the dimensionality of data with neural networks. Science **313**(5786), 504–507 (2006)
10. Hauser, M., Ray, A.: Principles of Riemannian geometry in neural networks. In: Advances in Neural Information Processing Systems, pp. 2807–2816 (2017)
11. Ham, J.H., Lee, D.D., Mika, S., Schölkopf, B.: A kernel view of the dimensionality reduction of manifolds. In: International Conference on Machine Learning (2004)
12. Ghojogh, B., Crowley, M.: Unsupervised and supervised principal component analysis: tutorial. arXiv preprint arXiv:1906.03148 (2019)
13. Karimi, A.H.: Exploring new forms of random projections for prediction and dimensionality reduction in big-data regimes. Master's thesis, University of Waterloo (2018)
14. Achlioptas, D.: Database-friendly random projections: Johnson-Lindenstrauss with binary coins. J. Comput. Syst. Sci. **66**(4), 671–687 (2003)
15. Parikh, N., Boyd, S.: Proximal algorithms. Found. Trends® Optim. **1**(3), 127–239 (2014)
16. Hofmann, T., Scholkopf, B., Smola, A.J.: Kernal methods in machine learning. Ann. Stat. **36**, 117–1220 (2008)
17. Ah-Pine, J.: Normalized kernels as similarity indices. In: Zaki, M.J., Yu, J.X., Ravindran, B., Pudi, V. (eds.) PAKDD 2010. LNCS (LNAI), vol. 6119, pp. 362–373. Springer, Heidelberg (2010). https://doi.org/10.1007/978-3-642-13672-6_36
18. Alperin, J.L.: Local Representation Theory: Modular Representations as an Introduction to the Local Representation Theory of Finite Groups, vol. 11. Cambridge University Press, Cambridge (1993)
19. Clevert, D.A., Unterthiner, T., Hochreiter, S.: Fast and accurate deep network learning by exponential linear units (ELUs). In: International Conference on Learning Representations (ICLR) (2016)
20. Krogh, A., Hertz, J.A.: A simple weight decay can improve generalization. In: Advances in Neural Information Processing Systems, pp. 950–957 (1992)
21. Ghojogh, B., Crowley, M.: The theory behind overfitting, cross validation, regularization, bagging, and boosting: tutorial. arXiv preprint arXiv:1905.12787 (2019)
22. Ioffe, S., Szegedy, C.: Batch normalization: Accelerating deep network training by reducing internal covariate shift. arXiv preprint arXiv:1502.03167 (2015)
23. LeCun, Y., Bottou, L., Bengio, Y., Haffner, P.: Gradient-based learning applied to document recognition. Proc. IEEE **86**(11), 2278–2324 (1998)

Parallel Implementation of the DRLSE Algorithm

Daniel Popp Coelho$^{(\boxtimes)}$ and Sérgio Shiguemi Furuie

School of Engineering, University of São Paulo, São Paulo, Brazil
1danielcoelho@usp.br

Abstract. The Distance-Regularized Level Set Evolution (DRLSE) algorithm solves many problems that plague the class of Level Set algorithms, but has a significant computational cost and is sensitive to its many parameters. Configuring these parameters is a time-intensive trial-and-error task that limits the usability of the algorithm. This is especially true in the field of Medical Imaging, where it would be otherwise highly suitable. The aim of this work is to develop a parallel implementation of the algorithm using the Compute-Unified Device Architecture (CUDA) for Graphics Processing Units (GPU), which would reduce the computational cost of the algorithm, bringing it to the interactive regime. This would lessen the burden of configuring its parameters and broaden its application. Using consumer-grade hardware, we observed performance gains between roughly 800% and 1700% when comparing against a purely serial C++ implementation we developed, and gains between roughly 180% and 500%, when comparing against the MATLAB reference implementation of DRLSE, both depending on input image resolution.

Keywords: CUDA · DRLSE · Level Sets

1 Introduction

The field of medical imaging possesses many different imaging modalities and techniques. For this field, image segmentation allows delineating organs and internal structures, information that is used for analysis, diagnosis, treatment planning, monitoring of medical conditions and more. The quality of the segmentation, then, directly affects the quality and effectiveness of these processes, again highlighting the importance of the correct choice of segmentation algorithm [1,2].

Level Set segmentation algorithms define a higher dimensional function on the domain of the image, and starting from a user-input initial state of this function, iterate it with some procedure, formulated in such a way that the set

---

The results published here are in part based upon data generated by the TCGA Research Network: http://cancergenome.nih.gov/. This study was financed in part by the Coordenação de Aperfeiçoamento de Pessoal de Nível Superior - Brazil (CAPES) - Finance Code 001.

A. Campilho et al. (Eds.): ICIAR 2020, LNCS 12132, pp. 25–35, 2020.
https://doi.org/10.1007/978-3-030-50516-5_3

of pixels for which the Level Set Function equals zero (called the zero-level set) converges on a target region. After a set number of iterations, the zero-level set represents the contour that is the result of the segmentation [3,4]. Those algorithms are capable of segmenting images with soft and hard edges, and the generated contour is also capable of splitting and merging during the iteration process, which is ideal for segmenting biological structures with complex shapes and varying intensity levels.

The Distance-Regularized Level-Set Evolution or DRLSE is an established Level Sets algorithm that effectively solves some of the usual problems with that class of algorithms, including the problem of reinitialization [4]. It has since been applied in many different contexts since its inception. The elevated computational cost and sensitivity to input parameters are still significant disadvantages of the class, however.

The configuration of DRLSE parameters is a manual task that usually involves trial and error and specialized knowledge, and depends on the specific image used and target object to segment. The situation is made worse by the elevated computational cost of the algorithm itself, that implies in slow iteration cycles. Fortunately, like most Level Sets algorithms, the DRLSE algorithm is highly parallelizable.

Graphics Processing Units (GPUs) are pieces of hardware originally designed for acceleration and parallelization of 3D rendering tasks. Through frameworks like CUDA (Compute Unified Device Architecture) [5], it is now possible to use GPUs for acceleration of General-Purpose algorithms (GPGPU) [5,6].

The specifics of GPU architecture vary greatly between manufacturers and generations. For the purposes of this paper, suffice to say that a GPU is composed of many streaming multiprocessors (SM), which are groups of individual processor cores. When a programmer dispatches a CUDA kernel (small GPU program) for execution on the GPU, he or she specifies how many blocks and threads these kernels should execute on, which are abstractions of SM and processor cores, respectively [5,6]. The abstraction allows CUDA code to be relatively independent of the exact GPU architecture it executes on. It also allows the hardware to manage kernel execution more freely. For example, when more blocks are requested at a time than are available on the hardware, they are placed in a queue and dispatched to SM as they become free. Additionally, CUDA allows multiple blocks to be executed concurrently on the same SM under some conditions [5].

The CUDA framework will be used in this paper to develop kernels for a parallel implementation of the DRLSE algorithm, allowing not only for faster usage, but faster iteration cycles when choosing the optimal segmentation parameters.

2   Materials

The program was developed and tested exclusively on a computer with an AMD Ryzen 7 2700X 3.70 GHz processor, 32 GB of DDR4 RAM, Windows 10 Enterprise N (64 bit) and an NVIDIA GTX 1060 6 GB GPU (Pascal architecture) [7].

The program was always executed out of a WD Blue 1 TB 7200 RPM SATA 6 Gb/s 3.5 in. hard disk drive.

The algorithm was developed and built for CUDA 9.1, using CMake 3.10.1 and Microsoft Visual C++ Compiler 14 (2015, v140). We also developed a purely serial implementation of DRLSE for C++ to be used as a point of reference for the performance comparisons. Additionally, the reference MATLAB implementation of DRLSE [8] was used for performance comparisons.

All of the medical images used in this work were retrieved from [9, 10].

NVIDIA's official profiler, nvprof was used for timing execution of all CUDA programs analyzed in this work. It is available bundled with the CUDA 9.1 SDK.

3 Methods

Special care was taken to guarantee the operations performed were as similar as possible between all implementations, as a way to isolate the performance impacts of parallelization. With that in mind, gradients were calculated via central differences, and second derivatives were implemented using discrete laplacian kernels, so as to mirror MATLAB's grad and del2 functions, respectively. Tests were then performed on large datasets to guarantee the results agreed for these isolated operations, between all three implementations.

We then implemented the DRLSE algorithm according to [4] and the MATLAB reference [8]. For more details and the derivation of the following equations, the reader is directed to the original DRLSE work at [4]. For the purposes of this work, suffice to describe the main iteration equation and its components.

Assuming that Ω describes the domain of an input image I, the Level Set function can be defined as $\Phi : \Omega \rightarrow \mathbb{R}$. The final zero-level set of Φ, the result of the segmentation, is then a contour given by a set S of image pixels, described by $S = \{x, y \in \Omega \mid \Phi_{x,y} = 0\}$.

For the bidimensional case (2D images, where $\Omega = \mathbb{R}^2$), (1) describes the main iteration equation for the DRLSE algorithm. In (1), $\Phi_{x,y}^k$ describes the value of Φ for pixel $[x, y]$ at iteration number k.

$$\Phi_{x,y}^{k+1} = \Phi_{x,y}^k + \tau \cdot [\mu \cdot div(\frac{p'(|\nabla \Phi_{x,y}^k|)}{|\nabla \Phi_{k,y}^k|} \cdot \nabla \Phi_{x,y}^k)$$

$$+ \lambda \cdot \delta_\varepsilon(\Phi_{x,y}^k) \cdot div(g(x,y) \cdot \frac{\nabla \Phi_{x,y}^k}{|\nabla \Phi_{x,y}^k|}) \tag{1}$$

$$+ \alpha \cdot g(x,y) \cdot \delta_\varepsilon(\Phi_{x,y}^k)]$$

At each time step and for every pixel, $\Phi_{x,y}^k$ is incremented by a term weighted by a time constant τ. The value incremented (enveloped in brackets in (1)) is a combination of the regularization term (weighted by the constant $\mu > 0$), the length term (weighted by the constant $\lambda > 0$), and the area term (weighted by the constant $\alpha \in \mathbb{R}$). These three terms are formulated in terms of $p(s)$, a double-well energy potential function, and its derivative, $p'(s)$, both described respectively in (2) and (3).

$$p(s) = \begin{cases} \frac{1}{(2\pi)^2}(1 - \cos(2\pi \cdot s)) & s \le 1 \\ \frac{1}{2}(s-1)^2 & s \ge 1 \end{cases} \tag{2}$$

$$p'(s) = \begin{cases} \frac{1}{2\pi}sin(2\pi \cdot s) & s \le 1 \\ s - 1 & s \ge 1 \end{cases} \tag{3}$$

The indicator image $g(x, y) : \Omega \to \mathbb{R}$, used in (1) and described in (4), is constructed in such a way as to have low values near the contours of interest. In our implementation and in [8], (4) uses the magnitude of the gradient of the input image I convolved with a gaussian kernel G_σ. The constant σ describes the standard deviation of G_σ, and is used as an additional paramater for this implementation. The reasoning behind this is that its optimal value depends on the hardness of the borders of the target object to segment.

$$g(x, y) = \frac{1}{1 + |\nabla G_\sigma \cdot I(x, y)|^2} \tag{4}$$

Finally, $\delta_\varepsilon(s)$ describes a simple approximation of Dirac's delta function, described by (5).

$$\delta_\varepsilon(s) = \begin{cases} \frac{1}{3}(1 + \cos\frac{\pi s}{1.5}) & |s| \le 1.5 \\ 0 & |s| > 1.5 \end{cases} \tag{5}$$

The main parameters of this DRLSE implementation, that need to be configured before each segmentation, are α, μ, λ, and σ. The DRLSE algorithm is sensitive to these parameters, and whereas we determined experimentally that a sensible range of potential values for α, μ, λ would be $[-15, 15]$ (respecting the fact that $\mu > 0$ and $\lambda > 0$), it is common to find scenarios where a change of 0.1 in any of these parameters drastically alters the final segmentation result.

The edge indicator image $g(x, y)$ can be constructed with Algorithm 1, executed on the CPU. Step 3 of Algorithm 1 involves dispatching the kernel *EdgeIndicatorKernel* to the GPU, which is described on Algorithm 2. For brevity, gaussian convolution and the gradient calculation kernel are omitted here.

Note that *edge_grad_image* has the dimensions of *input_image*, however it possesses two channels, x and y, describing the gradient in the x and y directions, respectively. The *gradient kernel* described on step 4 of Algorithm 1 fills both channels of this image based on the values of *edge_image*.

After the construction of the edge indicator image, the main DRLSE iteration procedure is executed on the CPU. That procedure, for an iteration process with 1000 iterations, is described on Algorithm 3.

The laplace kernel used at line 4 of Algorithm 3 is omitted for brevity, but involves a simple convolution with a 3×3 laplacian kernel.

It is worth noting that *phi_grad_image* has the dimensions of *input_image*, however it possesses four channels, unlike *edge_grad_image*. The first two channels describe the gradient of *phi_image* in the x and y directions, respectively.

Channels 3 and 4 of *phi_grad_image* contain the normalized gradient in the x and y direction, respectively, so that the magnitude of the gradient equals one. In the same manner as before, the normalized gradient kernel fills in all four channels of *phi_grad_image* with the gradient and normalized gradient of *phi_image*.

Algorithm 1: Edge Indicator Image $g(x, y)$ (CPU)

> **input** : *gaussian_kernel15x15*, *input_image*
> **output:** *edge_image*, *edge_grad_image*

1 Execute convolution kernel with *gaussian_kernel15x15* and *input_image* generating *temp_image*;
2 Define *edge_image* and *edge_grad_image*, both with same dimensions as *input_image*;
3 Execute *EdgeIndicatorKernel* with *temp_image* and *edge_image*, modifying *edge_image*;
4 Execute gradient kernel with *edge_image* and *edge_grad_image*, modifying *edge_grad_image*;
5 Return *edge_image* and *edge_grad_image*;

Algorithm 2: EdgeIndicatorKernel (GPU)

> **input** : *temp_image*, *edge_image*
> **output:** Nothing

1 Acquire position $[x, y]$ of the current pixel;
2 Define $plusX = temp\_image[x + 1, y]$;
3 Define $minX = temp\_image[x - 1, y]$;
4 Define $plusY = temp\_image[x, y + 1]$;
5 Define $minY = temp\_image[x, y - 1]$;
6 Define $gradX = plusX - minX$;
7 Define $gradY = plusY - minY$;
8 Define $g = 1.0f/(1.0f + 0.25f * (gradX * gradX + gradY * gradY))$;
9 Write $edge\_image[x, y] = g$;

Step 5 of Algorithm 3 involves dispatching the *LevelSetKernel*, which is described on Algorithm 4. This is the main kernel of this implementation, and is larger than usual so as to minimize the overhead of dispatching multiple kernel calls. The main DRLSE iteration equation described in (1) is implemented in step 12 of Algorithm 4.

The procedures *DistRegPre* and *DiracDelta* used in steps 11 and 12 of Algorithm 4, respectively, are identical to the ones used in [8], so are omitted for brevity. It is worth noting however that *DistRegPre* receives a 4-component value from *phi_grad_image* and returns a value with two components, x, and y. These x and y members of the returned results are accessed directly in step 11, for brevity.

Algorithm 3: Iterate DRLSE (CPU)

input : *input_image, phi_image, edge_image, edge_grad_image, mu, lambda, alpha*

output: *phi_image*

1 Define *phi_grad_image* and *laplace_image*, both with same dimensions as *input_image*;

2 **for** $i = 1$ *to* 1000 **do**

3   Execute the normalized gradient kernel with *phi_image* and *phi_grad_image*, modifying *phi_grad_image*;

4   Execute laplace kernel with *phi_image* and *laplace_image*, modifying *laplace_image*;

5   Execute *LevelSetKernel* with *mu, lambda, alpha, phi_image, edge_image, edge_grad_image, phi_grad_image, laplace_image*, modifying *phi_image*;

6 **end**

7 Return final *phi_image*;

Algorithm 4: LevelSetKernel (GPU)

input : *mu, lambda, alpha, tau, phi_image, edge_image, edge_grad_image, phi_grad_image, laplace_image*

output: Nothing

1 Acquire position *[x,y]* of the current pixel;

2 Define *phi = phi_image[x,y]*;

3 Define *phi_grad = phi_grad_image[x,y]*;

4 Define *edge = edge_image[x,y]*;

5 Define *edge_grad = edge_grad_image[x,y]*;

6 Define *plusX = phi_grad_image[x + 1, y]*;

7 Define *minX = phi_grad_image[x − 1, y]*;

8 Define *plusY = phi_grad_image[x, y + 1]*;

9 Define *minY = phi_grad_image[x, y − 1]*;

10 Define *curvature = 0.5f * (plusX.xnorm − minX.xnorm + plusY.ynorm − minY.ynorm)*;

11 Define *delta=DiracDelta(phi)*;

12 Define *increment = tau * (mu * dist_reg + lambda * delta * (edge_grad.x * phi_grad.xnorm + edge_grad.y * phi_grad.ynorm + edge * curvature) + alpha * delta * edge)*;

13 Write *phi[x,y] = phi + increment*;

4   Results and Discussion

4.1   LevelSetKernel's Occupancy

Occupancy is a common performance metric for CUDA programs, and it roughly measures how many of the GPU's cores are used when the kernel is dispatched [6]. 50% occupancy means that half of the processors were idle during execution. Ideally the programmer aims for 100% occupancy, as it usually means the hardware is being used to its full capacity.

We determined experimentally that the optimal number of threads per block for the LevelSetKernel on the NVIDIA GTX 1060 6GB is 64. In normal situations the GPU's block scheduler would have been able to just run more concurrent blocks on the same SM, which would still allow for 100% occupancy.

The LevelSetKernel, however, obtains a theoretical occupancy of 62.5%, and observed occupancy of 59%, according to nvprof. This is explained by the fact that the LevelSetKernel is bottlenecked by its usage of 48 registers per thread. With 64 threads per block, the kernel uses a total of 3072 registers per block. Given the hardware limit of 65536 threads per SM, the scheduler is restricted to dispatching a maximum of 20 concurrent blocks per SM. Those 20 blocks, each with 64 threads, lead to a maximum theoretical number of 1280 concurrent threads per SM, compared to the device maximum of 2048 threads per multiprocessor. The ratio of 1280 to 2048 corresponds to the 62.5% theoretical occupancy limit.

The standard recommendation for optimizing scenarios where register usage is preventing 100% occupancy is to limit the number of registers that the compiler may allocate to each thread. Restricting the number of registers used per thread would allow the dispatching of more concurrent blocks to each SM. It would also, on the other hand, lead to the occasional eviction of some of kernel's used data from fast registers into higher-latency memory (L1 cache). Whether the trade-off is worth it in terms of performance varies greatly with application. To investigate this experimentally, we performed test segmentations on a 512 × 512 pixel input image, where the maximum number of registers per thread were varied from the ideal 48 to 32. Three kernels dispatched from Algorithm 3 were timed, as well as the total time after 1000 complete iterations. The results are displayed on Table 1.

Note that for all trials performed in this work that record execution time, the time required to allocate the required memory, and the time required to transfer data to and from the GPU, when appropriate, has been ignored. Additionally, all data describing execution time of CUDA programs has been collected only after a proper period of "warm-up", where the CUDA driver performs just-in-time compilation of the kernels and any additional one-time setup procedures.

The data in Table 1 shows that even though occupancy does increase as predicted, the restriction of register count leads to an overall loss of performance, meaning it is not an effective method of optimizing this CUDA program. Also note the fact that the normalized gradient average execution time (NG time) and laplace gradient execution time (Laplace time) are not as affected from the

Table 1. Occupancy and performance as a function of register usage

mag reg[a]	used reg LSK[a]	Theor. Occ. (%)	Real Occ. (%)	LSK time[b] (μs)	NG time[b] (μs)	Laplace time[b] (μs)	1000 iter. total (ms)
48	48	62.5	59.2	89.081	37.029	25.772	180.27
46	46	62.5	59.2	96.633	40.095	27.240	179.71
44	40	75.0	70.3	102.96	38.621	23.661	181.52
42	40	75.0	70.5	104.75	38.026	26.425	183.60
40	40	75.0	70.3	111.40	39.729	24.736	189.69
38	38	75.0	70.7	110.43	40.055	26.626	191.60
36	32	100.0	90.4	130.84	39.842	26.606	206.55
34	32	100.0	90.9	128.31	40.649	26.801	208.94
32	32	100.0	91.5	129.60	40.557	26.607	204.58

[a] max reg describes the hard limit set on how many registers a thread could use, and used reg LSK indicates how many registers the LevelSetKernel actually used.
[b] LSK time indicates the average "wall clock" execution time for the LevelSetKernel after 1585000 executions. Data was retrieved from nvprov so standard deviation was inaccessible. NG time describes the analogue for the normalized gradient kernel, and Laplace time describes the analogue for the laplace kernel.

register count restriction as the *LevelSetKernel* average execution time (LSK time). This is due to the fact that the normalized gradient kernel and the laplace kernel use 22 and 20 registers per thread, respectively.

4.2   Implementation Verification

With the aim of ensuring that the overall DRLSE implementations in C++ and CUDA matched the reference in [8], a 512×512 pixel medical image was segmented with the exact same parameters and initial Level Set contour, and the results were compared. The final zero-level set segmentations after 1000 iterations each are shown in Fig. 1. Visual inspection suggests a near perfect match between the results obtained from all three implementations, shown in Fig. 1.

This mismatch can be explained by two discrepancies between the implementations. The first and greater discrepancy arises from how the three programs behave on calculations that depend on neighboring pixels (such as gradients and convolution results) at the extreme borders of the images. Given that the DRLSE algorithm uses many of such operations, the influence from this discrepancy spreads from the border pixels inwards as the Level Set is iterated.

The second, and lesser source of discrepancies between the results obtained by the three implementations is that the C++ and CUDA versions exclusively use 32-bit floating point numbers, while MATLAB exclusively uses 64-bit floating-point numbers. Despite of this drawback, we intentionally chose this approach for the CUDA implementation due to smaller memory usage and memory transfer time, and the difference in performance: The NVIDIA GTX 1060 GPU provides a theoretical performance of 4.375 TFLOPS (trillion floating point operations per second) on 32-bit floating point numbers, while it provides only 0.137 TFLOPS on 64-bit floating point numbers. This difference in performance alone can lead to a theoretical performance gain of nearly 32 times.

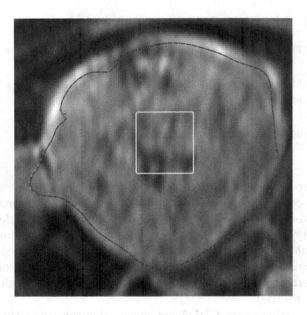

Fig. 1. Final zero-level set contours produced from the different implementations. Solid green: CUDA; Dashed red: MATLAB reference; Dotted blue: C++; Solid white: Initial Level Set function for all three scenarios (Color figure online)

4.3 Total Memory Usage

Given that the input data is also converted to 32-bit floating point numbers, and the fact that *cudaSurfaceObject_t* memory (main data structure we used to store image data) is aligned to 512 bytes on the NVIDIA GTX 1060 6 GB, the total memory consumption of the algorithm, in bytes, can be described by (6).

$$Total = 11 \cdot \max(width \cdot 4, \ 512) \cdot height \tag{6}$$

This means that for a 512×512 pixel image, the total memory consumption of the algorithm is roughly 11.5 MB, which corresponds to approximately 0.2% of the total global memory capacity of the NVIDIA GTX 1060 6 GB.

4.4 Performance and Time Complexity

For all performance tests, 10 segmentations of 1000 iterations each were performed for each DRLSE implementation, for images of varying sizes. The resulting "wall clock" execution times are displayed in Table 2.

Observing the results in Table 2, the performance advantage of the CUDA implementation over both C++ and MATLAB implementations is immediately apparent. For the larger image size of 1024×1024 pixels, the C++ program takes 1700 times as much time as the CUDA program takes to complete, while the MATLAB code takes over 500 times as long as the CUDA program. It should

Table 2. Performance comparisons

Image size (square)	MATLAB time (s)	C++ time (s)	CUDA time (s)	MATLAB time/ CUDA time	C++ time/ CUDA time
128	3.313 ± 0.048	14.465 ± 0.058	0.018 ± 0.001	184.1	803.6
256	8.655 ± 0.052	57.739 ± 0.274	0.044 ± 0.005	197.7	1312.3
512	70.540 ± 0.396	228.892 ± 2.254	0.147 ± 0.001	479.9	1557.1
1024	274.390 ± 0.0487	924.158 ± 5.376	0.547 ± 0.001	501.6	1690.0

be noted however, that these results are not strict performance benchmarks, as the MATLAB code has not necessarily been written with performance in mind.

It is also possible to note how the C++ program is almost perfectly linear in time complexity with respect to the number of pixels, meaning time complexity $O(width * height)$, as one would expect of a purely single-threaded serial implementation. The same cannot be observed from the data collected from MATLAB, as it likely internally uses parallelization mechanisms such as Single-Instruction, Multiple Data (SIMD).

The data referring to the CUDA implementation suggests a time complexity close to linear. This can also be observed on the execution time ratio when compared to the purely linear C++ implementation, as image size increases. This behavior is due to how even though the SM execute the kernel purely in parallel at first (conferring some level of independence to the dimensions of the input data), the behavior changes once the hardware is saturated. Once all SM are occupied, the remaining kernel blocks to execute are placed in a queue, and executed once SM become free at a constant rate. As the image size increases, the initial independence to the dimensions of the input data becomes progressively less significant, and the time complexity approaches linearity.

5   Conclusions and Further Work

The CUDA implementation developed presents significant performance advantages over a purely single-threaded C++ implementation, and the reference MATLAB implementation [8]. Additionally, the CUDA program developed is capable of performing 1000 iterations in under a second, enough to segment even 1024 × 1024 pixel images. This fact allows the segmentation of medical images with in an interactive regime, which is critical for the trial and error procedure involved in configuring the parameters of the DRLSE algorithm.

References

1. Nuruozi, A., et al.: Medical image segmentation methods, algorithms and applications. IETE Techn. Rev. **31**(3), 199–213 (2014)
2. Taha, A.A., Hanbury, A.: Metrics for evaluating 3D medical image segmentation: analysis, selection, and tool. BMC Med. Imaging **15**(1), 29 (2015)

3. Zhang, K., Zhang, L., Song, H., Zhang, D.: Reinitialization-free level set evolution via reaction diffusion. IEEE Trans. Image Process. **22**(1), 258–271 (2013)
4. Li, C., Xu, C., Gui, C., Fox, M.D.: Distance regularized level set evolution and its application to image segmentation. IEEE Trans. Image Process. **19**(12), 3243–3254 (2010)
5. NVIDIA: NVIDIA CUDA programming guide. Version: 10.1.2.243 (2019). https://docs.nvidia.com/cuda/cuda-c-programming-guide/. Accessed 26 Jan 2020
6. Cheng, J., Grossman, M., McKercher, T.: Professional CUDA C Programming. Wiley, Indianapolis (2014)
7. NVIDIA: NVIDIA Tesla P100 the most advanced datacenter accelerator ever built featuring Pascal GP100, the world's fastest GPU. Version: 01.1 (2014). https://images.nvidia.com/content/pdf/tesla/whitepaper/pascal-architecture-whitepaper.pdf. Accessed 26 Jan 2020
8. Li, C.: Reference implementation for the distance regularized level set evolution (DRLSE) algorithm. http://www.imagecomputing.org/~cmli/DRLSE/. Accessed 26 Jan 2020
9. Erickson, B.J., et al.: Radiology data from the cancer genome atlas liver hepatocellular carcinoma [TCGA-LIHC] collection. The Cancer Imaging Archive (2016). https://doi.org/10.7937/K9/TCIA.2016.IMMQW8UQ. Accessed 26 Jan 2020
10. Clark, K., et al.: The cancer imaging archive (TCIA): maintaining and operating a public information repository. J. Digit. Imaging **26**(6), 1045–1057 (2013)

A Multiscale Energy-Based Time-Domain Approach for Interference Detection in Non-stationary Signals

Vittoria Bruni[1,2], Lorenzo Della Cioppa[1(✉)], and Domenico Vitulano[1,2]

[1] SBAI Department, Sapienza – Rome University,
Via Antonio Scarpa 14, 00161 Rome, Italy
`lorenzo.dellacioppa@uniroma1.it`
[2] Istituto per le Applicazioni del Calcolo, Via dei Taurini 19, 00185 Rome, Italy

Abstract. Identification and extraction of individual modes in non-stationary multicomponent signals is a challenging task which is shared by several applications, like micro-doppler human gait analysis, surveillance or medical data analysis. State-of-the-art methods are not capable yet to correctly estimate individual modes if their instantaneous frequencies laws are not separable. The knowledge of time instants where modes interference occurs could represent a useful information to use in separation strategies. To this aim, a novel time-domain method that is capable of locating interferences is investigated in this paper. Its main property is the use of multiscale energy for selecting the best analysis scale without requiring either the use of time-frequency representations or imaging methods. The performance of the proposed method is evaluated through several numerical simulations and comparative studies with the state of the art Rényi entropy based method. Finally, an example concerning a potential application to simulated micro-doppler human gait data is provided.

Keywords: Multicomponent signal · Interferences detection · Multiscale energy · Time-scale analysis · Human gait

1 Introduction

Multicomponent non-stationary signals analysis is required in many real world applications, including radio communications, surveillance, human gait recognition, radar and sonar signals processing, gravitational waves and medical data analysis. One of the main issues is to extract the single modes composing the multicomponent signal. In order to better illustrate the problem and motivate the proposed work, we refer to micro-doppler human gait classification and recognition. It consists in analysing the fine time-frequency variations of a radar signal in order to gain knowledge not only about the general position of an incoming object, but also about its moving parts (see Fig. 1). It plays a crucial role for

A. Campilho et al. (Eds.): ICIAR 2020, LNCS 12132, pp. 36–47, 2020.
https://doi.org/10.1007/978-3-030-50516-5_4

pedestrian detection in self-driving cars where the individual modes composing a micro-doppler signal correspond to movements of the body and the limbs.

Several methods have been proposed in the literature to analyse multicomponent signals, either relying on time-frequency (TF) representations [3,11] or on empirical mode decomposition [9]. Individual modes appear as ridges in TF representation but, due to energy spread in the TF domain, post processing is required to enhance TF readability. Reassignment [1,13] is one of the most common enhancement methods. It consists of reallocating TF points toward the centroids of the TF distribution. Despite its simplicity, reassignment performs very poorly in the vicinity of two interfering modes, as shown in Fig. 2. On the other hand, empirical mode decomposition directly extracts the individual modes through a time domain process, but the extracted modes are themselves non-stationary, requiring Hilbert spectral analysis for instantaneous frequencies estimation. A more rigorous alternative to empirical mode decomposition is Daubechies' synchrosqueezing transform [7,13], which combines reassignment and modes extraction on a complex wavelet representation. More recent literature focuses on the reconstruction of the individual modes in TF interference regions [6], but the precision of the proposed methods strongly depends on interference localization.

TF representations and reassignment method are commonly used [8] for addressing human gait recognition and classification issues. However, the precision of such tools depends on the selection of a proper analysis window and on the robustness of reassignment to modes interference. With regard to the first issue, experiment-based heuristics are commonly used [8]. With regard to the second issue, if several gestures are present in a micro-doppler signal, it is crucial to avoid the extraction of segments of different components as a single mode. In this case, knowing where modes overlap (interference regions) allows to segment spectrogram portions that might be misreassigned.

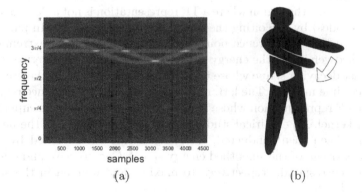

(a)   (b)

Fig. 1. Simulated micro-doppler data: spectrogram (a) of the swinging arms movement (b). The non stationary modes in (a) are generated by the movement of the arms.

To this aim, counting methods may be used. For example, in [17], a method based on short-time Rényi entropy is proposed to estimate the number of modes in a multicomponent signal. In fact, if the modes are locally similar, short-time Rényi entropy is capable of correctly estimating the number of components [2,10]. As a side effect, interferences represent drops in the counting signal. Despite the overall good performances, the short-time Rényi entropy suffers from the choice of the window length. In addition, the method does not distinguish between constructive interference, i.e. when two overlapping modes are in phase and the total energy increases, and destructive interference, i.e. when two overlapping modes are out of phase and the energy decreases.

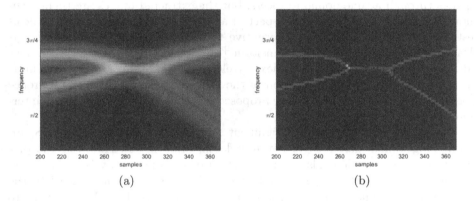

Fig. 2. Spectrogram (a) and reassigned spectrogram (b) of two interfering modes in a multicomponent signal. As it can be observed, wherever two instantaneous frequency laws are not separable, reassignment method is not able to distinguish them.

The method proposed in this paper aims at the detection of interference temporal intervals (i.e. the region where a TF representation is not able to distinguish individual modes) by exploiting the properties of signal energy. In particular, it will be shown that interference points correspond to the local extremal points of a smoothed version of the energy signal. Each of them characterizes the main lobe of a fast decaying shape whose first sidelobes correspond to the boundary of interference time interval. The latter corresponds to the interference region measured on a TF representation whose analysis window has the same support of the smoothing kernel. An empirical study concerning the selection of the best kernel support has been then conducted. Interferences are then located by selecting significant extrema of the smoothed energy signal. Constructive and destructive interference correspond, respectively, to maxima and minima in the smoothed energy signal.

The remainder of the paper is organized as follows. Next section introduces the proposed method, including the theoretical background and the algorithm. In Sect. 3 a performance analysis is presented through numerical experiments. Finally, conclusions are drawn in Sect. 4.

2   The Proposed Method

The proposed method aims at defining the optimal resolution scale at which interference analysis should be performed. To this aim, a cost function of a proper smoothed energy signal is evaluated at different scales. The latter is used to identify the optimal scale, i.e. the scale at which signal natural oscillations are removed while modes interference is emphasized.

Next sections provide details and theoretical bases.

2.1   Mathematical Formulation

Let $f(t) : \Omega \subset \mathbb{R} \to \mathbb{R}$ be a multicomponent signal, i.e. such that

$$f(t) = \sum_{k=1}^{n} A_k \sin(\varphi_k(t)), \tag{1}$$

where A_k are positive constant amplitudes, $\varphi_k(t)$ are phase functions, and Ω is a limited interval of the real line. The *smoothed energy signal* is defined as follows.

Definition 1. *Let $f(t)$ be defined as in eq. (1), $\lambda(t)$ be a compactly supported function such that $\int_{\mathbb{R}} \lambda(t)dt = 1$ and let $\lambda_s(t) = s^{-1}\lambda(t/s)$. The smoothed energy signal is defined as*

$$M_f(u, s) = \int_{\Omega} \lambda_s (u - t) \, |f(t)|^2 dt. \tag{2}$$

Since $\lambda(t)$ is the impulse response of a low pass filter by definition, $M_f(u, s)$ is expected to be constant if no interferences occur, and to vary abruptly where interference occurs.

In fact, following the same arguments[1] as in [6] and using Plancharel theorem, the *smoothed energy signal* itself is a multicomponent signal with specific instantaneous frequency laws and time dependent amplitudes. To better illustrate this fact, let us consider the two component signal $f(t) = f_1(t) + f_2(t)$, with $f_1(t) = \sin(\varphi_1(t))$ and $f_2(t) = \sin(\varphi_2(t))$. The energy signal $M_f(u, s)$ can be decomposed as

$$M_f(u, s) = (\lambda_s * |f_1|^2)(u) + (\lambda_s * |f_2|^2)(u) + 2(\lambda_s * f_1 f_2)(u), \tag{3}$$

where $*$ denotes the convolution product. Since the two modes have constant amplitude, the first two terms in the second member of Eq. (3) are constant, while the third term can be rewritten as

$$2(\lambda_s * f_1 f_2)(u) = (\lambda_s * \tilde{f}_1)(u) + (\lambda_s * \tilde{f}_2)(u), \tag{4}$$

---

[1] In [6] it has been shown that the energy of the spectrogram of a multicomponent signal computed with respect to the frequency axis is itself a multicomponent signal, whose frequencies depend on the sums and differences of the original frequencies.

where $\tilde{f}_1(t) = -\cos(\varphi_1(t) + \varphi_2(t))/2$ and $\tilde{f}_2(t) = \cos(\varphi_1(t) - \varphi_2(t))/2$. Hence, if no interference occurs, both $\tilde{f}_1$ and $\tilde{f}_2$ oscillate and thus they are filtered out by the low pass filter. Conversely, if there exists a point t_0 such that $\varphi_1(t_0) = \varphi_2(t_0)$, then $\tilde{f}_1(t)$ still is filtered out around t_0, while $\tilde{f}_2(t)$ is not, being almost constant.

It is evident that the size of the support of the kernel $\lambda_s(t)$ plays a crucial role. A shown in Fig. 3 (a), a support that is too small would not filter out enough high frequencies from the smoothed energy signal to enhance interference regions. On the contrary, a too large support would provide an excessive smoothing of the signal (see Fig. 3 (b)). The optimal scale is then the one at which all the small oscillations are sufficiently suppressed, but the low frequency information is not oversmoothed by the filtering process.

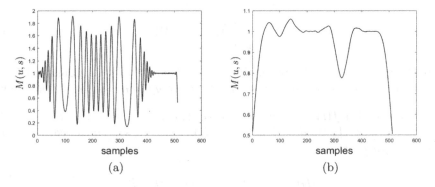

Fig. 3. Smoothed energy signal in case of too small kernel support (a) and too large kernel support (b)

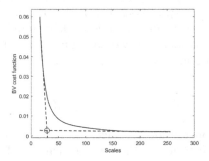

Fig. 4. Behaviour of $C_{BV}(M(u,s))$ with respect to the scale. It resembles the behaviour of rate/distortion curve. The dashed lines approximate the beginning and the end of the curve and their intersection, marked by a circle, is the optimal point according to the method in [14].

In order to select the optimal scale $\tilde{s}$ automatically, a rate/distortion curve is constructed using BV cost function, which is defined as

$$C_{BV}(g) = \int_{\Omega} |g'(t)| dt, \tag{5}$$

where g is a function $g(t) : \Omega \to \mathbb{R}$. The BV cost function measures how much the signal g oscillates. The typical profile of C_{BV} for the smoothed energy signal $M_f(u, s)$ with respect to the scales is shown in Fig. 4. It strongly resembles a rate/distortion curve, where C_{BV} is the distortion while s is the rate. In fact, as the scale increases the signal oscillates less, and the optimal scale $\tilde{s}$ is the point in the rate/distortion curve that gives the best compromise between oscillations and scale. Several methods have been proposed to estimate the optimal point in a rate distortion curve, such as methods based on Occam's Razor principle [14,15] or information-based methods [4,5]. The technique used in this paper, proposed in [14], approximates a given rate distortion curve with two straight lines, one estimated at the beginning of the curve, the other at the end. The optimal point is then found as the intersection of these two lines, as shown in Fig. 4. In our experiments the lines were estimated as the ones that best fit the curve, respectively, on the first five and on the last five values of $C_{BV}(M(u, s))$, in the least squares sense.

Extracting the Extrema. The extraction of local extrema at the optimal scale $\tilde{s}$ is performed, as they carry information concerning modes interference.

Let $P = \{t_1, \ldots, t_m\}$ the time positions of the local extrema of $M_f(u, \tilde{s})$. In order to discard the extremal points which are due to small numerical oscillations, the following measure of variation on the set P is defined:

$$DP(i) = \frac{1}{2} \Big[|M_f(t_{i+1}, \tilde{s}) - M_f(t_i, \tilde{s})| + |M_f(t_i, \tilde{s}) - M_f(t_{i-1}, \tilde{s})|. \Big] \tag{6}$$

The function DP measures the variation of the extremal point located in t_i with respect to its neighbours. Given a threshold τ, the set $\tilde{P}$ of the time positions of the interferences is the set of the t_i's such that $DP(i) \geq \tau$.

To determine the interference regions R_i, to each $t_i \in \tilde{P}$ the interval $[t_{i-1}, t_{i+1}]$ is assigned and interference regions with common boundaries are merged.

2.2 The Algorithm

The proposed algorithm consists of the following steps.

Input: Signal $f(t)$, scales S, threshold τ.
Output: Set of interference time positions $\tilde{P}$ and interference regions R_i's.
  1. For each scale $s \in S$, compute $M_f(u, s)$, as in Eq. (2);
  2. for each scale $s \in S$, compute $C_{BV}(M_f(u, s))$ as in Eq. (5);
  3. find the optimal level $\tilde{s}$ using the method in [14] applied to $C_{BV}(M_f(u, s))$;

4. extract the extrema from $M_f(u, \tilde{s})$ and collect their time positions in the set $P = \{t_1, \ldots, t_m\}$;
5. for each $t_i \in P$, if $DP(i) > \tau$, assign t_i to the set $\tilde{P}$, where DP is defined in eq. (6);
6. for each $t_i \in \tilde{P}$, assign the interference region as $R_i = [t_{i-1}, t_{i+1}]$;
7. for each R_i and R_j, if $t_{i+1} = t_{j-1}$ or $t_{i-1} = t_{j+1}$, merge R_i and R_j.

3 Performance Analysis

The proposed method has been tested on several signals. Results concerning two synthetic signals consisting of two chirps with different time-frequency laws will be presented and discussed in the sequel. Numerical simulations have been conducted taking into account the case of interference between a linear and a quadratic chirp and the case of interference between a linear and a sine-modulated chirp. The test signals are 2 s long at sampling frequency $\omega_s = 256$ Hz, resulting in a length of 512 samples. A 4-th order B-spline has been used as smoothing kernel $\lambda(t)$, $S = \{\lfloor 2^{4+i/8} \rfloor\}_{i=0,\ldots,32}$ and the threshold τ has been set as $\tau = 1/5$. The threshold value has been fixed empirically.

Test signals are obtained as a linear combination of the following chirps:

$$f_1(t) = \sin\left(\pi\omega_s \left[t^2/10 + 2t/5 + 1/5\right]\right); \tag{7}$$

$$f_2(t) = \sin\left(\pi\omega_s \left[t/2 + \sin\left(5\pi t/2\right)/20\right]\right); \tag{8}$$

$$f_3(t) = \sin\left(\pi\omega_s \left[t^3/5 + 6t^2/10 + t/10 + 1/2\right]\right). \tag{9}$$

These synthetic signals represent well real life applications, such as the responses of micro-doppler signals [12, 16].

For $f(t) = f_1(t) + f_2(t)$ (linear and sinusoidal), the selected scale is $\tilde{s} = 30$, depicted in Fig. 5 (c), and the corresponding smoothed energy signal $M_f(u, \tilde{s})$ is depicted in Fig. 5 (a). The significant extrema are correctly identified by the proposed method, thus correctly estimating interference centres and widths, as shown in Fig. 5 (b). The results for the test signal $f(t) = f_1(t) + f_3(t)$ (linear and quadratic) mimic those of the previous case. As it can be seen in Figs. 6 (a) and 6 (b) interference time positions and the interference regions are correctly estimated and the selected scale is $\tilde{s} = 46$ (Fig. 6 (c)). Comparing the smoothed energy signals in Figs. 5 (a), and 6 (a) with the relative spectrograms, it can be observed that constructive interferences correspond to maxima in $M_f(u, \tilde{s})$ while destructive interferences correspond to minima.

It is also worth highlighting that the method shows the same precision in detecting interference regions for signals whose components contribute with different amplitudes.

To further validate the proposed approach, in Fig. 7 (a) and (b) details of the reassigned spectrogram of both considered test signals are shown, along with the estimated interference points and interference regions. The spectrograms have been computed using a smoothing window of size equal to the corresponding

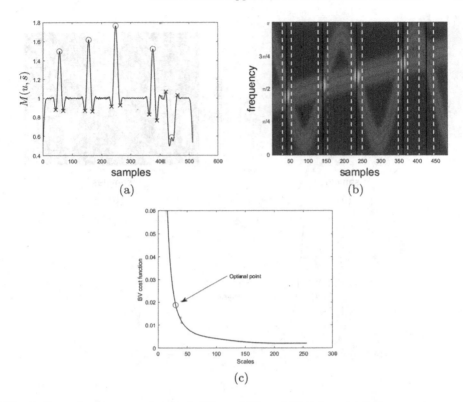

Fig. 5. Smoothed energy signal of $f(t) = f_1(t) + f_2(t)$ (a) and relative spectrogram (b) computed using a smoothing window of length $\tilde{s}$. Crosses (a) and black lines (b) represent interferences while circles (a) and dashed lines (b) represent the interference regions. The corresponding rate/distortion curve is shown in (c); the estimated optimal point is marked by a circle.

optimal scales $\tilde{s}$. As it can be observed, the detected regions exactly correspond to the regions where reassignment fails.

In Fig. 8 the same test signals are evaluated using short-time Rényi entropy based method proposed in [17]. The choice of the window length is crucial in this case. As shown in Figs. 8 (a) and (c), using a window of 16 samples, the counting signal tends to oscillate heavily. Conversely, if the window length is increased to 32, the small oscillations are dampened but the signal is less readable and the two rightmost interferences are barely distinguishable (Fig. 8 (d)). In addition, since interference regions are identified as drops in the counting signal, it is not possible to classify the type of interference (constructive or destructive). On the contrary, the proposed method takes advantage of the adaptive selection of the scale that defines the optimal window length and it is capable of classifying the type of interference.

Finally, the proposed method has been tested on synthetic data simulating micro-doppler signal of a man swinging both its arms. The simulation has

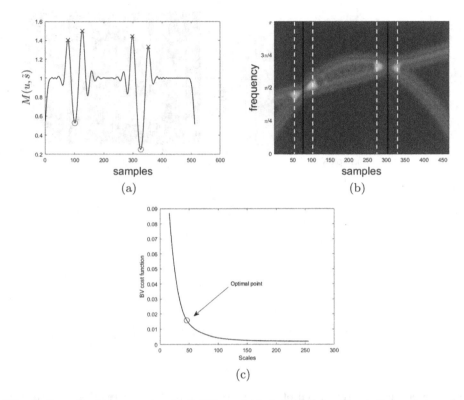

Fig. 6. Smoothed energy signal of $f(t) = f_1(t) + f_3(t)$ (a) and relative spectrogram (b) computed using a smoothing window of length $\tilde{s}$. Crosses (a) and black lines (b) represent interferences while circles (a) and dashed lines (b) represent the interference regions. The corresponding rate/distortion curve is shown in (c); the estimated optimal point is marked by a circle.

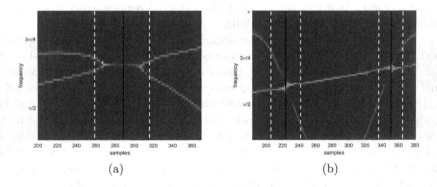

Fig. 7. Reassigned spectrograms: interference between linear and quadratic chirps (a) and between linear and sine-modulated chirps (b) (details). White dashed lines limit interference regions and correspond to regions where reassignment fails.

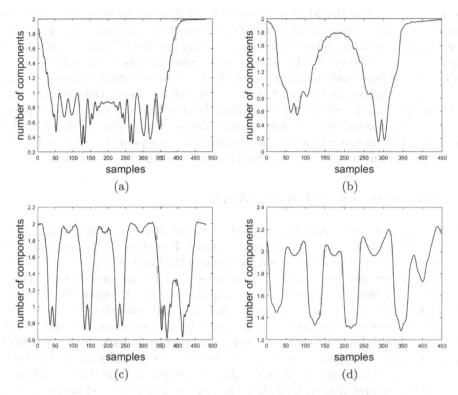

Fig. 8. Number of components estimated using short time Rényi entropy [17]. Interference between linear and quadratic chirps (a & b) and linear and sine-modulated chirps (c & d). The first column refers to a 16 taps window while the second column refers to 32 taps window

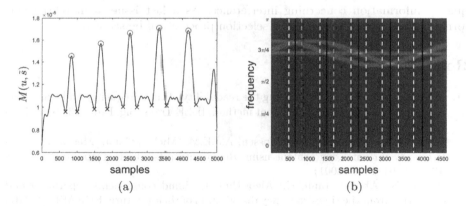

Fig. 9. Simulated micro-doppler data: smoothed energy signal (a) and spectrogram (b) using a window of size $\tilde{s}$. Crosses (a) and black lines (b) represent interferences while circles (a) and dashed lines (b) represent the interference regions.

been realized using the MatLab Phased Array System Toolbox. In Fig. 9 the smoothed energy signal (a) and the spectrogram of the micro-doppler signal (b) are depicted. As it can be observed, even if signal amplitude is not constant over time, the method correctly locates the interference points where reassignment would fail. It turns out the proposed method could be twofold advantageous in human gait micro-doppler signal analysis. On the one hand, the optimal scale selection process described in Sect. 2 might be implemented in existing separation and analysis strategies to automatize analysis window size selection; on the other hand the temporal interference occurrence can represent a further feature to be used in human gait the recognition and classification procedures.

4 Conclusion and Future Work

In this paper a novel time-domain method for interference detection in multicomponent chirp-like signals has been proposed. Even though the method does not rely on any time-frequency representation of the input signal, it is capable to correctly detect time locations and amplitudes of TF modes interference, thanks to the use of a smoothed energy signal computed at a specific scale. The method is automatic and robust to signals having not comparable amplitude. In addition, it results more stable and precise than some existing competing methods. These features make the method useful in several applications. In particular, the case of data human gait recognition and classification for micro-doppler signals has been considered in this paper through simulations using synthetic data. Although the results are preliminary, they show that the proposed method performs well on both synthetic data and simulated human micro-doppler data. This motivates future research concerning the actual use of the proposed method in this specific application, as mentioned at the end of the previous section.

For example, robustness to noise and to modes having different temporal duration will be investigated as well as the possibility of also extracting frequency information concerning interference. As a last issue, a more rigorous formalization of the optimal scale selection process will be studied.

References

1. Auger, F., Flandrin, P.: Improving the readability of time-frequency and time-scale representations by the reassignment method. IEEE Trans. Sig. Proc. **43**, 1068–1089 (1995)
2. Baraniuk, R.G., Flandrin, P., Janssen, A.J.E.M., Michel, O.J.J.: Measuring time-frequency informations content using Renyi entropies. IEEE Trans. Inf. Theory. **47**(4), 1391–1409 (2001)
3. Barkat, B., Abed-Meraim, K.: Algorithms for blind components separation and extraction from the time-frequency distribution of their mixture. EURASIP J. Adv. Signal Process. **2004**(13), 1–9 (2004). https://doi.org/10.1155/S1110865704404193
4. Bruni, V., Della Cioppa, L., Vitulano, D.: An automatic and parameter-free information-based method for sparse representation in wavelet bases. Mathematics and Computers in Simulation, in press (2019)

5. Bruni, V., Della Cioppa, L., Vitulano, D.: An entropy-based approach for shape description. In: Proceedings of EUSIPCO 2018, 2018-September, pp. 603–607 (2018). https://doi.org/10.23919/EUSIPCO.2018.8553507
6. Bruni, V., Tartaglione M., Vitulano, D.: On the time-frequency reassignment of interfering modes in multicomponent FM signals. In: Proceedings of EUSIPCO 2018, 2018, pp. 722–726 (2018). https://doi.org/10.23919/EUSIPCO.2018.8553498
7. Daubechies, I., Lu, J., Wu, H.T.: Synchrosqueezed wavelet transforms: an empirical mode decomposition-like tool. Appl. Comput. Harmon. Anal. **30**(2), 243–261 (2011)
8. Ding, Y., Tang, J.: Micro-Doppler trajectory estimation of pedestrians using a continuous-wave Radar. IEEE Trans. Geosci. Remote Sens. **52**(9), 5807–5819 (2014)
9. Huang, N.E., et al.: The empirical mode decomposition and the Hilbert spectrum for nonlinear and non-stationary time series analysis. Proc. R. Soc. Lond. A **454**, 903–995 (1998). https://doi.org/10.1098/rspa.1998.0193
10. Lerga, J., Saulig, N., Lerga, R., Milanović, Ž.: Effects of TFD thresholding on EEG signal analysis based on the local Rényi entropy. In: 2017 2nd International Multidisciplinary Conference on Computer and Energy Science (SpliTech) (2017)
11. Li, Z., Martin, N.: A time-frequency based method for the detection and tracking of multiple non-linearly modulated components with birth and deaths. IEEE Trans. on Sig. Proc. **64**, 1132–1146 (2016)
12. Lyonnet B., Ioana, C., Amin, M.G.: Human gait classification using microDoppler time-frequency signal representations. In: 2010 IEEE Radar Conference, Washington, DC, pp. 915-919 (2010). https://doi.org/10.1109/RADAR.2010.5494489
13. Meignen, S., Oberlin, T., McLaughlin, S.: A new algorithm for multicomponent signal analysis based on Synchrosqueezing: with an application to signal sampling and denoising. IEEE Trans. Signal Process. **60**, 11 (2012)
14. Natarajan, B.K.: Filtering random noise from deterministic signals via data compression. IEEE Trans. Signal Process. **43**(11), 2595–2605 (1995)
15. Rissanen, J.: Modeling by shortest data description. Automatica **14**, 465–471 (1978)
16. Sakamoto, T., Sato, T., Aubry, P.J., Yarovoy, A.G.: Texture-based automatic separation of echoes from distributed moving targets in uwb radar signals. IEEE Trans Geos. Rem. Sen. **53**, 1 (2015)
17. Sucic, V., Saulig, N., Boashash, B.: Analysis of local time-frequency entropy features for nonstationary signal components time support detection. Dig. Sig. Proc. **34**, 56–66 (2014)

SMAT: Smart Multiple Affinity Metrics
for Multiple Object Tracking

Nicolas Franco Gonzalez[✉], Andres Ospina[✉], and Philippe Calvez[✉]

CSAI Laboratory at ENGIE, Paris, France
nicolas.franco@uao.edu.co, andres.ospina@external.engie.com,
philippe.calvez1@engie.com

Abstract. This research introduces a novel multiple object tracking
algorithm called SMAT (Smart Multiple Affinity Metric Tracking) that
works as an online tracking-by-detection approach. The use of various
characteristics from observation is established as a critical factor for
improving tracking performance. By using the position, motion, appearance, and a correction component, our approach achieves an accuracy
comparable to state of the art trackers. We use the optical flow to
track the motion of the objects, we show that tracking accuracy can
be improved by using a neural network to select key points to be tracked
by the optical flow. The proposed algorithm is evaluated by using the
KITTI Tracking Benchmark for the class CAR.

Keywords: Online multiple object tracking · Tracking by detection

1 Introduction

Multiple object tracking or MOT is an important problematic in computer
vision. This problematic has many potential applications, such as tracking and
analyzing the movement of vehicles and pedestrians on the road, helping self-
driving cars to make decisions, tracking and analyzing the movement of cells
or organisms in time-lapse microscopy images or helping robots to pick up and
place things in environments such as farms or industries. The broad area of
application reflects the importance of developing accurate objects trackers.

MOT can be explained as the task of locating and tracking multiple objects
of interest in video footage, identifying their position in every frame, and maintaining the identity (ID) of each target through its trajectory. There are many
challenges in tracking multiple objects, such as the random motion of objects,
crowded scenes, partial or full occlusion of objects, objects and camera viewpoint
variations, illumination changes, background appearance changes, and non-ideal
weather conditions.

This paper introduces an online MOT based on the tracking-by-detection
paradigm. In an online approach, uniquely the previous tracked objects and the
current frame are available to the algorithm. Tracking-by-detection means that
in every frame, the objects are detected and considered as targets. The proposed

© Springer Nature Switzerland AG 2020
A. Campilho et al. (Eds.): ICIAR 2020, LNCS 12132, pp. 48–62, 2020.
https://doi.org/10.1007/978-3-030-50516-5_5

pipeline is composed of two main modules: detection and tracking. For detection, we test our system with two different detectors Faster-RCNN [24] and RRC [23]. The tracking algorithm is composed of three elements: Affinity metrics, data association and past corrector. for multiple object tracking The affinity metric outputs the probability of two observations from different frames being of the same target. For this we rely on three factors estimated from an observation: position, appearance, and motion. We use three affinity metrics inspired by state of the art trackers such as [4,5,30]: Intersection over Union (IoU) score, appearance distance, and optical flow affinity. The scores generated by the affinities are analyzed by the data association component with the objective of linking the current observations to the past observations, by giving the same ID in the cases that the target is the same. These process results are then passed into the corrector component, called tubulet interpolation which aims to fill empty spaces in the trajectories produced by detection failures.

There are multiple challenges and benchmarks for Multiple object tracking as MOT Challenge [7], KITTI Tracking Challenge [8], DETRAC [20] between others. This work uses to train, test and experiment using the KITTI Tracking Challenge [8] dataset for the car category. This limitation is due that we don't want to concentrate our efforts training the detector. The main idea is to concentrate on the tracking.

The main contributions of this paper are:

- The development of a novel tracking algorithm called Smart Multiple Affinity Metrics Tracker (SMAT). This algorithm combines three affinity metrics that evaluate the position, appearance, and motion of the targets.
- We tested the algorithm on the KITTI Tracking Benchmark and our approach produces competitive results. It was ranked 12th in this challenge (01/2020). Having the best multiple object tracking precision (MOTP). Also, in the subset of the top 12 submissions: we have the least identity switches (IDs) and the second best trajectory fragmentation (FRAG).
- Our experiments showed that the proposed affinity metrics complement each other to reduce errors produced along the tracking-by-detection framework.
- In near online [5], an affinity metric is used which is based on optical flow. We propose an improvement on the way the interest points are chosen for this metric by using an neural network called "hourglass"[26], instead of popular corner detectors as [27] and [25]. Better tracking accuracy results are obtained with the use of this network.
- A tubulet interpolation method was used to fill the empty spaces in a trajectory produced by detection failures. This technique allowed us to correct the past observations relying on the information provided by the motion model.

2   Related Work

Due to the rapid advancement in object detection thanks to convolutional neural network (CNN), tracking-by-detection has become a popular framework for addressing the multiple target tracking problem. These methods depend on an object detector to generate object candidates to track. Then, based on the information extracted from detections (for example the position or appearance), the tracking is done by associating the detections.

A MOT approach can treat the association of the detections either as an online or offline problem. Global trackers [1,18,21] assume detections from all frames are available to process. In recent global trackers, the association is done by popular approaches as multiple hypotheses tracking (MHT) [13] and Bayesian filtering based tracking [15]. These methods achieve higher data association accuracy than online trackers as they consider all the detections from all frames. Contrary to global trackers, online trackers [12,30,31] do not use any data from future frames. They use the data available up to the current instance to tackle the association problem. Such trackers often solve this via the Hungarian algorithm [14]. Their advantage is that online methods can be applied in real-time applications such as autonomous cars. In these methods, a key factor for having an excellent performance is to use a relevant affinity metric. The affinity metric estimates how much similarity exists between 2 detections across time. Then, based on this information the association between these detections is done or not. For the affinity metric some trackers such as [3,4] rely in the information provided by the Intersection over Union (IoU) score. Other trackers such as [30] rely on the appearance information.

Recently, near online trackers proposed an Aggregated Local Flow Descriptor (ALFD) [5] to be used as affinity metric. The ALFD applies the optical flow to estimate the relative motion pattern between a pair of temporally distant detections. For doing that, the ALDF computes long term interest point trajectories (IPTs). If two observations have many IPTs in common this means that they are more likely to represent the same target. In [5] they use the algorithm FAST [25] for computing the interest points to be tracked by the optical flow. In contrast to these trackers, we propose a novel architecture that uses IoU score, optical flow affinity and appearance distance to infer if two observations correspond to the same target. The data association is done by the Hungarian algorithm.

3   Smart Multiple Affinity Tracker

The proposed algorithm (SMAT) is shown in Fig. 1. The inputs of the algorithm are the current frame and the identified tracked objects from the previous frame. In first place, the objects are detected in the current frame.

Then, three different algorithms compute the affinity or probability of being the same object between the detections and the previous tracked objects. These algorithms rely on three factors estimated from the detections and frames: position, appearance and motion. We use three affinity metrics: IoU score, appearance distance and optical flow affinity. The generated affinities are used by the data association component to link the current observations with the past observations by assigning the same ID in the cases that the object is the same. At each iteration, the corrector analyses if there is a re-identification of a lost target.

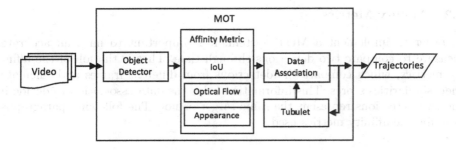

Fig. 1. SMAT overview

In that case, the corrector component will fill the empty spaces in the trajectories produced by detection failures using interpolation.

3.1   Object Detection

The proposed tracker is an online tracking-by-detection approach, where at each frame the objects of interest are detected and then associated to the targets of the past frames.

In this research we had two different stages:

- In the first experiment, we used the detector Faster R-CNN [24] with a ResNET-101 [6] as a backbone. This was selected based on the work developed by [10]. They evaluate the speed/memory/accuracy balance of different feature extractors and meta architectures. This configuration was chosen for the good trade-off between computing time and accuracy. We use this detector for the first experiment related to the affinity metrics.
- In the second experiment, aiming to improve our results, we used the detector RRC [23] due to its strong accuracy in the detection task.

The results obtained for those detectors in the KITTI Object Detection Benchmark [8] were:

Table 1. KITTI Object detection benchmark results

Method	Easy	Moderate	Hard
TuSimple [32] (Best)	94.47	95.12	86.45
RRC [23] (Used)	93.40	95.68	87.37
Faster-RCNN [24] (Used)	79.11	87.90	70.19

3.2   Affinity Metrics

In order to implement a MOT System, it is important to have an accurate measure to compare two detections through time. That is the job of the affinity metrics, which compare the detections from different frames and calculate their similarities scores. This information helps the data association to decide if the two detections represent the same target or not. The following paragraphs describe the affinity metrics used.

Intersection over Union Score: IoU is computed by dividing the area of overlap by the area of union between two bounding boxes that represent the detections. To use it as an affinity metric, the detections from frame t are compared with the tracked objects from frame $t-1$. The results are registered in a cost matrix named C_{IoU} that will be used by the data association. This process is summarized in Fig. 2. Where D_t and D_{t-1} are the predicted bounding boxes for the current frame and previous tracks.

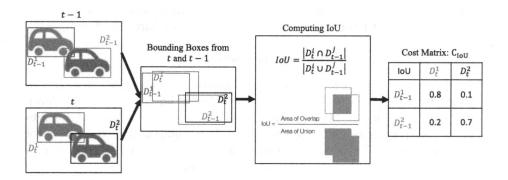

Fig. 2. Computing IoU cost matrix

Optical Flow Affinity: The Aggregated Local Flow Descriptor (ALFD) was introduced by [5] in 2015. The ALDF robustly measure the similarity between two detections across time using optical flow. Inspired by them, we developed a simplified version of the ALDF called optical flow affinity.

The optical flow affinity uses the Lucas-Kanade sparse optical motion algorithm [11]. This algorithm starts by identifying interest points (IPs) in the detections. Now, the optical flow algorithm tracks this points regardless of the detections, this track is called interest point trajectory (IPTs). Also, an ID is given to each trajectory.

To compute the affinity, the detection bounding box is divided in 4 sectors (as proposed in [5]) and a description of the detection is made based on the locations of the IPTs with respect to the sectors. Then, each tracked object from $t-1$ is compared with each detection from frame t. The number of IPTs that are common per sector (number of matches) are counted and divided by the total number of matches and the number of IPTs of the target using Eq. 1.

$$score_i = \frac{\sum_{i=0}^{N_{sector}} matches(t_i, d_i)}{t_{i\,IDs}} \qquad (1)$$

where d_i is the observation from current frame, t_i is the target from past frame, $t_{i\,IDs}$ is the total number of IPTs contained by the target and N_{sector} is the total number of sectors per target (4 in this case). The results are stored in a cost matrix C_{of} (see Fig. 3).

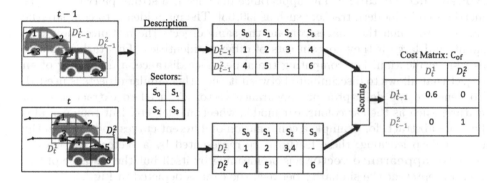

Fig. 3. Optical flow affinity

When working with sparse optical flow, the interest points (IPs) should be easy to be re-identified in the subsequent frames. Approaches as Shi Tomasi Corner Detector [27] or FAST [25] are used as a way of correctly choosing these IPs. However, these approaches do not have any notion of the shape of the target and therefore they find interest points in objects that are not of interest. These interest points are called outliers. The outliers reduce the accuracy of the optical flow affinity metric because it introduces wrong information to the evaluation. For reducing these outliers, we propose the use of a method that computes key-points for a given target.

In [26] they trained a stacked hourglass for the task of detecting key-points in vehicles. They obtained 93.4 of percentage of Correct Key-points (PCK) in the class car for the Pascal 3D dataset. The hourglass [22] can produce 36 points as shown in Fig. 4. In this work we selected only 8 points: (1, 2, 14, 15, 17, 32, 33, 35). These points were strategically chosen due to their position (easily identified in different views of a car), and the information that they provide (they are well tracked, and provide useful information on many components of the car to calculate the affinity metric).

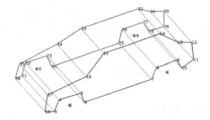

Fig. 4. Keypoints generated by hourglass

Appearance Distance: The appearance distance is a strong pairwise affinity metric used in modern trackers such as [30,33]. The main idea is to compare the car images, when the images contain the same objects the distance should be small, and large if they contain cars of different identities. This task is known as re-identification. For computing the appearance distance, a descriptor of an object that allows to discriminate between it and other similar objects is needed. To generate the descriptor or appearance vector we need to extract a set of features such as the car colour, car model, wheels model, etc. But they can also be more abstract, for example a combination of different curves and lines. In the case of deep learning, these features are represented by a vector. The feature vector or **appearance vector** has no meaning by itself but the distance of two vectors represent the similarity between the cars as depicted in Fig. 5).

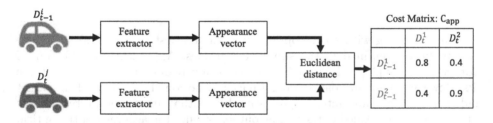

Fig. 5. Computing appearance distance (Color figure online)

For computing the appearance vector we use the Multiple Granularity Network (MGN) [29]. MGN was chosen because of its great performance in person re-identification datasets such as CUHK03 [16]. We trained the model similar to [29], with the main difference being the input size: 384×384. The model is trained in the VERI dataset [19]. For training, many mini-batch are created by randomly selecting 20 identities and randomly sampled 4 images for each identity. In result we get 80% mean average precision (mAP) for the test set in the VERI dataset.

We should note that, the result of this algorithm for similar cars the affinity distance is small. However, for the data association the affinity for similar cars should be large. Then, knowing that the maximum distance is one. For the data association, the value of the appearance affinity is corrected by computing one minus the affinity distance.

3.3   Data Association

For the data association the Hungarian Algorithm [14] is used. The affinity metrics described previously are used when comparing the current detection with the previous tracked objects. Each affinity metric produces a cost matrix by comparing every detection with every tracked object. Then, we sum each cost matrix multiplied by a weight, as shown in Eq. 2.

$$C_{total} = w_{IoU}C_{IoU} + w_{of}C_{of} + w_{app}C_{app} \qquad (2)$$

where C_{total} is the total cost matrix. w_{IoU}, w_{of} and w_{app} are the weights. This multiplication is done to prioritize or balance the costs because of the nature of the affinity, i.e. the values of the optical flow affinity have a mean value lower than the other two affinities. Therefore, we choose to multiply the optical flow cost matrix by 1.4. The other affinities are multiplied by 1.

Then using the total cost matrix C_{total}, the Hungarian Algorithm will assign which detections represent the same target by maximizing the cost assignment. Associations with a score lower than 0.3 are deleted. Then, new identities are created with the unmatched detections. The information of the terminated tracked objects or tracklets (identities that were not found in the present frame) is stored. Then, in subsequent frames the algorithm will look for reappearances of these terminated tracklets. This means that, in every frame we will first compare the detections from the present frame with the detections from the past frame. The detections not associated will be compared with the terminated tracklets. If there are some matches, the old IDs will be assigned. Otherwise, new ids will be generated. In practice, the system stores the information of terminated tracklets for 13 frames. If during that period the ID does not reappear then this will be definitely deleted.

3.4   Estimating Trajectories for Partially Lost Objects

When the object detection fails, as depicted in Fig. 6 at the time $t-1$, a fragmentation in the estimation of a target's trajectory is produced. The fragmentation

Fig. 6. Detection failure

happens when it is unknown the position of the target over a period of time. To reduce fragmentation we propose a technique called tubelet interpolation.

Although the object is not detected, the optical flow still follows the target. Relying on this information an interpolation of the bounding box is done for filling the empty spaces in the trajectories (see Fig. 7).

Fig. 7. Tubulet interpolation

Procedure: The correction of the fragmentation starts; the information of the matched bounding boxes is known. The current bounding box in time t and the last known bounding box $t - n$ are assessed. Therefore, the object was lost for n frames. The bounding boxes coordinates are defined as $[x_1, y_1, x_2, y_2]$, where (x_1, y_1) is the top left corner and (x_2, y_2) is the bottom right corner of the rectangle. It is assumed that velocity between frames is linear. Equation 3 computes the velocities v_x and v_y of the targets between frame t and $t - \Delta t$. Also, the width w_r and height h_r change ratio are calculated using the Eq. 4 (for v_y and h_r replace x by y in Eqs. 3 and 4). Finally, to reproduce the bounding box coordinates between the frames, Eq. 5 and 6 (for y_1 and y_2 replace x by y and w_r by h_r) is used.

$$v_x = \frac{x_{1(t)} - x_{1(t-n)}}{n} \tag{3}$$

$$w_r = \frac{(x_{2(t)} - x_{1(t)}) - (x_{2(t-n)} - x_{1(t-n)})}{n} \tag{4}$$

$$x_{1(t-n+k)} = x_{1(t-n)} + v_x * k \tag{5}$$

$$x_{2(t-n+k)} = x_{1(t-n)} + (x_{2(t-n)} - x_{1(t-n)}) + w_r * k \tag{6}$$

where k is a number between 0 and n.

4   Experimentation

Our approach was evaluated in the KITTI Tracking Benchmark [8] on training and testing dataset. Different configurations of SMAT were proposed. The best performing configuration in the training set was used to report the result on the benchmark on the Car class.

4.1   Metrics

The metrics used to evaluate the multi-target tracking performance are defined in [17], along with the widely used CLEAR MOT metrics [2]. Some of these metrics are explained below, where ($\uparrow$) means the higher the better and ($\downarrow$) the lower the better:

- MOTA($\uparrow$): Multi-object tracking accuracy
- MOTP($\uparrow$): Multi-object tracking precision
- MT($\uparrow$): Ratio of ground truth trajectories successfully tracked for at least 80 % of their life span.
- ML($\downarrow$): Mostly lost trajectories. Trajectories tracked for less than 20% of its total length.
- PT($\downarrow$): The ratio of partially tracked trajectories, i.e., MT - ML
- FP($\downarrow$): Total number of wrong detections
- FN($\downarrow$): Total number of missed detections
- ID sw($\downarrow$): Number of times the ID of a tracker switches to a different previously tracked target
- Frag($\downarrow$): Number of times a trajectory is interrupted during tracking.

Table 2. Results in the KITTI tracking training set using different configurations

Config	MOTA	MOTP	Recall	Precision	MT	PT	ML	TP	FP	FN	IDS	FRAG
IoU	0.8776	0.9107	0.9164	0.9867	0.8422	0.1525	0.0053	24797	334	2261	350	926
IoU+Sh	0.8870	0.9107	0.9164	0.9867	0.8422	0.1525	0.0053	24797	334	2261	126	705
IoU+FS	0.8853	0.9107	0.9164	0.9867	0.8422	0.1525	0.0053	24797	334	2261	167	744
IoU+Hg	**0.8878**	0.9107	0.9164	0.9867	0.8422	0.1525	0.0053	24797	334	2261	**105**	**687**
IoU+Ap	0.8882	0.9107	0.9164	0.9867	0.8422	0.1525	0.0053	24797	334	2261	96	685
IoU+Sh+Ap	0.8895	0.9107	0.9164	0.9867	0.8422	0.1525	0.0053	24797	334	2261	64	650
IoU+Hg+Ap	**0.8899**	0.9107	0.9164	0.9867	0.8422	0.1525	0.0053	24797	334	2261	**55**	**642**
IoU+FS+Ap	0.8896	0.9107	0.9164	0.9867	0.8422	0.1525	0.0053	24797	334	2261	63	650
IoU+Sh+Ap+Tb	0.9151	0.9069	0.9509	**0.9768**	0.9238	0.0727	0.0035	25982	**618**	1342	84	306
IoU+Hg+Ap+Tb	**0.9201**	0.9062	**0.9563**	0.9762	**0.9326**	**0.0638**	0.0035	**26158**	639	**1195**	89	**266**
IoU+FS+Ap+Tb	0.9160	**0.9066**	0.9535	0.9753	0.9255	0.0709	0.0035	26050	659	1269	93	293

4.2   Experiments

Different configurations were tested with the training set of KITTI Tracking. In the first experiment is evaluated the performance of using a tracker with IoU as an affinity metric. Surprisingly, a MOTA of 87.76% was obtained. However, there were many id-switches. For reducing the number of ID switches to improve the accuracy we added the optical flow affinity to the tracker formulation. As the optical flow affinity highly depends on the IPs, different interest points detectors were used to see which could give better results. The configurations proposed were:

- **IoU:** Tracking using intersection over union as affinity
- **IoU+Sh:** IoU and optical flow with Shi Tomasi
- **IoU+FS:** IoU and optical flow with FAST
- **IoU+Hg:** IoU and optical flow with Hourglass

The results are shown in Table 2. In all cases, using optical flow improves the MOTA by reducing the ID switches. This is because in some situations there are large movements of vehicles from one frame to another generating low IoU scores. However, in these situations the optical flows can still provides relevant information to associated the ids. Also, in the situation were some objects are moving very close to each other, the optical flow affinity helps to discriminate well between these. The configuration with less ID switches was the one that uses an hourglass as interest point detector. This is because the hourglass was trained to find key-points on vehicles so it presents less outliers points than the others. In the case of FAST and Shi Tomasi they are looking for finding corners in the image. In many cases, bounding boxes contains not only the vehicle, also other pieces of objects. That causes that these corner detectors produce points in zones that are not interesting, generating wrong points to track.

The IoU score and the optical flow affinity fails in cases were the objects are occluded. In order to make our tracker robust in these situations, the appearance distance metric (+Ap) was added to the configurations aforementioned. By doing this we managed to obtain a MOTA of 88.99% and reduce the id-switches from 350 to 55 in the best method (see IoU+Hg+Ap from Table 2).

Although the ID switches were greatly reduced and the MOTA was increased from 87.76% to 88.99%, the model was not still good enough to be ranked in the first 20 positions of the challenge. This was partly because there were many false negatives (2261 in all methods shown). Due to the false negatives the models were also presenting many fragmentations (642 in the best case). When evaluating MOT system, each trajectory had a unique start and end and it was assumed that there was no fragmentation in the trajectories [2]. However, the object detectors present failures between frames. This increases the number of false negatives and fragmentation. To deal with this we added a tubulet interpolation (+Tb) to the tracking formulation as it was explained before. From Table 2 we concluded that the positive aspects of the interpolation are: the number of fragmentation and false negatives are reduced by filling the empty spaces of the trajectories, the recall is increased by generating more bounding boxes, the mostly tracked (MT) metric is increased, and the ratio of partially tracked trajectories is reduced along with the mostly lost (ML) metric. The negative aspects are: the false positives and the id-switches increase because sometimes the corrections of past frames are wrong and the precision decreases because in some cases the created bounding box does not match completely well the objects. Although the tubulet interpolation has negative aspects, the MOTA increased more than 2% in all the configurations, proving that the effect of positive aspects outweighted the negatives. The Table 3 shows the effect in percentage of adding different components to a basic tracker that uses only IoU. Green values mean the result is improved while red values means the result gets worse.

Table 3. Components contributions in % training set

Config	MOTA%	MOTP%	Recall%	Precision%	MT%	PT%	ML%	TP%	FP%	FN%	IDS%	FRAG%
+OF	+ 0.99±0.21	=	=	=	=	=	=	=	=	=	- 61±9	- 23±3
+Ap	+ 1.6	=	=	=	=	=	=	=	=	=	- 73	- 30
+OF+Ap	+ 2±0.2	=	=	=	=	=	=	=	=	=	- 83±1	- 30
+OF+Ap+Tb	+ 4±0.2	- 0.42±0.3	+ 3.7±0.3	- 1	+ 8.9±0.6	- 8	- 0.2	+ 5.2±0.3	+ 91±6	- 43.5±3.5	- 75±1	- 70.5±0.5

During the writing of this paper, we saw the opportunity of using a better detector called RRC in the place of the Faster R-CNN. Therefore, using the best performing model in the previous experiment (IoU + Hg + Ap + Tb), the model is tested with the RRC detector. The results are shown in the next section.

4.3 Results

Two submission were done for the KITTI Tracking Benchmark. One using a Faster R-CNN as object detector and other employing and Accurate Single Stage Detector Using Recurrent Rolling Convolution RRC. Both of them using the best performing model (IoU + Hg + Ap + Tb)]. In the Table 4 our approach (SMAT) with the best models of the challenge is compared. Due to the fact that the RRC presents better object detection accuracy than the Faster-RCNN, the architecture that use RRC is the best performing. SMAT+RRC is ranked 12th while SMAT+F-RCNN is ranked 20th in the challenge. Models that used other sensors different to the camera were not included in the Table 4. As shown, SMAT has competitive results in comparison with state of the art trackers.

Table 4. Results in the KITTI Tracking Benchmark

Config	MOTA%	MOTP%	MT%	ML%	IDS	FRAG
MASS [12]	85.04	85.53	74.31	2.77	301	744
SMAT+RRC (ours)	84.27	86.09	63.08	5.38	28	341
MOTBeyPix [26]	84.24	85.73	73.23	2.77	468	944
IMMDP [31]	83.04	82.74	60.62	11.38	172	365
JCSTD [28]	80.57	81.81	56.77	7.38	61	643
extraCK [9]	79.99	82.46	62.15	5.54	343	938
SMAT+F-RCNN (ours)	78.93	84.29	63.85	4.77	160	679

5 Conclusion

In this paper we propose a novel tracker architecture that uses the position, motion and appearance as characteristics for associating the targets to the observations. Based on these characteristics three affinity metrics were implemented: IoU score, optical flow affinity and appearance distance. Our experiments showed that for tracking the motion using optical flow the results are highly dependent on the selection of the interest points. A neural network called "hourglass" is used in order to compute interest points to follow. By using this instead of classical interest point detectors the tracking accuracy is improved. Through experiments we showed that the affinity metric complement each other to reduce mistakes committed in the tracking-by-detection framework. An analysis of the contributions generated for adding each affinity to the tracking formulation is done. A method called tubelet interpolation was proposed in order to reduce the fragmentation generated by detections failures. This method relies on the information provided by the optical flow. Finally, the proposed algorithm presents competitive results as it was ranked 12th in the KITTI Tracking Benchmark for the class Car.

In future work, we will see the performance difference between using a segmentation network plus a classic interest point detector, instead of the detection network plus the hourglass network in order to compute key points. The segmentation will avoid the points outside of the object. Therefore, the difference in time and performance could be studied. In the other hand, we will experiment different position models as a Kalman filter and how we can joint the information of the optical flow. Also, we will study other data association algorithms.

References

1. Berclaz, J., Fleuret, F., Turetken, E., Fua, P.: Multiple object tracking using k-shortest paths optimization. IEEE Trans. Pattern Anal. Mach. Intell. **33**(9), 1806–1819 (2011). https://doi.org/10.1109/TPAMI.2011.21
2. Bernardin, K., Stiefelhagen, R.: Evaluating multiple object tracking performance: the clear mot metrics. J. Image Video Process. **2008**, 1 (2008)
3. Bewley, A., Ge, Z., Ott, L., Ramos, F., Upcroft, B.: Simple online and realtime tracking. In: 2016 IEEE International Conference on Image Processing (ICIP), pp. 3464–3468. IEEE (2016)
4. Bochinski, E., Eiselein, V., Sikora, T.: High-speed tracking-by-detection without using image information. In: 2017 14th IEEE International Conference on Advanced Video and Signal Based Surveillance (AVSS), pp. 1–6. IEEE (2017)
5. Choi, W.: Near-online multi-target tracking with aggregated local flow descriptor. In: 2015 IEEE International Conference on Computer Vision (ICCV), December 2015. https://doi.org/10.1109/iccv.2015.347, http://dx.doi.org/10.1109/iccv.2015.347
6. Dai, J., Li, Y., He, K., Sun, J.: R-FCN: Object detection via region-based fully convolutional networks. In: Advances in Neural Information Processing Systems, pp. 379–387 (2016)

7. Dendorfer, P., et al.: CVPR19 tracking and detection challenge: how crowded can it get? arXiv:1906.04567 [cs], June 2019, arXiv: 1906.04567
8. Geiger, A., Lenz, P., Urtasun, R.: Are we ready for autonomous driving? the kitti vision benchmark suite. In: Conference on Computer Vision and Pattern Recognition (CVPR) (2012)
9. Gündüz, G., Acarman, T.: A lightweight online multiple object vehicle tracking method. In: 2018 IEEE Intelligent Vehicles Symposium (IV). pp. 427–432, June 2018. https://doi.org/10.1109/IVS.2018.8500386
10. Huang, J., et al.: Speed/accuracy trade-offs for modern convolutional object detectors. In: Proceedings of the IEEE Conference on Computer Vision and Pattern Recognition, pp. 7310–7311 (2017)
11. Kanade, L.: An iterative image registration technique with an application to stereo vision
12. Karunasekera, H., Wang, H., Zhang, H.: Multiple object tracking with attention to appearance, structure, motion and size. IEEE Access **7**, 104423–104434 (2019). https://doi.org/10.1109/ACCESS.2019.2932301
13. Kim, C., Li, F., Ciptadi, A., Rehg, J.M.: Multiple hypothesis tracking revisited. In: 2015 IEEE International Conference on Computer Vision (ICCV), pp. 4696–4704, December 2015. https://doi.org/10.1109/ICCV.2015.533
14. Kuhn, H.W.: The hungarian method for the assignment problem. Naval Res. Logistics Quarterly **2**(1–2), 83–97 (1955)
15. Lee, B., Erdenee, E., Jin, S., Nam, M.Y., Jung, Y.G., Rhee, P.K.: Multi-class multi-object tracking using changing point detection. In: Hua, G., Jégou, H. (eds.) ECCV 2016. LNCS, vol. 9914, pp. 68–83. Springer, Cham (2016). https://doi.org/10.1007/978-3-319-48881-3_6
16. Li, W., Zhao, R., Xiao, T., Wang, X.: Deepreid: deep filter pairing neural network for person re-identification. In: Proceedings of the IEEE Conference on Computer Vision and Pattern Recognition, pp. 152–159 (2014)
17. Li, Y., Huang, C., Nevatia, R.: Learning to associate: hybridboosted multi-target tracker for crowded scene. In: 2009 IEEE Conference on Computer Vision and Pattern Recognition, pp. 2953–2960, June 2009. https://doi.org/10.1109/CVPR.2009.5206735
18. Zhang, L., Li, Y., Nevatia, R.: Global data association for multi-object tracking using network flows. In: 2008 IEEE Conference on Computer Vision and Pattern Recognition, pp. 1–8, June 2008. https://doi.org/10.1109/CVPR.2008.4587584
19. Liu, X., Liu, W., Ma, H., Fu, H.: Large-scale vehicle re-identification in urban surveillance videos. In: 2016 IEEE International Conference on Multimedia and Expo (ICME), pp. 1–6, July 2016
20. Lyu, S., et al.: UA-DETRAC 2017: Report of AVSS2017 & IWT4S challenge on advanced traffic monitoring. In: 2017 14th IEEE International Conference on Advanced Video and Signal Based Surveillance (AVSS), pp. 1–7. IEEE (2017)
21. Milan, A., Roth, S., Schindler, K.: Continuous energy minimization for multitarget tracking. IEEE Trans. Pattern Anal. Mach. Intell. **36**(1), 58–72 (2014). https://doi.org/10.1109/TPAMI.2013.103
22. Murthy, J.K., Sharma, S., Krishna, K.M.: Shape priors for real-time monocular object localization in dynamic environments. In: 2017 IEEE/RSJ International Conference on Intelligent Robots and Systems (IROS), pp. 1768–1774. IEEE (2017)
23. Ren, J., et al.: Accurate single stage detector using recurrent rolling convolution. http://arxiv.org/abs/1704.05776

24. Ren, S., He, K., Girshick, R., Sun, J.: Faster R-CNN: towards real-time object detection with region proposal networks. Adv. Neural Inf. Process. Syst. **36**, 91–99 (2015)
25. Rosten, E., Porter, R., Drummond, T.: Faster and better: a machine learning approach to corner detection. IEEE Trans. Pattern Anal. Mach. Intell. **32**(1), 105–119 (2008)
26. Sharma, S., Ansari, J.A., Murthy, J.K., Krishna, K.M.: Beyond pixels: leveraging geometry and shape cues for online multi-object tracking. In: 2018 IEEE International Conference on Robotics and Automation (ICRA), pp. 3508–3515. IEEE (2018)
27. Shi, J., et al.: Good features to track. In: 1994 Proceedings of IEEE Conference on Computer Vision and Pattern Recognition, pp. 593–600. IEEE (1994)
28. Tian, W., Lauer, M., Chen, L.: Online multi-object tracking using joint domain information in traffic scenarios. IEEE Trans. Intell. Transport. Syst. **39**, 1–11 (2019)
29. Wang, G., Yuan, Y., Chen, X., Li, J., Zhou, X.: Learning discriminative features with multiple granularities for person re-identification. In: 2018 ACM Multimedia Conference on Multimedia Conference, pp. 274–282. ACM (2018)
30. Wojke, N., Bewley, A., Paulus, D.: Simple online and realtime tracking with a deep association metric. In: 2017 IEEE International Conference on Image Processing (ICIP), pp. 3645–3649. IEEE (2017)
31. Xiang, Y., Alahi, A., Savarese, S.: Learning to track: online multi-object tracking by decision making. In: The IEEE International Conference on Computer Vision (ICCV), December 2015
32. Yang, F., Choi, W., Lin, Y.: Exploit all the layers: fast and accurate CNN object detector with scale dependent pooling and cascaded rejection classifiers. In: The IEEE Conference on Computer Vision and Pattern Recognition (CVPR), June 2016
33. Yu, F., Li, W., Li, Q., Liu, Y., Shi, X., Yan, J.: POI: multiple object tracking with high performance detection and appearance feature. In: Hua, G., Jégou, H. (eds.) ECCV 2016. LNCS, vol. 9914, pp. 36–42. Springer, Cham (2016). https://doi.org/10.1007/978-3-319-48881-3_3

Combining Mixture Models and Spectral Clustering for Data Partitioning

Julien Muzeau(✉) , Maria Oliver-Parera , Patricia Ladret ,
and Pascal Bertolino

Université Grenoble Alpes, CNRS, Grenoble INP, GIPSA-lab, 38000 Grenoble, France
{julien.muzeau,maria.oliver-parera,patricia.ladret,
pascal.bertolino}@gipsa-lab.grenoble-inp.fr

Abstract. Gaussian Mixture Models are widely used nowadays, thanks to the simplicity and efficiency of the Expectation-Maximization algorithm. However, determining the optimal number of components is tricky and, in the context of data partitioning, may differ from the actual number of clusters. We propose to apply a post-processing step by means of Spectral Clustering: it allows a clever merging of similar Gaussians thanks to the Bhattacharyya distance so that clusters of any shape are automatically discovered. The proposed method shows a significant improvement compared to the classical Gaussian Mixture clustering approach and promising results against well-known partitioning algorithms with respect to the number of parameters.

Keywords: Gaussian Mixture model · Spectral Clustering ·
Bhattacharyya coefficient · Bayesian Information Criterion

1 Introduction

Cluster analysis is a fundamental task in data science as it allows to gather individuals that show similar features. Clustering belongs to the set of unsupervised learning methods: they are among the most challenging ones in machine learning as they aim at blindly determining the label of each point. In other words, the objective is to find out which group each point belongs to, without having any ground-truth available. Clustering has been applied to a variety of fields such as community detection, segmentation and natural language processing [9].

Due to the amount of topics clustering can be applied to, numerous techniques have been proposed to tackle this problem. They can be classified as: hierarchical (agglomerative or divisive), centroid-based, density-based, graph-based or distribution-based [17]. This work focuses on distribution-based cluster analysis algorithms, which address the problem from a statistical point of view by considering the probability density of the data. In particular, we study the case of Gaussian Mixture (GM) Clustering which models the data by a combination of normal distributions: each Gaussian represents one cluster of points.

Supported by Auvergne-Rhône-Alpes region.

A. Campilho et al. (Eds.): ICIAR 2020, LNCS 12132, pp. 63–75, 2020.
https://doi.org/10.1007/978-3-030-50516-5_6

Determining the right number of GM components only from the data is still a current topic of research. Three strategies exist regarding this challenge. The first one consists in initializing the Gaussian Mixture Model (GMM) with a low number of components and increasing it until convergence [21]. The second approach aims at iteratively merging components until a stopping criterion is met [6]. Finally, the third group of methods applies an optimization process to minimize a criterion over possible numbers of components [12]. Many criteria have been proposed such as the Bayesian Information Criterion (BIC) [19], the Minimum Message Length (MML) [20] or the Akaike Information Criterion (AIC) [1].

Nevertheless, the aforementioned approaches may suffer from overfitting. In fact, model selection is based on the minimization of $\mathcal{L}$, the likelihood function. Yet, adding more components leads to a decrease in $\mathcal{L}$. As a consequence, the resulting GMM often ends up with too many components: the model accurately represents the data density but overestimates the number of actual clusters. A fusion step can subsequently be added to reduce the number of components by merging similar Gaussians. In most cases, the GMM generates strong overlapping components and gathering them allows a simplification of the model without any loss of information. Hence, several methods have been developed.

In order to automatically detect the correct number of clusters, no matter their distribution, we propose a three-fold process. First, the data is approximated by a GMM, optimally selected through the minimization of the BIC, leading to an overfitted model with too many components compared to the number of clusters. Then, to decide if two components belong to the same group or not, the Bhattacharyya coefficients are computed to estimate the similarity between each pair of components of the GMM. Lastly, the final clusters are determined by merging similar Gaussians thanks to Spectral Clustering (SC).

The remainder of this paper is organized as follows: the idea of Gaussian Mixture is explained and detailed in Sect. 2. Section 3 describes our proposal and results are given in Sect. 4. Finally, Sect. 5 draws conclusions and gives some prospects about future work.

2   Gaussian Mixture Model

2.1   Principle

A mixture model is a probabilistic model that approximates the density of a dataset by a weighted sum of probability distributions of the same kind but differently parameterized. This article is focused on Gaussian Mixture Models, *i.e.* the aforementioned distributions are assumed to be normal in any dimension.

A Gaussian Mixture Model $\mathcal{M}$ with K components can be defined as

$$\mathcal{M} = \sum_{i=1}^{K} \pi_i \, \mathcal{N}(\boldsymbol{\mu}_i; \boldsymbol{\Sigma}_i), \tag{1}$$

where π_i is the weight associated to the ith component with $\sum_{i=1}^{K} \pi_i = 1$, $\mathcal{N}(\boldsymbol{\mu}_i; \boldsymbol{\Sigma}_i)$ is the multivariate normal distribution with mean $\boldsymbol{\mu}_i \in \mathbb{R}^n$ and covariance matrix $\boldsymbol{\Sigma}_i \in \mathbb{R}^{n \times n}$, and n represents the dimensionality of the data to be modeled. In other words, for any vector $\boldsymbol{x}$ in $\mathbb{R}^n$,

$$p(\boldsymbol{x}|\boldsymbol{\mu}_i, \boldsymbol{\Sigma}_i) = \frac{1}{(2\pi)^{n/2}|\boldsymbol{\Sigma}_i|^{1/2}} \exp\left(-\frac{C_i}{2}\right), \tag{2}$$

with $C_i = (\boldsymbol{x} - \boldsymbol{\mu}_i)^T \boldsymbol{\Sigma}_i^{-1} (\boldsymbol{x} - \boldsymbol{\mu}_i)$.

2.2 Expectation-Maximization Algorithm

When the number of components K is known, it is possible to determine such a mixture model only from the data $\{\boldsymbol{x}_i \in \mathbb{R}^n, i = 1, \ldots, N\}$, as defined in Eq. (1). To do this, one makes use of the Expectation-Maximization (EM) algorithm developed by Dempster *et al.* in 1977 [4]. It consists in determining the parameters of the Gaussian Mixture, namely $\pi_1, \ldots, \pi_K$, $\boldsymbol{\mu}_1, \ldots, \boldsymbol{\mu}_K$, $\boldsymbol{\Sigma}_1, \ldots, \boldsymbol{\Sigma}_K$, that maximize the likelihood function associated to $\mathcal{M}$, in an iterative manner.

In real terms, the algorithm can be broken down into four steps.

1. **Parameters initialization**
   The classical approaches set $\pi_1 = \cdots = \pi_K = \frac{1}{K}$ and $\boldsymbol{\Sigma}_1 = \cdots = \boldsymbol{\Sigma}_K = \boldsymbol{\Sigma}$ (the whole data covariance matrix). Regarding the means, it is common to initialize them with randomly chosen data points.
2. **E(xpectation) step**
   One computes here the probability γ_{ik} of data at index i being generated by component k for all i in $\{1, \ldots, N\}$ and for all k in $\{1, \ldots, K\}$. That is

$$\gamma_{ik} = \frac{\pi_k \, \mathcal{N}(\boldsymbol{x}_i|\boldsymbol{\mu}_k; \boldsymbol{\Sigma}_k)}{\sum_{j=1}^{K} \pi_j \, \mathcal{N}(\boldsymbol{x}_i|\boldsymbol{\mu}_j; \boldsymbol{\Sigma}_j)}. \tag{3}$$

3. **M(aximization) step**
   The parameters are updated thanks to the previously computed probabilities. For each component $k \in [\![1; K]\!]$:

$$\pi_k = \frac{1}{N} \sum_{i=1}^{N} \gamma_{ik}, \tag{4}$$

$$\boldsymbol{\mu}_k = \frac{\sum_{i=1}^{N} \gamma_{ik} \boldsymbol{x}_i}{\sum_{i=1}^{N} \gamma_{ik}}, \tag{5}$$

$$\boldsymbol{\Sigma}_k = \frac{\sum_{i=1}^{N} \gamma_{ik}(\boldsymbol{x}_i - \boldsymbol{\mu}_k)^T(\boldsymbol{x}_i - \boldsymbol{\mu}_k)}{\sum_{i=1}^{N} \gamma_{ik}}. \tag{6}$$

4. Steps 2 and 3 are repeated until convergence.

2.3 High-Dimensional Gaussian Mixture Model

Even though the EM algorithm returns accurate outputs in most cases, issues occur when working in high-dimensional spaces (when the dimensionality is greater than five). As a matter of fact, what is called the "curse of dimensionality" causes a breakdown of any kind of distances as the number of dimensions increases. A small value, repeated by the dimensionality, leads to a huge error.

This issue appears in the Gaussian Mixture models and the EM algorithm. In particular, the Mahalanobis distance C_i of Eq. (2) suffers from this phenomenon, especially due to the covariance matrix. In fact, the number of elements in such a matrix increases quadratically with respect to the dimensionality, enforcing small errors to accumulate, thus leading to an unrealistic Mahalanobis distance.

To overcome this challenge, the algorithm developed in [16] is used. The idea is to add a regularization parameter in the maximization step of the EM algorithm (see Subsect. 2.2). More precisely, after the covariance matrix of each component is computed, the Graphical Lasso method, proposed by Friedman et al. [8], is applied to each of those matrices. The idea is as follows: an l_1-norm penalization term imposes the covariance matrices to become sparse, that means with a maximum of zeros outside of the diagonal. Eventually, the elements close to zero, that would yield small errors, are set to zero: an immediate consequence is the vanishing of the errors that disturbs the final computed distance.

2.4 Clustering via GMM

In addition to its ability to approximate data set distributions, clustering by Gaussian Mixture model is also possible. For instance, the three blobs pictured by black crosses in Fig. 1a can readily be partitioned thanks to a GMM with 3 components whose covariances are represented by red ellipses. Each point is associated to one Gaussian only and the three clusters are retrieved.

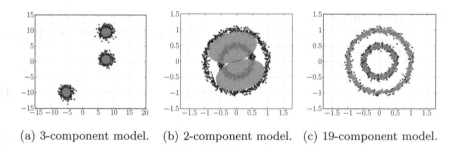

(a) 3-component model.   (b) 2-component model.   (c) 19-component model.

Fig. 1. 2-D data sets and associated Gaussian Mixture models. (Color figure online)

However, this kind of clustering fails in cases where the different data groups are not spherical. As a case of point, when applied to two concentric circles of different radii (see Fig. 1b), the method completely misses the data distribution and influences the subsequent partitioning in the wrong direction.

Several remarks have to be done at this point. First, this issue occurs in this case because the true clusters (the two rings) share the same mean. Moreover, the number of components is chosen equal to the number of clusters (2 here). It implies that the number of groups is available prior to the execution, which is unrealistic in most practical situations. It also shows that setting the number of components equal to the number of clusters may be inappropriate.

3   Gaussian Spectral Clustering

In this section, we propose a parameter-free Gaussian Mixture based clustering method: Gaussian Spectral Clustering (GSC). The idea is twofold: first, the input data is modeled by a GMM, whose number of components exceeds the actual number of clusters. Then, those Gaussians are merged in a smart manner, thanks to Spectral Clustering [13], to discover the real clusters.

3.1   Assess Mixture Model Quality

As pointed out in Subsect. 2.4, the choice of the number of components of a mixture model is critical and is often not related to the number of distinct data groups. We propose an exhaustive search for the optimal data modeling: more precisely, we try several models on the input data (or equivalently, several number of components) and select the one that fits the data the best. We remind that the number of components may mismatch the true number of clusters.

A question which rightfully arises next is the choice of the criterion that allows model selection. Many techniques have been developed through the years to address this issue [12]. We propose to use the Bayesian Information Criterion (BIC) [19] because it does not underestimate (asymptotically) the number of true components [11] and the resulting density estimate is consistent with the groundtruth [10,15]. It is defined as:

$$BIC = t \ln N - 2 \ln \mathcal{L}, \tag{7}$$

where t is the number of parameters to be estimated in the model, N the number of data points and $\mathcal{L}$ the likelihood function associed to the model $\mathcal{M}_K$.

The model leading to the lowest BIC value is assumed to be optimal, as it perfectly fits the original data. In particular, in the situation shown in Fig. 1b and c, the minimum is reached for 19 components and such a model adapts better to the data. According to definition (7), one can understand that this criterion is a trade-off between how good the model fits the data, represented by the likelihood function $\mathcal{L}$, and its complexity, embodied in t (the number of parameters).

3.2   Determine Gaussians Similarity

The main consequence of determining the optimal GM model through the minimization of the BIC value lies in the fact that the final number of components
may not be indicative of the actual number of clusters present in the data. Due to
the fact that a normal distribution is only able to accurately model an elliptical
data group, one can see a Gaussian Mixture model as an approximation of a data
set distribution by Gaussians. It follows that the number of components of the
optimal mixture model is necessarily greater than or equal to the true number of
clusters, as non-elliptical data clusters are decomposed into several Gaussians.
It is also clear that normal distributions which belong to the same cluster show
similarities among themselves, contrary to those from distant clusters.

Consider for example the 19-component GMM (see Fig. 1c) of the two concentric circles depicted in Fig. 1. This model fits the data better than the 2-
component one, although the data is composed of 2 clusters. Moreover, the
Gaussians which belong to the outer ring are similar to each other, pair by pair,
but differ from the ones of the inner ring, and vice-versa. Merging them into one
only set leads to a better data partitioning.

In order to measure similarity between Gaussians, we introduce in this
subsection the Bhattacharyya distance and coefficient [3]. The Bhattacharyya
distance d_B and coefficient c_B between two multivariate normal distributions
$p \sim \mathcal{N}(\boldsymbol{\mu_p}, \boldsymbol{\Sigma_p})$ and $q \sim \mathcal{N}(\boldsymbol{\mu_q}, \boldsymbol{\Sigma_q})$ are defined as:

$$d_B(p,q) = \frac{1}{8}(\boldsymbol{\mu_p} - \boldsymbol{\mu_q})^T \boldsymbol{\Sigma}^{-1}(\boldsymbol{\mu_p} - \boldsymbol{\mu_q}) + \frac{1}{2}\ln\frac{|\boldsymbol{\Sigma}|}{\sqrt{|\boldsymbol{\Sigma_p}||\boldsymbol{\Sigma_q}|}}, \tag{8}$$

$$c_B(p,q) = \exp\left(-d_B(p,q)\right), \tag{9}$$

with $\boldsymbol{\Sigma} = (\boldsymbol{\Sigma_p} + \boldsymbol{\Sigma_q})/2$ and $|.|$ the determinant of a (square) matrix.

A geometrical interpretation is to be drawn from this coefficient: it actually
approximates the overlap ratio between two statistical distributions (normal
in our case). It approaches 1 when the two compared distributions are quasi-
identical and tends towards 0 in the case of two dissimilar ones.

Let us now suppose that, from a specific data set, an optimal Gaussian Mix-
ture model with C components $\mathcal{N}_1, \ldots, \mathcal{N}_C$ is determined, as explained in Subsect. 3.1. Consequently, we can build the similarity matrix $\boldsymbol{S} = (S_{ij})_{i=1...C,j=1...C}$
whose elements are equal to the pairwise Bhattacharyya coefficients. Precisely:

$$S_{ij} = \begin{cases} c_B(\mathcal{N}_i, \mathcal{N}_j) & \text{if } i \neq j \\ 0 & \text{if } i = j \end{cases} \quad \forall (i,j) \in [\![1; C]\!]^2. \tag{10}$$

This gives a matrix of correspondences, where each coefficient reflects the
similarity between the two Gaussians involved. One can also highlight that this
matrix indeed is symmetric. The next subsection details the method used to
decide when Gaussians are overlapping enough to be considered as belonging to
the same cluster and to be merged.

3.3   Apply Spectral Clustering

The similarity matrix defined in the previous subsection (Eq. (10)) embeds the similarity between each pair of Gaussians.

In the perspective of clustering, the idea that comes next consists in partitioning this matrix so that sets of significant overlapping normal distributions are discovered. In other words, we want to determine clusters among which a "path" from one Gaussian to another is readily available, either directly or through other Gaussians from the same cluster.

This objective can be achieved by various means, we propose in this paper to make use of the spectral clustering approach proposed in [13]. Assuming the number N_c of clusters to be discovered, five steps have to be executed.

1. Normalize S by $L = D^{-1/2} S D^{-1/2}$ where D is the row-wise sum of S (its non-diagonal elements equal 0).
2. Obtain $V \in \mathbb{R}^{C \times C}$ the eigenvectors of L. We assume they are sorted in descending order of the eigenvalues.
3. Crop V as to keep the N_c largest eigenvectors: X is then of size $C \times N_c$.
4. Obtain Y by normalizing the rows of X. In the end, the norm of each row of Y is equal to 1.
5. Apply K-means algorithm [14] to Y, considering each row as a data point.

From this point, the rows of Y, which correspond to the normal distributions of the Gaussian Mixture model, are partitionned into N_c clusters. Consequently, the label of each GMM component is modified according to the K-means clustering output. Each original data point label is affected the same way.

The central disadvantage of spectral clustering is the need to specify the number of clusters. It implies an interaction with the user and the prior knowledge of how many groups are hidden in the data: this last piece of information is surrealist in most practical applications, especially in higher dimensions.

To overcome this issue, many ideas have been proposed. We propose to iterate over steps 3–5 from the spectral clustering algorithm: the idea is to provide an exhaustive search for the number of clusters, from 1 to C. At each iteration, the distortion of the K-means output (*i.e.* the sum of squared distances from each data point to the centroid of its cluster) is computed: the number of clusters yielding the lowest distortion value is assumed to be the actual number of groups.

At this point, the authors would like to highlight two elements from the exhaustive search step. First of all, the idea seems to show similarities with internal clustering validation measures (as a reminder, such a metric aims at comparing two partitionning of the same data set, possibly computed by two different algorithms, without any ground-truth). However, in our case, the data is evolving at each iteration, the dimensionality is increasing as well. Secondly, as was said earlier, it is almost impossible in real life scenarios to have a guess about the number of clusters. Nonetheless, a range of possible values seems more reasonable. This algorithm allows the inclusion of such prior knowledge which leads to a process acceleration.

3.4   Summary

We summarize in this subsection the proposed method. Given as input the data set $\{\boldsymbol{x}_i \in \mathbb{R}^n, i = 1 \ldots N\}$, our algorithm is made of the three following steps:

1. Determine several Gaussian Mixture models of the data with an increasing number of components. Keep the model yielding the lowest BIC.
2. Compute the similarity matrix $\boldsymbol{S}$ (Bhattacharyya coefficient between each pair of Gaussians).
3. Apply spectral clustering on $\boldsymbol{S}$ for different number of clusters. Keep the value which leads to the lowest K-means distortion.

The proposal is fully non-parametric: it is however possible to add, if available, constraints about the number of components in the mixture model or about the number of data clusters.

4   Experiments

This section is devoted to the comparison of our approach with other methods. Subsection 4.2 is divided in three groups. First, we compare GSC with GMM, as we want to show that our method outperforms GMM. Secondly, as most techniques need as input the number of clusters, we devise an extension of our method in this direction and compare the performance of GSC in both situations. Finally, we compare our method against well-known clustering algorithms.

4.1   Databases and Assessment Metric

To evaluate our method, we use more than 100 datasets from the `Clustering benchmark`[1]. We then compare our predicted partitioning with the available ground-truth. More precisely, we use the Fowlkes-Mallows (FM) score [7]:

$$FM = \sqrt{\frac{TP}{TP + FP} \times \frac{TP}{TP + FN}}, \tag{11}$$

where TP is the number of true positives, FP the number of false positives and FN the number of false negatives. It equals 1 in the case of a predicted clustering similar to the ground-truth, 0 otherwise.

For the following experiments, the range of the number of components in which to seek the optimal GMM is fixed to $[1; 75]$. Moreover, the SC stage of our algorithm tries all numbers of clusters from 1 to C, where C is the number of components of the optimal mixture model determined in the first stage.

---

[1] https://github.com/deric/clustering-benchmark.

4.2   Results

Comparison Against GM Clustering. As a preliminary, Fig. 2 displays a visual comparison between the proposed algorithm and the classical Gaussian Mixture (GM) approach on a few bidimensional synthetic datasets. One can observe identical results for the second and fourth columns. However, our method (top row) outperforms GM clustering (bottom row) in the other two cases: the latter is indeed unable to retrieve non-elliptical data clusters. We also highlight that, unlike GM clustering for which the number of clusters has to be specified by the user, GSC accurately discovers the right number of groups autonomously.

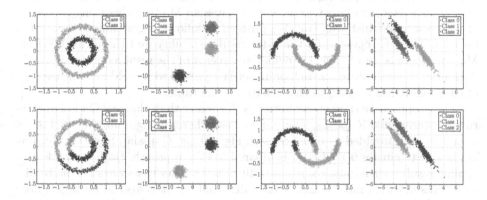

Fig. 2. Comparison of our approach (top row) with the classical GM clustering (bottom row) on 4 datasets. Each marker/color represents a different class. (Color figure online)

Comparison of Two Variants of GSC. We compare in this section the results given by the proposed approach in two situations: (1) when the number of clusters is given as input, denoted as GSC_gt, and (2) when it is totally parameter-free, denoted as GSC_free. For this purpose, we make use of the histogram of the Fowlkes-Mallows scores computed over all the datasets. Fig. 3 depicts the superposition of both histograms and both means (dashed lines).

As expected, GSC_gt outperforms GSC_free. Actually, the mean over all the datasets obtained by GSC_gt is 0.8604 while the one for GSC_free equals 0.7783. This difference stems from the K-Means iterations within Spectral Clustering (step 3 of Subsect. 3.4) and can be explained according to three factors:

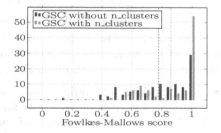

Fig. 3. Histograms of GSC_gt (in orange) and GSC_free (in blue). (Color figure online)

1. K-Means is applied on a small number of data points, namely the number of components in the optimal Gaussian Mixture Model.
2. The distortion only takes into account the distances between the points and their associated centroid. The number of clusters and the dimensionality are not included in this computation.
3. The limit case where K-Means algorithm is launched with the same number of clusters than the number of data points leads to a zero distortion, it is then considered the best clustering for any data set.

Comparison Against Other Clustering Algorithms. In order to provide a fair comparison, we put ourselves in conditions where the number of clusters is known, as it is a mandatory argument in Gaussian Mixture clustering, K-means [2] and Spectral Clustering [13]. We proceed as follows: algorithms are executed for each dataset, repeated 50 times for results stability and the mean Fowlkes-Mallows score is kept. Table 1 provides an excerpt of the results obtained over 23 datasets. The best score for each point set is represented in bold and the average of each algorithm is given in the last row of the table.

The proposed method shows better performance than classical Gaussian Mixture clustering and K-means. Thus, in the case where the right number of clusters is known and provided as input to both algorithms, Gaussian Spectral Clustering is able to cluster 9.1% better compared to GM approach. Both methods give similar results when data groups are elliptical, in other cases ours is able to discover clusters with more complex data distribution. It is also to be highlighted that GM clustering performs better than K-means as the latter is a specific case of the former for spherical clusters. In order to show the outperformance of GSC over GM, we perform a z-test, as we know the variance and the mean of the distributions and both of them follow a gaussian. Given the means, $\mu_{GSC} = 0.860$ and $\mu_{GM} = 0.798$, and the standard deviations, $\sigma_{GSC} = 0.171$ and $\sigma_{GM} = 0.197$, of both models computed on $n = 101$ datasets, we consider the null hypothesis $\mathcal{H}_0$: $\mu_{GM} = \mu_{GSC}$ and the alternative hypothesis $\mathcal{H}_1$: $\mu_{GM} < \mu_{GSC}$. We want to show with a 99% confidence ($u_{0.01} = 2.33$) that $\mathcal{H}_1$ holds. Then,

$$z = \frac{\mu_{GM} - \mu_{GSC}}{\sqrt{\sigma_{GM}^2/n + \sigma_{GSC}^2/n}} = -2.735 < -2.33 = -u_{0.01}. \tag{12}$$

Thus, we can reject the null hypothesis $\mathcal{H}_0$.

GSC shows better performance than DBSCAN [5,18] but is outpaced by Spectral Clustering. It is however important to keep in mind that several parameters have to be set for those methods:

- In DBSCAN, the radius ε of a spherical neighborhood and the number $MinPts$ of data points in order for such a neighborhood to be valid.
- In SC, the standard-deviation of the RBF kernel and the number of clusters.

A quasi-exhaustive search over all the parameters is conducted, only the greatest FM score is kept in memory. Such a process is usually unpractical in most real-world clustering applications. We remind that our method requires, at this point, only the number of clusters and is automatic except from this parameter.

Table 1. Fowlkes-Mallows score of our method and four others on 23 datasets. The number of GMM components determined by our method is displayed in the first column. The average score for each datasets is reported in the last row.

Dataset		GSC (ours)	DBSCAN	GM	Kmeans	SC
2d-20c-no0	(22)	0.987	0.9705	0.9479	0.9658	**0.9916**
2d-3c-no123	(4)	**0.9814**	0.8897	0.937	0.8206	0.9405
2d-4c-no4	(6)	**0.9999**	0.9625	0.8985	0.9819	0.9922
2d-4c-no9	(5)	1	0.8961	0.9915	0.9241	0.9752
2d-4c	(4)	1	1	1	1	1
curves1	(14)	1	1	0.499	0.499	1
curves2	(13)	**0.9065**	0.3763	0.9047	0.9465	0.952
dartboard2	(63)	0.5698	**0.9246**	0.547	0.5465	0.5484
donut1	(21)	**0.9958**	0.991	0.5263	0.5046	0.992
elly-2d10c13s	(8)	0.9367	0.707	**0.9368**	0.9076	0.929
engytime	(2)	0.6141	0.732	0.6895	0.736	**0.9696**
pathbased	(7)	0.6923	**0.9797**	0.8556	0.7984	0.9794
pmf	(6)	0.9916	0.9251	0.9759	**0.9932**	**0.9932**
spherical_5_2	(4)	1	1	0.9891	1	1
spherical_6_2	(6)	1	1	0.5281	0.5162	1
square1	(4)	0.9411	0.4992	0.9411	0.9372	**0.9469**
square2	(4)	0.8925	0.4992	0.8922	0.8943	**0.8978**
tetra	(4)	0.6499	**0.7068**	0.5956	0.574	0.7053
twenty	(20)	0.9999	0.7067	1	1	1
twodiamonds	(6)	0.8651	0.947	0.9315	0.9297	1
wingnut	(5)	0.9953	0.9421	0.9953	0.9953	**0.9993**
zelnik3	(6)	1	1	1	0.6467	1
zelnik5	(4)	0.8098	1	0.7889	0.7907	0.8078
Average		0.8604	0.8516	0.7887	0.76	**0.884**

In summary, our model gives competitive results in most datasets but three, namely DARTBOARD2, ENGYTIME and TETRA. Those bad scores are mainly caused by inaccurate GM models due to strongly overlapping clusters or data groups with too few points.

5   Conclusion and Prospects

We propose in this article an improvement of GM clustering. Our method combines a modelling of a specific data set by a mixture of weighted normal distributions. Then, similar Gaussians are merged using the SC algorithm. Our model, named Gaussian Spectral Clustering (GSC), is able to retrieve clusters with complex shape, contrary to GM partitioning which is limited to elliptical data groups. Constraints on the number of components, the number of clusters or both, can be applied in order to speed GSC up and obtain realistic results regarding to specific applications. A non-parametric algorithm is also derived.

Several leads for improvement will be considered for future work. First, the selection of the optimal GMM, *i.e.* trying all models within a range of possible ones, is naive and expensive. Moreover, the random initialization of the EM algorithm makes it difficult to obtain stable results. Finally, since the estimation of the number of clusters in the SC step is prone to errors, an auto-determination technique or another evaluation criterion for K-Means may be used.

References

1. Akaike, H.: A new look at the statistical model identification. IEEE Trans. Auto. Control **19**(6), 716–723 (1974)
2. Arthur, D., Vassilvitskii, S.: K-means++: the advantages of careful seeding. In: ACM-SIAM Symposium on Discrete Algorithms, January 2007
3. Bhattacharyya, A.: On a measure of divergence between two statistical populations defined by their probability distributions. Bull. Calcutta Math. Soc. **7**, 99–109 (1943)
4. Dempster, A.P., Laird, N.M., Rubin, D.B.: Maximum likelihood from incomplete data via the em algorithm. J. Royal Stat. Soc. **39**(1), 1–38 (1977)
5. Ester, M., Hans-Peter, K., Sander, J., Xu, X.: A density-based algorithm for discovering clusters in large spatial databases with noise. In: International Conference on Knowledge Discovery and Data Mining, pp. 226–231, December 1997
6. Figueiredo, M., Jain, A.: Unsupervised learning of finite mixture models. IEEE Trans. Pattern Anal. Mach. Intell. **24**(3), 381–396 (2002)
7. Fowlkes, E.B., Mallows, C.L.: A method for comparing two hierarchical clusterings. J. Am. Stat. Assoc. **78**(383), 553–569 (1983)
8. Friedman, J., Hastie, T., Tibshirani, R.: Sparse inverse covariance estimation with the graphical lasso. Biostatistics **9**(3), 432–441 (2008)
9. Ghosal, A., Nandy, A., Das, A.K., Goswami, S., Panday, M.: A short review on different clustering techniques and their applications. In: Mandal, J., Bhattacharya, D. (eds.) Emerging Technology in Modelling and Graphics, pp. 69–83. Springer, Singapore (2020). https://doi.org/10.1007/978-981-13-7403-6_9
10. Keribin, C.: Consistent estimation of the order of mixture models. Sankhyā: The Indian Journal of Statistics, Series A, pp. 49–66 (2000)
11. Leroux, B.G.: Consistent estimation of a mixing distribution. Ann. Stat. **20**, 1350–1360 (1992)
12. McLachlan, G.J., Rathnayake, S.: On the number of components in a gaussian mixture model. Wiley Interdisciplinary Rev. Data Min. Knowl. Disc. **4**(5), 341–355 (2014)
13. Ng, A.Y., Jordan, M.I., Weiss, Y.: On spectral clustering: analysis and an algorithm. In: Advances in Neural Information Processing Systems, pp. 849–856 (2001)
14. Pelleg, D., Moore, A.: X-means: extending k-means with efficient estimation of the number of clusters. In: International Conference on Machine Learning, pp. 727–734 (2000)
15. Roeder, K., Wasserman, L.: Practical bayesian density estimation using mixtures of normals. J. Am. Stat. Assoc. **92**(439), 894–902 (1997)

16. Ruan, L., Yuan, M., Zou, H.: Regularized parameter estimation in high-dimensional gaussian mixture models. Neural Comput. **23**(6), 1605–1622 (2011)
17. Saxena, A., et al.: A review of clustering techniques and developments. Neurocomputing **267**, 664–681 (2017)
18. Schubert, E., Sander, J., Ester, M., Kriegel, H.P., Xu, X.: Dbscan revisited, revisited: why and how you should (still) use dbscan. ACM Trans. Database Syst. **42**(3), 1–21 (2017)
19. Schwarz, G.: Estimating the dimension of a model. Ann. Stat. **6**(2), 461–464 (1978)
20. Wallace, C.S.: Statistical and Inductive Inference by Minimum Message Length. Springer, New York (2005). https://doi.org/10.1007/0-387-27656-4
21. Zhang, Z., Chen, C., Sun, J., Chan, K.L.: EM algorithms for gaussian mixtures with split-and-merge operation. Pattern Recogn. **36**(9), 1973–1983 (2003)

MSPNet: Multi-level Semantic Pyramid Network for Real-Time Object Detection

Ji Li and Yingdong Ma[✉]

Inner Mongolia University, The Inner Mongolia Autonomous Region, College Road no. 235, Hohhot, China
csmyd@imu.edu.cn

Abstract. With increasing demand of running Convolutional Neural Networks (CNNs) on mobile devices, real-time object detection has made great progress in recent years. However, modern approaches usually compromise detection accuracy to achieve real-time inference speed. Some light weight top-down CNN detectors suffer from problems of spatial information loss and lack of multi-level semantic information. In this paper, we introduce an efficient CNN architecture, the Multi-level Semantic Pyramid Network (MSPNet), for real-time object detection on devices with limited resource and computational power. The proposed MSPNet consists of two main modules to enhance spatial details and multi-level semantic information. The multi-scale feature fusion module integrates different level features to tackle the problem of spatial information loss. Meanwhile, a light weight multi-level semantic enhancement module is developed which transforms multiple layer features to strengthen semantic information. The proposed light weight object detection framework has been evaluated on CIFAR-100, PASCAL VOC and MS COCO datasets. Experimental results demonstrate that our method achieves state-of-the-art results while maintains a compact structure for real-time object detection.

Keywords: Real-time object detection · Multi-scale feature fusion · Multi-level semantic information

1 Introduction

Real-time object detection is a fundamental computer vision task. With rapid development of mobile devices, there are increasing interests in designing Convolutional Neural Network models (CNNs) for speed sensitive applications, such as robotics, video surveillance, autonomous driving and augmented reality. Real-time object detection on mobile devices is a challenging task due to state-of-the-art CNNs require high computational resources beyond the capabilities of many mobile and embedded devices.

To tackle the problem, some light weight networks adopt small backbone and simple structure that compromise detection accuracy to inference speed. For example, the Light-head R-CNN [1] implemented real-time detection by using a small backbone. However, small backbone makes the network prone to overfitting. Iandola et al. proposed the

A. Campilho et al. (Eds.): ICIAR 2020, LNCS 12132, pp. 76–88, 2020.
https://doi.org/10.1007/978-3-030-50516-5_7

SqueezeNet [2] which uses a fire module to reduce parameters and computational cost. Though the method is simple and effective, the lack of effective spatial details leads to accuracy degradation. Meanwhile, it is difficult for light weight networks with small backbone to provide feature maps with large receptive field. Global Convolution Network [3] utilizes "large kernel" to enlarge the receptive field, while it leads to a sharp increase in computational cost. ICNet [4] adopts a multi-branch framework, in which coarse prediction map obtained from deep feature maps are refined by medium- and high-resolution features. This method enhances spatial details but the semantic information is computed mainly from deep feature maps. In fact, feature maps in previous layers not only contain spatial detail but have different level semantic information. With these observations, we aim to implement a CNN to achieve accurate object detection while maintain compact architecture. The proposed multi-level semantic pyramid network consists of two modules to integrate multiple layer features. The overall framework is shown in Fig. 1. The main contributions of this paper are summarized as follows:

1. We propose a light weight network architecture that consists of a multi-scale feature fusion (MFF) module to preserve spatial information and a multi-level semantic enhancement (MSE) module to extract different level semantic features. The new model enhances network representation ability for both fine-level spatial details and high-level semantic information. Meanwhile, the proposed model maintains a compact structure for real-time object detection.
2. The multi-scale feature fusion module integrates different scale features to enrich spatial information. In the light weight multi-level semantic enhancement module, features of various layers are transformed to shallow, medium and deep features. Shallow features and deep features are further combined to compute global semantic clues. These features are aggregated to generate semantic segmentation maps which are used as semantic guidance to improve detection performance.
3. With 304 × 304 input images, MSPNet achieves 78.2% mAP on the PASCAL VOC 07/12 and 30.1% AP on the MS COCO datasets, outperforming state-of-the-art light weight object detectors. Furthermore, experiments demonstrate the efficiency of MSPNet, e.g. our module operates at 126 frames per second (FPS) on PASCAL VOC 2007 with a single GTX1080Ti GPU.

2 Related Work

2.1 Light Weight Deep Neural Network

To construct a compact detector, some models either compress and prune typical CNNs or adopt a light weight structure. As an example, the MobileNet [5, 6] utilizes depthwise separable convolution to build a light weight deep neural network. Other light weight networks, such as ShuffleNet [7, 8] reduces computational cost by using the pointwise group convolution and adopt the channel shuffle operation to obtain feature maps from different groups. The ThunderNet adopts a light weight backbone and a compressed RPN network with discriminative feature representation to realize effective two-stage detector [9]. In [10], Wang et al. proposed the PeleeNet which consists of a stem block and a set of dense blocks. Different from other light weight networks, the PeleeNet adopts conventional convolutions to achieve efficient architecture.

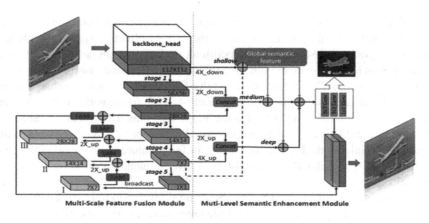

Fig. 1. The multi-level semantic pyramid network. The main structure is composed of the feature fusion module (left half part) and the semantic enhancement module (right half part). LAM: light weight attention module.

2.2 Multi-layer Feature Fusion

Combination of multi-layer feature maps is a common method to make better use of different level features. Lin et al. developed a top-down feature pyramid architecture with lateral connections to construct the FPN model [11]. The feature pyramid enhances semantic information by combing semantically strong deep feature maps with high-resolution shallow feature maps. In the SSD network [12], additional layers are added to the baseline model to enrich semantic clues. The network makes prediction from multiple layer feature maps with different resolutions. DSSD adds deconvolutional layers to the top of SSD network to build a U-shape structure [13]. These feature maps are then combined with different scale feature maps to enhance different level features. The STDN [14] adopts multiple layer features with different resolutions and the scale-transfer layers to generate a large size feature map. In [15], Kong et al. proposed a light weight reconfiguration architecture to combine multiple level features. The architecture employs global attention to extract global semantic features which followed by a local reconfiguration to model low-level features.

2.3 Attention Mechanism

The attention mechanism has been successfully applied in many computer vision tasks to boost the representational power of CNNs. The residual attention network [16] is built by stacking multiple attention modules to generate attention-aware features. The trunk branch performs feature processing and the soft mask branch learns weight for output features. In [17], Hu et al. proposed a compact squeeze and excitation structure to adjust output response by modeling the relationship between channel features. The channel-wise attention learned by SENet is used to select important feature channels. In CBAM [18] and BAM [19], attention modules are integrated with CNNs to compute attention maps in both channel and spatial dimensions. These attention modules sequentially transmit input feature map to channel attention module and spatial attention module for feature refinement. Li et al. propose the pyramid attention network which combines attention mechanism with spatial pyramid to provide pixel-level attention for high-level features extraction [20].

Table 1. Architecture of the multi-level semantic pyramid network

Stage	Output size	Operation	
Input	224 × 224 × 3		
PeleeNet_head			
Multi-scale feature fusion module			
MFF feature I	7 × 7 × 704	Broadcasting	
		Element-wise sum	
MFF feature II	14 × 14 × 704	2× up sampling	
		Element-wise sum	
MFF feature III	28 × 28 × 704	2× up sampling	
		Element-wise sum	
Global semantic feature			
Multi-level semantic enhancement module			
MSE deep feature	28 × 28 × 704	2/4× up sampling	Sum
		Concatenate	
MSE medium feature	28 × 28 × 704	2x down sampling	
		Concatenate	
MSE shallow feature	28 × 28 × 704	4× down sampling	
Classification			

2.4 Light Weight Semantic Segmentation

Wu et al. introduced a light weight context guided network for semantic segmentation [21]. The CGNet contains multiple context guided blocks which learns joint features by using a local feature extractor, a surrounding context extractor, and a global context extractor. The Light weight RefineNet implements real-time segmentation by using light weight residual convolutional units and light weight chained residual pooling [22]. By replacing 3 × 3 convolutions with 1 × 1 convolutions, the method reduces model parameters while achieving similar performance to the original RefineNet [23]. Zhang et al. proposed the detector with enriched semantics network which consists of a detection branch and a segmentation branch [24].

Different to these works, we aim to improve the discriminative power of light weight network by enriching spatial details and high-level semantic information. The proposed network extracts multi-scale spatial details and different level semantic information in two modules simultaneously. These enriched features are further integrated to generate semantic segmentation map which is used to guide object detection.

3   Multi-level Semantic Pyramid Network

In the task of object detection, both spatial details and object-level semantic information are crucial to achieve high accuracy. However, it is difficult to meet these demands simultaneously in a light weight top-down CNN structure. In this work, we introduce the multi-level semantic pyramid network to solve the problem. The proposed MSP-Net applies a feature fusion module to strengthen multi-scale spatial features. A light weight multi-level semantic enhancement module is developed, which aggregates global semantic features with different features to enhance multiple level semantic information. The overall structure of the MSPNet is shown in Table 1.

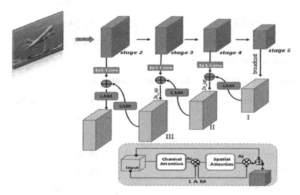

Fig. 2. The multi-scale feature fusion module.

3.1   Multi-scale Feature Fusion Module

Most light weight networks make prediction mainly on deep feature maps. While deep features provide rich semantic information, the top-down CNN structure suffers from fine-level spatial information loss. Thus, lack of suitable strategy to preserve multi-scale spatial details is one of the main issues of light weight CNNs.

We present a feature fusion module to tackle this problem. Specifically, a multi-level feature pyramid is built in the feature fusion module. The proposed MFF module refines features by aggregating multiple scale spatial features from different pyramid layers. The structure of the MFF module is shown in Fig. 2. In the module, we build a four-stage feature pyramid to fuse multiple level features. Firstly, the broadcasting is applied to the global average pooling layer. Secondly, feature maps with smaller sizes are $2\times$ upsampled by the bilinear interpolation to match spatial dimensions of previous layer feature maps. Then feature maps of two levels are merged by using element-wise add.

The Light weight Attention Module. In the proposed multi-scale feature fusion module, a light weight attention module (LAM) is applied to learn weight for multiple level features. The LAM consists of a channel attention module (CAM) and a spatial attention module (SAM) as shown in Fig. 2. The LAM sequentially transforms input features to the channel attention module and the spatial attention module. The channel and spatial attention are computed as:

$$Ac = Sigmoid(MaxPool(F)) \tag{1}$$

$$Ac' = F \otimes Ac \tag{2}$$

$$As = Sigmoid\left\{Conv^1\left(Concat\left(M^3, M^6\right)\right)\right\} \tag{3}$$

$$M^3 = MaxPool\left\{Conv^{d3}\left(Conv^1(Ac')\right)\right\} \tag{4}$$

$$M^6 = MaxPool\left\{Conv^{d6}\left(Conv^1(Ac')\right)\right\} \tag{5}$$

$$As' = Ac' \otimes As \tag{6}$$

Where, $Conv^1$ is 1×1 convolution, $Conv^{d3}$ and $Conv^{d6}$ are dilated convolution with dilation rate of 3 and 6 respectively. For the given input feature F, the refined feature map F' is computed as:

$$F' = F \otimes As' \tag{7}$$

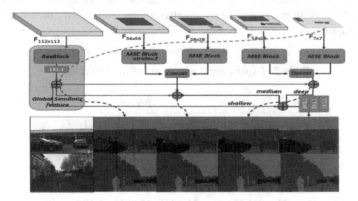

Fig. 3. The multi-level semantic enhancement module.

3.2 Multi-level Semantic Enhancement Module

We develop a light weight multi-level semantic enhancement module to further improve detection accuracy. In the semantic module, multi-level features are combined to form shallow, medium and deep features. Figure 3 shows the architecture. Specifically, features in different levels are resized to 28×28 pixels by downsampling and upsampling. The upsampled $F_{7\times7}$ and $F_{14\times14}$ are concatenated to get the deep features. Likewise, the medium features are obtained by integrating $F_{28\times28}$ features and $F_{56\times56}$ features. $F_{112\times112}$ is used as the shallow features. However, only deep features are not enough as segmentation also require object boundaries information to facilitate different scale objects localization. In addition, when feature maps propagate from top level to low level, progressively upsampling might cause information dilution. we combine the $F_{7\times7}$ features and the $F_{112\times112}$ features to yield global semantic features. These global semantic features not only contain high-level semantic cues, but also provide fine-level object location information. The deep features and medium features are then combined with global semantic features to form output features.

In the MSE module, we apply a two-path MSE block to reduce training time. For each level feature maps, the input is transformed by a stack of three convolutional layers. The two 1×1 convolutions reduces and then restores dimensions that makes the 3×3 DW_Conv [5] running on features with small dimensions. MSE Block has a shortcut path which downsamples input features by using the max pooling. The shortcut path is designed to enhance semantic information of different level features.

4   Experiments

In this section, we evaluate the effectiveness of MSPNet on CIFAR-100 [25], PASCAL VOC [26] and MS COCO [27] benchmarks. The CIFAR-100 consists of 60,000 32×32 color images in 100 classes where the training and testing sets contain 50,000 and 10,000 images respectively. We use VOC 2007 trainval and VOC 2012 trainval as the training data, and use VOC 2007 testval as the test data. For MS COCO, we use a popular split which takes trainval35k for training, minival for validation, and we report results on test-dev 2017. Ablation experiments are also conducted to verify the effectiveness of different components.

Table 2. Experimental results on CIFAR-100 dataset.

Module	Params	Error (%)
ResNet 50 [28]	23.71M	21.49
ResNet 101 [28]	42.70M	20.00
WideResNet 28 [29]	23.40M	20.40
ResNeXt 29 [30]	34.52M	18.18
DenseNet-BC-250 (k = 24) [31]	15.30M	19.64
MobileNet [5]	3.30M	18.30
ShuffleNet [7]	2.50M	16.60
PreResNet 110 [32]	1.73M	22.2
Pelee [10]	1.60M	15.90
MSPNet	**2.20M**	**10.3**

4.1 Experiments on CIFAR-100

The MSPNet is trained on the public platform TensorFlow with batch size of 128. We set the initial learning rate to 10^{-3}. The learning rate changes to 10^{-4} at 60k iterations and 10^{-5} at 100k iterations. Experimental results on CIFAR-100 dataset are shown in Table 2. The MSPNet has 10.3% error rate, about 5.6% lower than the baseline model with slight network parameter increase of 0.6M. Our model outperforms other light weight networks, e.g. compared to MobileNet and ShuffleNet, the MSPNet achieves 8.0% and 6.3% performance improvement with fewer model parameters.

Table 3. Experimental results on PASCAL VOC.

Module	Backbone	Input dimension	MFLOPs	Data	mAP (%)
R-FCN [35]	ResNet-101	600 × 1000	58900	12	77.4
HyperNet [34]	VGG-16	600 × 1000	–	07 + 12	76.3
RON384 [36]	VGG-16	384 × 384	–	12	73.0
SSD300 [12]	VGG-16	300 × 300	31750	07 + 12	77.5
SSD321 [12]	ResNet-101	321 × 321	15400	12	77.1
DSSD321 [13]	ResNet-101	321 × 321	21200	07 + 12	78.6
DES300 [24]	VGG16	300 × 300	–	12	77.1
RefineDet320 [37]	VGG-16	320 × 320	–	12	78.1
DSOD300 [38]	DenseNet	300 × 300	–	07 + 12	77.7
YOLOv2 [33]	Darknet-19	416 × 416	17400	07 + 12	76.8
YOLOv2 [33]	Darknet-19	288 × 288	8360	07 + 12	69.0
PFPNet-R320 [39]	VGG-16	320 × 320	–	12	77.7
Tiny-YOLOv2 [33]	DarkNet-19	416 × 416	3490	07 + 12	57.1
MobileNet-SSD [5]	MobileNet	300 × 300	1150	07 + 12	68.0
MobileNet-SSD [5]	MobileNet	300 × 300	1150	07 + 12 + coco	72.7
Tiny-DSOD [40]	–	300 × 300	1060	07	72.1
Pelee [10]	DenseNet-41	304 × 304	1210	07 + 12 + coco	76.4
Pelee [10]	DenseNet-41	304 × 304	1210	07 + 12	70.9
ThunderNet [9]	SNet146	320 × 320	–	07 + 12	75.1
MSPNet	**Pelee**	**304 × 304**	**1370**	**07 + 12**	**78.2**
MSPNet	**Pelee**	**512 × 512**	**1370**	**07 + 12**	**79.4**

4.2 Experiments on PASCAL VOC

We train the model with an initial learning rate of 0.05 and batch size is 32. The learning rate reduces to 0.005 at 80k iterations. Table 3 lists the experimental results on Pascal VOC dataset. We use the standard mean average precision (mAP) scores with IoU thresholds of 0.5 as evaluation metric. Our model achieves 78.2% mAP, higher than the baseline model by 7.3%. Compared to other light weight models, the proposed MSPNet also has competitive results. For example, we observe 3.1% performance improvement than the ThunderNet [9] with SNet146. Furthermore, our model has better performance than some state-of-the-art CNNs, such as SSD300 [12], YOLOv2 [33], HyperNet [34] and R-FCN [35] with significant computational cost reduction.

Table 4. Experimental results on MS COCO

Module	Backbone	Input dimension	AP	AP_{50}	AP_{75}
ResNet-50 [28]	–	320×320	26.5	47.6	–
YOLOv2 [33]	DarkNet-19	416×416	21.6	44.0	19.2
SSD300 [12]	VGG-16	300×300	25.1	43.1	25.8
SSD512 [12]	VGG-16	512×512	28.8	48.5	30.3
DSSD321 [13]	ResNet-101	321×321	28.0	46.1	29.2
MDSSD300 [41]	VGG-16	300×300	26.8	46.0	27.7
Light-head r-cnn [1]	ShuffleNetv2	800×1200	23.7	–	–
RefineDet320 [37]	VGG-16	320×320	29.4	49.2	31.3
EFIPNet [42]	VGG-16	300×300	30.0	48.8	31.7
DES300 [24]	VGG-16	300×300	28.3	47.3	29.4
DSOD300 [38]	DenseNet	300×300	29.3	47.3	30.6
RON384++ [36]	VGG-16	384×384	27.4	40.5	27.1
PFPNet-S300 [39]	VGG-16	300×300	29.6	49.6	31.1
MobileNet-SSD [5]	MobileNet	300×300	19.3	–	–
MobileNetv2-SSDLite [6]	MobileNet	320×320	22.1	–	–
MobileNetv2 [6]	–	320×320	22.7	–	–
Tiny-DSOD [40]	–	300×300	23.2	40.4	22.8
Pelee [10]	DenseNet-41	304×304	22.4	38.3	22.9
ThunderNet [9]	SNet146	320×320	23.6	40.2	24.5
ShuffleNetv1 [7]	–	320×320	20.8	–	–
ShuffleNetv2 [8]	–	320×320	22.7	–	–
MSPNet	**Pelee**	$\mathbf{304 \times 304}$	**30.1**	**48.9**	**31.5**
MSPNet	**Pelee**	$\mathbf{512 \times 512}$	**35.2**	**52.1**	**36.7**

4.3 Experiments on MS COCO

We train the MSPNet with an initial learning rate of 0.05 and the batch size is 32. The learning rate changes to 0.01 at 40k iterations and 0.001 at 45k iterations. The evaluation metric of MS COCO is the average precision (AP) scores, which includes AP_{50} and AP_{75}, with IoU thresholds of 0.5 and 0.75 respectively. As shown in Table 4, MSPNet achieves 30.1% AP, surpasses most light weight networks, such as MobileNet-SSD [5] and PeleeNet [10] with the mostly same computational cost.

4.4 Ablation Experiments

Table 5 shows the results of ablation experiments. Compared to Pelee (70.9% mAP), utilizing feature fusion module alone obtains 3.4% accuracy improvement. Similarly, the third and the fourth row show that using attention module and semantic enhancement module separately increases performance by 1.2% and 3.7%, respectively. In the attention experiment, the feature fusion module and semantic enhancement module are removed and the attention module is applied to refine each stage backbone feature maps. It can be seen from these experiments that multi-scale feature fusion module is necessary to object detection as it enhances both spatial information and high-level semantic features. Utilization of the multi-scale feature fusion module and the multi-level semantic enhancement module separately yields 75.6% mAP (the sixth row) and 75.1% mAP (the seventh row), respectively. The last two experiments show that our model achieves 77.7% mAP on the PASCAL VOC dataset by utilizing the feature fusion module, the attention module, and the semantic enhancement module (without global semantic features). Integrating the global semantic features further improve detection performance to 78.2%.

Table 5. Ablation study of different modules on PASCAL VOC.

| | Multi-level semantic pyramid network | | | | |
| | MFF | | MSE | | |
	Feature fusion module	Attention module	Semantic enhancement module	Global semantic feature	mAP (%)
1	Pelee [10]				70.9
2	✓				74.3
3		✓			72.1
4			✓		74.6
5				✓	71.8
6	✓	✓			75.6
7			✓	✓	75.1
8	✓	✓	✓		77.7
9	✓	✓	✓	✓	**78.2**

5   Conclusion

In this paper, we propose a light weight network, the multi-level semantic pyramid network, to implement real-time object detection. The MSPNet consists of a multi-scale feature fusion module and a multi-level semantic enhancement module to improve network representation ability for both spatial details and multi-level semantic information. Specifically, the MFF integrates different level features to collect multi-scale spatial information. The light weight MSE transforms different level features to shallow, medium and deep features. These features are combined with global semantic features to generate semantic segmentation maps which are used as semantic guidance to improve detection performance. Experiments on different datasets demonstrate superior object detection performance of the proposed method as compared with state-of-the-art works.

References

1. Li, Z., Peng, C., Yu, G., Zhang, X., Deng, Y., Sun, J.: Light-head r-cnn: In defense of two-stage object detector. arXiv preprint arXiv:1711.07264 (2017)
2. Iandola, F.N., Han, S., Moskewicz, M.W., et al.: SqueezeNet: AlexNet-level accuracy with 50x fewer parameters and <0.5 MB model size. arXiv preprint arXiv:1602.07360 (2016)
3. Peng, C., Zhang, X., Yu, G., Luo, G., Sun, J.: Large kernel matters–improve semantic segmentation by global convolutional network. In: Proceedings of the IEEE Conference on Computer Vision and Pattern Recognition, pp. 4353–4361 (2017)
4. Zhao, H., Qi, X., Shen, X., Shi, J., Jia, J.: ICNet for real-time semantic segmentation on high-resolution images. In: Ferrari, V., Hebert, M., Sminchisescu, C., Weiss, Y. (eds.) ECCV 2018. LNCS, vol. 11207, pp. 418–434. Springer, Cham (2018). https://doi.org/10.1007/978-3-030-01219-9_25
5. Howard, A.G., Zhu, M., Chen, B., Kalenichenko, D., et al.: Mobilenets: efficient convolutional neural networks for mobile vision applications. arXiv preprint arXiv:1704.04861 (2017)
6. Sandler, M., Howard, A., Zhu, M., Zhmoginov, A., Chen, L.C.: Mobilenetv2: inverted residuals and linear bottlenecks. In: Proceedings of the IEEE Conference on Computer Vision and Pattern Recognition, pp. 4510–4520 (2018)
7. Zhang, X., Zhou, X., Lin, M., Sun, J.: Shufflenet: an extremely efficient convolutional neural network for mobile devices. In: Proceedings of the IEEE Conference on Computer Vision and Pattern Recognition, pp. 6848–6856 (2018)
8. Ma, N., Zhang, X., Zheng, H.-T., Sun, J.: ShuffleNet V2: practical guidelines for efficient CNN architecture design. In: Ferrari, V., Hebert, M., Sminchisescu, C., Weiss, Y. (eds.) Computer Vision – ECCV 2018. LNCS, vol. 11218, pp. 122–138. Springer, Cham (2018). https://doi.org/10.1007/978-3-030-01264-9_8
9. Qin, Z., et al.: ThunderNet: towards real-time generic object detection on mobile devices. In: Proceedings of the IEEE International Conference on Computer Vision, pp. 6718–6727 (2019)
10. Wang, R.J., Li, X., Ling, C.X.: Pelee: a real-time object detection system on mobile devices. In: Advances in Neural Information Processing Systems, pp. 1963–1972 (2018)
11. Lin, T.Y., Dollár, P., et al.: Feature pyramid networks for object detection. In: Proceedings of the IEEE Conference on Computer Vision and Pattern Recognition, pp. 2117–2125 (2017)
12. Liu, W., et al.: SSD: single shot multibox detector. In: Leibe, B., Matas, J., Sebe, N., Welling, M. (eds.) ECCV 2016. LNCS, vol. 9905, pp. 21–37. Springer, Cham (2016). https://doi.org/10.1007/978-3-319-46448-0_2

13. Fu, C.Y., Liu, W., Ranga, A., Tyagi, A., Berg, A.C.: Dssd: Deconvolutional single shot detector. arXiv preprint arXiv:1701.06659 (2017)
14. Tang, Y., et al.: Visual and semantic knowledge transfer for large scale semi-supervised object detection. IEEE Trans. Pattern Anal. Mach. Intell. **40**(12), 3045–3058 (2017)
15. Kong, T., Sun, F., Huang, W., Liu, H.: Deep feature pyramid reconfiguration for object detection. In: Ferrari, V., Hebert, M., Sminchisescu, C., Weiss, Y. (eds.) ECCV 2018. LNCS, vol. 11209, pp. 172–188. Springer, Cham (2018). https://doi.org/10.1007/978-3-030-01228-1_11
16. Wang, F., et al.: Residual attention network for image classification. In: Proceedings of the IEEE Conference on Computer Vision and Pattern Recognition, pp. 3156–3164 (2017)
17. Hu, J., Shen, L., Sun, G.: Squeeze-and-excitation networks. In: Proceedings of the IEEE Conference on Computer Vision and Pattern Recognition, pp. 7132–7141 (2018)
18. Woo, S., Park, J., Lee, J.-Y., Kweon, I.S.: CBAM: convolutional block attention module. In: Ferrari, V., Hebert, M., Sminchisescu, C., Weiss, Y. (eds.) ECCV 2018. LNCS, vol. 11211, pp. 3–19. Springer, Cham (2018). https://doi.org/10.1007/978-3-030-01234-2_1
19. Park, J., Woo, S., Lee, J.Y., Kweon, I.S.: Bam: Bottleneck attention module. arXiv preprint arXiv:1807.06514. (2018)
20. Li, H., Xiong, P., An, J., Wang, L.: Pyramid attention network for semantic segmentation. arXiv preprint arXiv:1805.10180 (2018)
21. Wu, T., Tang, S., Zhang, R., Zhang, Y.: CGNET: a light-weight context guided network for semantic segmentation. arXiv preprint arXiv:1811.08201 (2018)
22. Nekrasov, V., Shen, C., Reid, I.: Light-weight refinenet for real-time semantic segmentation. arXiv preprint arXiv:1810.03272 (2018)
23. Lin, G., Milan, A., Shen, C., Reid, I.: RefineNet: multi-path refinement networks for high-resolution semantic segmentation. In: Proceedings of the IEEE Conference on Computer Vision and Pattern Recognition, pp. 1925–1934 (2017)
24. Zhang, Z., Qiao, S., Xie, C., Shen, W., Wang, B., Yuille, A.L.: Single-shot object detection with enriched semantics. In: Proceedings of the IEEE Conference on Computer Vision and Pattern Recognition, pp. 5813–5821 (2018)
25. Krizhevsky, A., Hinton, G.: Learning multiple layers of features from tiny images. Computer Science Department, University of Toronto, vol. 1, no. 4, pp. 7 (2009)
26. Everingham, M., Van Gool, L., Williams, C.K., Winn, J., et al.: The pascal visual object classes (Voc) challenge. Int. J. Comput. Vis. **88**(2), 303–338 (2010). https://doi.org/10.1007/s11263-009-0275-4
27. Lin, T.-Y., et al.: Microsoft COCO: common objects in context. In: Fleet, D., Pajdla, T., Schiele, B., Tuytelaars, T. (eds.) ECCV 2014. LNCS, vol. 8693, pp. 740–755. Springer, Cham (2014). https://doi.org/10.1007/978-3-319-10602-1_48
28. He, K., Zhang, X., et al.: Deep residual learning for image recognition. In: Proceedings of the IEEE Conference on Computer Vision and Pattern Recognition, pp. 770–778 (2016)
29. Zagoruyko, S., Komodakis, N.: Wide residual networks. arXiv preprint arXiv:1605.07146 (2016)
30. Xie, S., Girshick, R., Dollár, P., Tu, Z., He, K.: Aggregated residual transformations for deep neural networks. In: Proceedings of the IEEE Conference on Computer Vision and Pattern Recognition, pp. 1492–1500 (2017)
31. Huang, G., Liu, Z., et al.: Densely connected convolutional networks. In: Proceedings of the IEEE Conference on Computer Vision and Pattern Recognition, pp. 4700–4708 (2017)
32. He, K., Zhang, X., Ren, S., Sun, J.: Identity mappings in deep residual networks. In: Leibe, B., Matas, J., Sebe, N., Welling, M. (eds.) ECCV 2016. LNCS, vol. 9908, pp. 630–645. Springer, Cham (2016). https://doi.org/10.1007/978-3-319-46493-0_38
33. Redmon, J., Farhadi, A.: YOLO9000: better, faster, stronger. In: Proceedings of the IEEE Conference on Computer Vision and Pattern Recognition, pp. 7263–7271 (2017)

34. Kong, T., Yao, A., Chen, Y., Sun, F.: HyperNet: towards accurate region proposal generation and joint object detection. In: Proceedings of the IEEE Conference on Computer Vision and Pattern Recognition, pp. 845–853 (2016)
35. Dai, J., Li, Y., He, K., Sun, J.: R-FCN: object detection via region-based fully convolutional networks. In: Advances in Neural Information Processing Systems, pp. 379–387 (2016)
36. Kong, T., Sun, F., Yao, A., Liu, H., Lu, M., Chen, Y.: RON: reverse connection with objectness prior networks for object detection. In: Proceedings of the IEEE Conference on Computer Vision and Pattern Recognition, pp. 5936–5944 (2017)
37. Zhang, S., Wen, L., Bian, X., Lei, Z., Li, S.Z.: Single-shot refinement neural network for object detection. In: Proceedings of the IEEE Conference on Computer Vision and Pattern Recognition, pp. 4203–4212 (2018)
38. Shen, Z., Liu, Z., Li, J., Jiang, Y.G., Chen, Y., Xue, X.: DSOD: learning deeply supervised object detectors from scratch. In: Proceedings of the IEEE International Conference on Computer Vision, pp. 1919–1927 (2017)
39. Kim, S.-W., Kook, H.-K., Sun, J.-Y., Kang, M.-C., Ko, S.-J.: Parallel feature pyramid network for object detection. In: Ferrari, V., Hebert, M., Sminchisescu, C., Weiss, Y. (eds.) ECCV 2018. LNCS, vol. 11209, pp. 239–256. Springer, Cham (2018). https://doi.org/10.1007/978-3-030-01228-1_15
40. Li, Y., Li, J., Lin, W., Li, J.: Tiny-DSOD: lightweight object detection for resource-restricted usages. arXiv preprint arXiv:1807.11013 (2018)
41. Xu, M., et al.: MDSSD: multi-scale deconvolutional single shot detector for small objects. arXiv preprint arXiv:1805.07009 (2018)
42. Pang, Y., Wang, T., Anwer, R.M., Khan, F.S., Shao, L.: Efficient featurized image pyramid network for single shot detector. In: Proceedings of the IEEE Conference on Computer Vision and Pattern Recognition, pp. 7336–7344 (2019)

Multi-domain Document Layout Understanding Using Few-Shot Object Detection

Pranaydeep Singh, Srikrishna Varadarajan$^{(\boxtimes)}$, Ankit Narayan Singh,
and Muktabh Mayank Srivastava

ParallelDots, Inc., Lewes, USA
{pranaydeep,srikrishna,ankit,muktabh}@paralleldots.com

Abstract. We try to address the problem of document layout understanding using a simple algorithm which generalizes across multiple domains while training on just few examples per domain. We approach this problem via supervised object detection method and propose a methodology to overcome the requirement of large datasets. We use the concept of transfer learning by pre-training our object detector on a simple artificial (source) dataset and fine-tuning it on a tiny domain specific (target) dataset. We show that this methodology works for multiple domains with training samples as less as 10 documents. We demonstrate the effect of each component of the methodology in the end result and show the superiority of this methodology over simple object detectors. We will open-source the code, trained models, source and target datasets upon acceptance.

Keywords: Object detection · Few-shot · Transfer learning · Domain-invariant · Document layout detection

1 Introduction

The understanding of document layout in terms of finding logical components such as title, paragraphs etc. is a preliminary step towards retrieving information from images of documents. The amount of variability in real-world data coming from multiple domains e.g., documents, invoices etc. makes it a challenging computer vision problem that has intrigued researchers for decades.

Various image processing methodologies [1,7,8] have approached the problem of understanding general documents as well as digitizing historical documents. With the onset of deep learning and data driven approaches, the problem was approached as a pixel-wise segmentation task [12], where each pixel is assigned a class based on its surrounding pixels. In this paper, we explore a new tangent, where the problem is approached as a few-shot object detection problem to identify relevant areas in a document. The motivation is to understand document

---

P. Singh—Contributed equally.

© Springer Nature Switzerland AG 2020
A. Campilho et al. (Eds.): ICIAR 2020, LNCS 12132, pp. 89–99, 2020.
https://doi.org/10.1007/978-3-030-50516-5_8

structure with as less as 10 tagged examples since digitization tasks generally don't have an abundance of tagged data at hand. However, understanding documents is a complicated task and a dataset consisting of just 10 examples is not enough to train an object detector especially (as they're fully supervised networks requiring large amounts of training data) to understand various structures, like tables or lists.

Hence, we use a transfer learning based approach where we give the network a general understanding of what basic features and structures are contained in a document and then proceed to train on a few-shot task for understanding of specific document types like invoices, resumes, academic papers, journals etc. A few-shot task is described widely as training the model using just a handful of tagged examples.

The initial network which is to be later used for fine-tuning needs to have a wide understanding of document structures and substructures and needs to be trained extensively for it to yield good results when fine-tuned with very less samples. There was no relevant dataset which accommodated these needs and hence, we artificially generated a simple dataset using HTML. We refer to this dataset as Source Dataset. We then proceed to train the described model on this dataset. This trained model now serves as the backbone of all future models we fine-tuned. Using as little as 10, and up to 50 images, we demonstrate that the obtained model learns to understand document structures. We also show that the methodology can be extended to any number of domains with few examples from each. In this paper, we demonstrate the methodology and its application to Invoices and Resume images. We call these domains as Target Domains and the datasets as Target Datasets.

Our contributions consists of the following points

- Applying state of the art object detection techniques for Document Layout Understanding
- Introducing a generalized algorithm which can perform Layout Understanding in multiple domains using just few tagged images (eg: 10).

2   Related Work

There are two sub-parts to the Document Layout Analysis problem

- Geometric Layout Analysis
- Logical Layout Analysis

Geometric Layout Analysis (GLA) is centred around understanding the basic geometric layout of a document, such as skew, page decomposition, text detection etc. Logical Layout Analysis (LLA) focuses on understanding the implied semantic labels in a document, like captions, subheading, table headings etc. GLA has been addressed mainly by image processing methods like Hough Transforms and Binarization. While the GLA problem is as old as Image Processing itself, LLA is a more recent problem and the one which we attempt to solve.

Approaches employed in LLA mainly follow the bottom-up approach. Bottom-up approaches work by finding the smallest entities like words or characters and attempt to aggregate them using a distance metric and an aggregation algorithm like K-Nearest Neighbors or K-D Trees. These approaches [1,7,8] have the advantage of being mostly unsupervised but involve tuning a lot of heuristics. They are also not scalable to document layouts which are different from those the algorithm is tuned on. Comparisons of such approaches are also covered by [6,11]. The most popular and widely used of these approaches is the Docstrum [8] algorithm. While deep learning approaches to LLA also exist, these approaches [3,12] require vast amounts of training data and only learn a fixed set of labels and are thus not useful for few-shot tasks with a wide variety of different labels. We explore an object detection based approach to LLA, which can be fine-tuned on as less as 10 images to understand semantic labels like address, total bill amount, skills, education etc.

Few shot object detection is a task where the tagged training set is very small (say 1–50 images total). Previous work has been explored on the PASCAL VOC/COCO/ImageNet dataset. [2] introduce a Low-shot Object Detector (LSTD) model which is pretrained on a huge Source Dataset and fine-tuned on a small (low-shot) target dataset. The LSTD model is based on Single Shot Detector (SSD) [5] and Faster-RCNN (FRCNN) [10]. Broadly, they use the SSD network to detect foreground segments and a classifier which takes ROIPooled features from the SSD feature maps to classify the detected regions. There are two regularizations introduced by [2], Background Regularization (BGR) and Tk-Regularization (Tk-R) which helps them in learning from just few examples in the target dataset. We use BGR to make the learning of Target domain easier and faster. This is achieved by making the learning of background part in the Target domain easier through this constraint. Tk-R tries to bridge the gap between predictions of the classifier on Source and Target domain. The Source dataset in our case is more basic while [2] assume the Source dataset to be very huge and comprehensive.

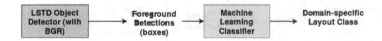

Fig. 1. Overview of the proposed method

3   Architecture

Our architecture is a two-step object detector. The first step is the detector (inspired from LSTD) which detects the foreground regions and the second step is the ML classifier which predicts the domain-specific layout class.

For the first step, we leverage a better feature extractor for the object detector. We use the Feature Pyramid Networks [4] as our feature extractor. This

(FPN based SSD) achieves state-of-the-art performance for a single model on PASCAL VOC dataset (object detection) as shown here[1].

On the Target dataset, many of the target classes cannot be distinguished by visual features alone. Hence we resorted to using a separate classifier (as opposed to the FRCNN based LSTD classifier) for the detected boxes. This involves taking text based features. Hence, while fine-tuning, a better alternative to this classifier is used in our system. The learning of target domain is made easier and faster by making use of the background regularization constraint.

4   Methodology

The task can be described as few shot document layout understanding. Our methodology consists of the following parts

1. Creating the artificial (Source) dataset.
2. Pretraining the model on the Source dataset.
3. Finetuning the model on the domain-specific (Target) dataset.
4. Training the ML classifier on the Target dataset (is combined with Step 3).

4.1   Dataset Generation

Our artifical dataset contains 160,000 images spanning multiple scales and sizes, accommodating for asymmetrically placed structures and elements. The dataset contained 8 basic layout classes: Title, Heading. Sub-Heading, Text Block, List, Table, Image Content, Image/Table Caption.

The textual content in the dataset was taken from a text dump consisting of a variety of online sources. The images were taken from a small dataset collected from Google Images. Apart from random images, the image dataset contained specific images collected using relevant keywords like graphs, tables, charts etc.

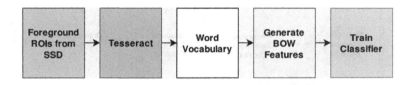

Fig. 2. Overview of the ML Classifier

4.2   Training

We train the LSTD model as it is on the Source Dataset. Once our model is trained on the Source Dataset, we move to fine-tune the model on the Target Datasets. Here we apply BGR. As mentioned earlier, we found that the performance of the inbuilt classifier in LSTD was not performing to our satisfaction,

---

[1] https://github.com/kuangliu/torchcv.

hence we decided to pass the foreground detections from the network through a seperate classifier.

Target Classification: To tackle the domain specific layout classes, we employed few ways to extract the best features so that we can train a classifier. We extracted the text from the detected box and used bag-of-words approach for getting the textual features. We also used other features related to the spatial configuration of the detected box. We use these features to train a machine learning algorithm to classify the detected bounding box to one of the classes. This is described in Fig. 2.

4.3   Implementation Details

For creating the artificial dataset, we generated HTML files which correspond to web documents and exported them into images using a webdriver. For the layout detection step, we implemented the LSTD network in PyTorch library. We use the FPNSSD from torchcv library (see footnote 1). For all experiments, we use SGD optimizer with learning rate of 0.0001 and momentum 0.9. We use L2 penalty of 0.0005. For the **layout classification** step, to extract text from a detected box we use the open-source LSTM-based Tesseract 4.0. We get our classifier using the tpot toolkit [9], which uses genetic programming to optimize machine learning pipelines. While reporting the results, we take the IoU threshold for evaluating object detection metrics as 0.5.

5   Invoice Dataset

We collected 170 invoices which includes variations in structure, domain and template. We refer to this as the Invoice Dataset. We manually tag this dataset into layouts of 5 main categories: Logo, Address, Bill/Invoice Information, Tables, Amount Information (Total). We use a fixed set of 100 images as our test set. We train our model on different (incremental) number of training images (k) and report the results correspondingly.

Table 1. LSTD end to end performance on **Invoice Dataset**

No. of training images (k)	Mean precision	Mean recall	Mean F1 score
10	0.4721	0.5188	0.4943
20	0.4962	0.5444	0.5192
30	0.5012	0.5791	0.5373
40	0.5244	0.601	0.5601
50	0.5316	0.6101	0.5682
60	0.5599	0.6214	0.589
70	0.56	0.6354	0.5953

6   Resume Dataset

The resume dataset is a set of 100 images collected from various sources containing resumes from different domains and layouts. As with the invoice dataset, this was manually tagged into 6 main categories: Education, Experience, Bio, Skills, Summary, Other. A fixed set of 50 images is used as the test set and training is done on an incremental number of training images ranging from 10 to 50.

Table 2. LSTD End to End performance on **Resume Dataset**

No. of training images (k)	Mean precision	Mean recall	Mean F1 score
10	0.6144	0.5888	0.6013
20	0.6398	0.6011	0.6198
30	0.6587	0.6218	0.6397
40	0.6712	0.6325	0.6513
50	0.6946	0.634	0.6629

7   Baselines

Table 3. Baseline (Docstrum) performance

Dataset	Precision	Recall	F1 score
Invoice	0.0547	0.1935	**0.0853**
Resume	0.2415	0.2559	**0.2485**

The Docstrum algorithm [8] serves as our baseline. The algorithm finds the connected components and their centroids. It then looks for the K-nearest neighbours (K = 5) of each component. Vectors are plotted from each centroid to its neighbours and these angles help in skew correction. The nearest-neighbor distance histogram has several peaks and these peaks typically represent between-character spacing, between-word spacing and between-line spacing. These values are then used to construct lines, words and text blocks with some predetermined tolerance for each spacing value.

We use Docstrum to construct blocks and then evaluate the outputs using the manually annotated ground truth boxes on both the target datasets ie. Invoices and Resumes. The results are reported in Table 3 while sample outputs of the method are shown in Fig. 3. Results of Docstrum can be compared with the foreground detection results (Table 4, Table 5) as end to end layout detection uses textual features.

8   Results

We perform multiple experiments to evaluate the performance of the proposed approach. We first show the effectiveness of source pretraining (SP) for few shot layout detection. Later, we evaluate our pipeline in 3 ways.

1. Evaluation of Foreground detection task
2. Evaluation of ML Classifier
3. Evaluation of end to end layout detection task

We evaluate the foreground detection performance of two types of models. In Tables 4, 5, *scratch* refers to the model which was trained from scratch, while SP refers to the model which was finetuned from the Source Pretraining (Step 2 in Sect. 4). One can notice an improvement of at least **40%** on F1 scores of Target Domain Layout Detection task. This signifies the importance of Source Pretraining in our proposed pipeline.

The evaluation of the ML Classifier on the foreground ROIs is shown in Table 6. The performance of our pipeline on the end to end layout detection task is shown in Table 1, 2 for the Invoice and Resume datasets respectively. The end to end pipeline consists of both the foreground detection and ML Classifier. We are able to obtain satisfactory performance even with 10 training images.

Table 4. LSTD foreground detection performance on **Invoice Dataset**. SP denotes Source Pretraining.

No. of training images	Precision		Recall		F1 Score	
	SP	Scratch	SP	Scratch	SP	Scratch
0	0.144	NA	0.4214	NA	0.2147	NA
10	0.5992	0.1078	0.6212	0.1991	**0.61**	0.1399
20	0.611	0.1377	0.7062	0.235	0.655	0.1736
30	0.6203	0.1744	0.7755	0.2768	0.6893	0.214
40	0.6767	0.1957	0.7901	0.2998	0.729	0.2368
50	0.6742	0.3018	0.7992	0.3036	0.7314	0.3027
60	0.7017	0.3738	0.8001	0.315	0.7484	0.3419
70	0.7292	0.3888	0.8132	0.3445	**0.7689**	0.3653

Table 5. LSTD foreground detection performance on **Resume Dataset**. SP denotes Source Pretraining.

No. of training images	Precision		Recall		F1 Score	
	SP	Scratch	SP	Scratch	SP	Scratch
0	0.035	NA	0.4311	NA	0.06	NA
10	0.8228	0.3797	0.821	0.3571	**0.8219**	0.368
20	0.8542	0.3859	0.8224	0.3928	0.838	0.3893
30	0.8655	0.5238	0.8291	0.5238	0.8469	0.5238
40	0.9123	0.5178	0.8363	0.7532	0.8726	0.6137
50	0.8977	0.6094	0.8343	0.61309	**0.8659**	0.6103

Table 6. Evaluation of ML Classifier on **Invoice** and **Resume** Datasets

Dataset	No. of training images (k)	Precision	Recall	F1 score
Invoice	70	0.7718	0.8135	0.7921
Resume	50	0.804	0.8946	0.8469

9   Conclusion

In this work, we have shown that object detection techniques can be used for Document Layout understanding. We have also shown that the proposed methodology can be scaled across multiple domains with just need of few tagged examples. The results also demonstrate the superiority of the methodology over existing object detection techniques. Document Layout analysis techniques assumes great importance in the information age as more and more documents are digitized and needs to be retrieved by understanding their content similar to digital content. Such techniques are useful in automating manually intensive business processes such as processing KYC documents or invoices. Document Layout analysis techniques also opens up the possibilities for businesses to mine documents such as paper receipts and extract valuable insights from them for market research purposes. Getting a large annotated corpus of data can be time-consuming and expensive for practical use-cases which further demonstrates the practical utility of our approach.

10 Qualitative Outputs

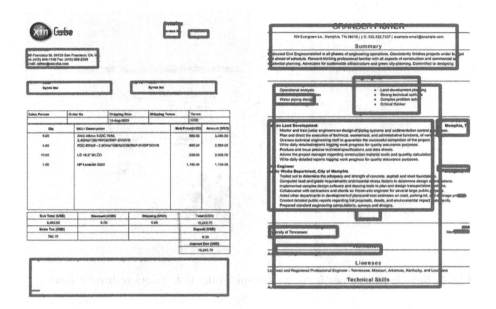

Fig. 3. Sample predictions of the baseline method on both Datasets

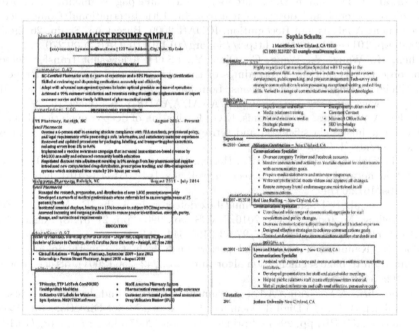

Fig. 4. Sample predictions from our system on the test images of Resume Dataset

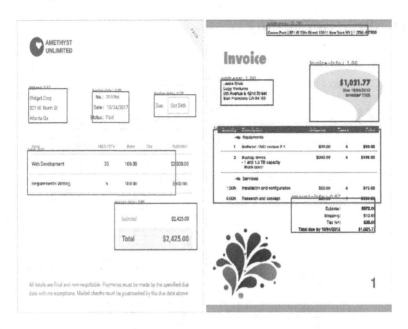

Fig. 5. Sample predictions from our system on the test images of Invoice Dataset

References

1. Agrawal, M., Doermann, D.S.: Voronoi++: a dynamic page segmentation approach based on voronoi and docstrum features. In: 10th International Conference on Document Analysis and Recognition, ICDAR 2009, Barcelona, Spain, 26–29 July 2009, pp. 1011–1015 (2009). https://doi.org/10.1109/ICDAR.2009.270
2. Chen, H., Wang, Y., Wang, G., Qiao, Y.: LSTD: a low-shot transfer detector for object detection. In: Proceedings of the Thirty-Second AAAI Conference on Artificial Intelligence, New Orleans, Louisiana, USA, 2–7 February 2018 (2018). https://www.aaai.org/ocs/index.php/AAAI/AAAI18/paper/view/16778
3. Harley, A.W., Ufkes, A., Derpanis, K.G.: Evaluation of deep convolutional nets for document image classification and retrieval, vol. abs/1502.07058 (2015). http://arxiv.org/abs/1502.07058
4. Lin, T., Dollár, P., Girshick, R.B., He, K., Hariharan, B., Belongie, S.J.: Feature pyramid networks for object detection, vol. abs/1612.03144 (2016). http://arxiv.org/abs/1612.03144
5. Liu, W., et al.: SSD: single shot multibox detector. In: Leibe, B., Matas, J., Sebe, N., Welling, M. (eds.) ECCV 2016. LNCS, vol. 9905, pp. 21–37. Springer, Cham (2016). https://doi.org/10.1007/978-3-319-46448-0_2
6. Mao, S., Kanungo, T.: Empirical performance evaluation methodology and its application to page segmentation algorithms. **23**, 242–256 (2001). https://doi.org/10.1109/34.910877
7. Namboodiri, A.M., Jain, A.K.: Document structure and layout analysis. In: Chaudhuri, B.B. (ed.) Digital Document Processing: Major Directions and Recent Advances. ACVPR, pp. 29–48. Springer, London (2007). https://doi.org/10.1007/978-1-84628-726-8_2

8. O'Gorman, L.: The document spectrum for page layout analysis. **15**, 1162–1173 (1993). https://doi.org/10.1109/34.244677
9. Olson, R.S., Bartley, N., Urbanowicz, R.J., Moore, J.H.: Evaluation of a tree-based pipeline optimization tool for automating data science. In: Proceedings of the 2016 on Genetic and Evolutionary Computation Conference, Denver, CO, USA, 20–24 July 2016, pp. 485–492 (2016). https://doi.org/10.1145/2908812.2908918
10. Ren, S., He, K., Girshick, R.B., Sun, J.: Faster R-CNN: towards real-time object detection with region proposal networks, vol. abs/1506.01497 (2015). http://arxiv.org/abs/1506.01497
11. Shafait, F., Keysers, D., Breuel, T.M.: Performance comparison of six algorithms for page segmentation. In: Bunke, H., Spitz, A.L. (eds.) DAS 2006. LNCS, vol. 3872, pp. 368–379. Springer, Heidelberg (2006). https://doi.org/10.1007/11669487_33
12. Yang, X., Yumer, E., Asente, P., Kraley, M., Kifer, D., Giles, C.L.: Learning to extract semantic structure from documents using multimodal fully convolutional neural networks. In: 2017 IEEE Conference on Computer Vision and Pattern Recognition, CVPR 2017, Honolulu, HI, USA, 21–26 July 2017, pp. 4342–4351 (2017). https://doi.org/10.1109/CVPR.2017.462

Object Tracking Through Residual and Dense LSTMs

Fabio Garcea$^{(\boxtimes)}$ (iD), Alessandro Cucco, Lia Morra (iD), and Fabrizio Lamberti (iD)

Dipartimento di Automatica e Informatica, Politecnico di Torino, Turin, Italy
{fabio.garcea,lia.morra,fabrizio.lamberti}@polito.it,
alessandro.cucco@studenti.polito.it

Abstract. Visual object tracking task is constantly gaining importance in several fields of application as traffic monitoring, robotics, and surveillance, to name a few. Dealing with changes in the appearance of the tracked object is paramount to achieve high tracking accuracy, and is usually achieved by continually learning features. Recently, deep learning-based trackers based on LSTMs (Long Short-Term Memory) recurrent neural networks have emerged as a powerful alternative, bypassing the need to retrain the feature extraction in an online fashion. Inspired by the success of residual and dense networks in image recognition, we propose here to enhance the capabilities of hybrid trackers using residual and/or dense LSTMs. By introducing skip connections, it is possible to increase the depth of the architecture while ensuring a fast convergence. Experimental results on the Re^3 tracker show that DenseLSTMs outperform Residual and regular LSTM, and offer a higher resilience to nuisances such as occlusions and out-of-view objects. Our case study supports the adoption of residual-based RNNs for enhancing the robustness of other trackers.

Keywords: Object tracking · Recurrent neural networks · Residual networks

1 Introduction

Visual object tracking plays a fundamental role in many applications including, e.g., robotics and video-surveillance. In this paper, we specifically focus on the problem of *generic* object tracking, which can be concisely phrased as follows: "given a bounding box enclosing an arbitrary object at time t, produce bounding boxes for that object in all future frames" [1].

Current generic 2D image tracking systems predominantly rely on training a tracker online according to the *tracking-by-detection* paradigm: an object-specific detector is continuously updated with the new object's aspect at every frame, to cope with changes in shape and appearance, as well as with occlusions, while the object moves. Compared with trackers trained completely offline, this approach is more robust and flexible, but these advantages are paid with a decrease in

© Springer Nature Switzerland AG 2020
A. Campilho et al. (Eds.): ICIAR 2020, LNCS 12132, pp. 100–111, 2020.
https://doi.org/10.1007/978-3-030-50516-5_9

the frame rate that can be achieved. In recent years, hybrid trackers that combine convolutional neural networks (CNNs) for visual feature extraction with Long Short-Term Memory (LSTM) recurrent neural networks have been widely adopted. An example is represented by the Re^3 tracker [1]: the CNN is trained completely offline, thus reducing the computational load at inference time, and the LSTM is trained to update and store an object-specific model. This method has shown increased accuracy and robustness against comparable trackers, especially during occlusions, but is still sensitive to changes in the object's appearance due to occlusions or partially out-of-view targets, to proximity with similar objects, as well as to the presence of background clutter.

A possible way to improve the performance of LSTM-based trackers is to increase the complexity of the recurrent module, e.g., by stacking several LSTMs. This approach, however, can make it harder for the training procedure to converge, due to the increased network depth. Inspired by the success of Residual Networks [2] and Dense Networks [3] in image recognition, few works in literature have explored the use of residual connections in LSTMs, mostly for speech and text analysis [4–8].

In principle, using deeper and more complex LSTM modules should improve the capability of the tracker to model long-term change sequences. Our contribution is thus the design and experimental validation of Dense and Residual LSTM modules for visual object tracking. To assess, *ceteris paribus*, the added benefit of residual connections in object tracking, we modified the established architecture of the Re^3 tracker [1]. Our experimental evaluation on the OTB50 and OTB100 benchmarks shows that Dense LSTM modules achieve higher robustness to occlusion and out-of-view targets while maintaining a similar parameter count compared to solutions adopting plain, non-residual layouts.

The rest of the paper is organized as follows. In Sect. 2, related work related to object tracking and residual networks is presented. Afterwards, in Sect. 3, we examine the tracker selected as the baseline and propose two different variations of the original layout involving residual connections in the recurrent module. In Sect. 4, we present the performance obtained by the proposed architectures on different benchmarks and compare them with state-of-the-art trackers. Finally, in Sect. 5, we discuss the main findings of our experiments and give some directions for future works.

2 Related Work

2.1 Object Tracking

Modern trackers can be roughly divided in *offline-trained*, *online-trained*, and *hybrid* [1,9]. Online trackers operate online, continually learning features to update the object's appearance during tracking: trackers adopting the well-known tracking-by-detection paradigm belong to this category. This type of tracker must carefully balance adaptation with real-time response abilities.

Recent works have exploited the capabilities of deep neural networks (DNNs) to learn from massive amounts of data by training CNN-based trackers completely offline [10]. These solutions rely on pre-trained CNNs for feature extraction and can operate at faster than real-time speed, but are intrinsically limited in coping with changes in objects' appearance due to movements, occlusions, blurring, etc.

Hybrid solutions like MDNet [11] and Re^3 [1] represent an attempt to merge best qualities from both offline- and online-trained solutions. In the Re^3 architecture, a CNN is trained offline to perform feature extraction, coupled with an LSTM module that keeps track of the object history over time. A multiresolution approach is used by combining high-level features derived from the full CNN to low-level features learned by the previous layers, thus increasing the robustness of the feature extraction. This architecture represents a good trade-off between fast, real-time tracking (it achieves speeds of 150 frames per second) and robustness against nuisances such as occlusions.

2.2   Residual Networks

Residual networks and, in more recent times, densely connected networks have shown superior accuracy and training properties than traditional sequential CNNs, and have consistently achieved state-of-the-art results in image classification and other visual tasks [12,13]. Densely connected CNNs, or DenseNets, represent an extension of the concept of skip connections: the output from each layer is passed as input to all subsequent layers and, as a consequence of the greater flexibility, have proven more effective than ResNets on a variety of visual tasks. A question that naturally arises is whether residual connections can prove as beneficial also for LSTM networks and, by extension, if the performance of hybrid trackers can be improved as well.

The idea of stacking LSTMs in a residual fashion has already been adopted in other fields of study such as distant speech recognition [5], sentiment intensity prediction [8], and object tracking [4]. Dense LSTMs stacking has been recently explored for sentence classification [6] and speech enhancement [7]. In [4], a rule-based residual RLSTM has been applied to tracking, achieving good results compared to other state-of-the-art trackers. However, given the complexity of object tracking networks and considering the role played by the feature extraction part (based on convolutional layers), by the training algorithm and by the training set, we believe that only by conducting controlled experiments the impact of residual connections can be fully appreciated. To the best of our knowledge, the role of Dense LSTMs in the context of tracking has not been investigated yet.

3   ResidualRe^3 and DenseRe^3

For our experiments, we selected the Re^3 tracker as baseline architecture, and propose two alternative LSTM modules: a ResidualLSTM block consisting of two cascaded LSTMs, and a DenseLSTM block in which four sequential LSTMs are densely connected by applying the same intuition used in DenseNets. The blocks, as well as their position in the overall architecture, are illustrated in Fig. 1. The number and size of layers were carefully chosen to keep the parameter count and the combined depth as similar as possible to the original Re^3 architecture. In the following sub-sections, the main characteristics of the two solutions will be illustrated.

3.1 Re^3

The Re^3 tracker was firstly proposed in [1]. It represents a hybrid solution to the problem of generic object tracking. The layout of this network can be mainly split into three modules solving different tasks. The first module is a stack of convolutional layers used to extract the embeddings from the object being tracked; a concatenation layer is fed with both low-level and high-level information to obtain a more complete representation of the object. In the second module, a recurrent block consisting of a stack of two LSTM layers, each one receiving the features extracted at the previous stage, can keep track of subsequent object's positions and transformations. Finally, a regression layer is used to predict the bounding box of the object in the current frame. The full model is fed with two frame crops from the sequence at each time step; one of them is centered at the object's position in the previous frame, whereas the other is centered at the same position but in the current frame. Both the crops are still large enough to carry some information about the background.

3.2   ResidualLSTM-Based RNN

In the ResidualRe^3 version of the tracker, a different architecture has been adopted for the recurrent module. In particular, a sequence of two LSTMs connected in series has been added to the input of the first LSTM module in a residual block fashion. We will refer to this structure as the ResidualLSTM block. The full layout of the recurrent module consists of a stack of three ResidualLSTM blocks. Since the outputs from both the convolutional module and the LSTMs are summed through a merge layer, they need to share the same number of units. In the original version of the tracker, the CNN output is set to use 1024 units, but we decided to downscale it to 768 units to keep the parameter count of the complete network comparable to that in the original version of the tracker. Moreover, a batch normalization layer has been added after the fully connected layer of the CNN to make its output comparable to that of the first ResidualLSTM block.

3.3   DenseLSTM-Based RNN

In the DenseLSTM version of the tracker, we decided to replace the recurrent module with a different structure exploiting dense residual connections; a stack of four LSTMs has been densely connected through skip connections, thus allowing each subsequent module to be fed from the output of the previous ones. We will refer to this structure as the DenseLSTM block. In this case, the fully connected layer on top of the CNN has been set to use 900 units instead of the original 1024 units to keep a low parameter count for the following recurrent module; moreover, a batch normalization layer has been added on top of this layer to speed up model convergence. In the full network layout, we used a single DenseLSTM block composed of four LSTMs with 512 units each. With these constraints, we were able to maintain a complexity similar to that of the original Re^3 tracker.

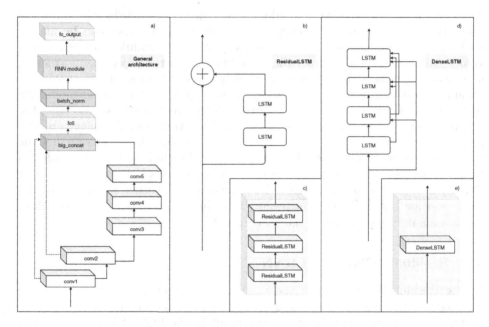

Fig. 1. A visual comparison of the proposed LSTM-based blocks. a) General structure of the Re^3 tracker; the main difference between Re^3, ResidualRe^3 and DenseRe^3 is the RNN module (in green). b) Basic structure of the ResidualLSTM tracker, consisting of a series of two LSTMs; in the ResidualRe^3 alternative shown in c), a stack of three ResidualLSTM blocks has been deployed as the recurrent module. d) Basic structure of the DenseRe^3 tracker where four LSTMs have been densely connected through skip connections; resulting structure used as recurrent module is reported in e). (Color figure online)

3.4   Training and Implementation Details

The training procedure is the same as in the original Re^3 paper [1]. We here summarize the most important steps. Before starting the training, synthetic data are produced with several augmentation techniques such as horizontal flipping and random noise generation, and weights from AlexNet are loaded in the CNN; LSTM states are initialized to zero, whereas other weights are set using MSRA initialization. The adopted optimizer is Adam with momentum and weight decay set to default values, and a learning rate decreasing from 10^{-5} to 10^{-6} after 10.000 iterations. Finally, the loss function is the Mean Absolute Error (MAE), and the number of iterations is 200.000. The training is initialized with 64 batch size, 2 unrolls and 1 probability of using the ground truth bounding boxes as a reference to crop the frame at the following time step; as soon as the loss plateaus, the batch size is halved and the unrolls are doubled (up to 32 unrolls). Moreover, the probability (initially 0) of mixing the predicted bounding boxes with the ground truth is increased using steps of 0.25; in this way, the network can learn from its errors during training thus being able to partly recover from errors at test time.

Since reproducibility of deep learning models is notoriously difficult to achieve, being the training procedure inherently random and affected by several factors including the training environment [14,15], the original model has been retrained following the steps in the original publication on the ILSVRC2014 DET and ILSVRC2017 VID datasets, starting from the original code provided by the authors.

The training procedure for the modified networks followed the same used for training the Re^3 original version with some minor changes. For the ResidualRe^3 tracker, a faster learning rate of 10^{-4} was initially set, then reduced to 10^{-5} and 10^{-6} when noticing that the loss function starts to plateau; moreover, a faster learning rate scaling of 10^{-1} was adopted for the finetuning of the CNN module weights. For the DenseRe^3 network, the procedure was similar, but we started to increase the probability of using the network prediction only after 32 unrolls and each time the loss function showed a plateau.

All the experiments were performed on a system configured with an i7 2600 CPU, 8 GB DDR3 1333 MHz RAM and an NVIDIA GTX 1060 3 GB GPU.

4 Experimental Results

First, we report the results of training and testing all the architectures on the ILSVRC2014 DET and ILSVRC2017 VID datasets, considering also the retrained Re^3 model provided by [1]. Secondly, we compare results obtained by the original published Re^3 model on the challenging OTB50 and OTB100 benchmarks [16] along with several state-of-the-art trackers. The OTB100 benchmark consists of 100 different image sequences reporting objects from different classes and assignable to different attributes (occlusion, motion blur, out-of-view, etc.).

4.1 Training ResidualRe^3 and DenseRe^3

The results for the architectures under test are reported in terms of two different metrics, namely, the number of targets lost by the tracker, and the Mean Intersection Over Union (IOU) between the predicted and the ground truth bounding boxes. The parameters count is reported as well, to highlight how the new architectures remain comparable to the model from the original paper.

Our results show an improvement in the Mean IOU score and a lower number of lost targets compared to the retrained version of Re^3, while keeping a comparable parameter count (Table 1). The evaluation of training results for DenseRe^3 showed significant improvements compared to the Retrained version of Re^3 and ResidualRe^3 (Table 1), thus demonstrating the advantages brought by the DenseLSTM blocks.

Table 1. Training results for different Re^3 architectures. It should be noticed how residual and dense LSTM improve performance with minimal increase in parameter count.

Tracker	Lost targets	Mean IOU	Parameter count
Re^3 (retrained)	350	0.64	85.699.686
ResidualRe^3	303	0.66	87.716.712
DenseRe^3	258	0.68	87.031.408

Concerning the original Re^3 architecture, it is worth noticing that we did not achieve the same performance of the model released by the authors even though, to the best of our knowledge, we followed the same training curriculum for Re^3. Specifically, our retrained model achieves 350 lost targets with a Mean IOU of 0.64 versus the 243 lost targets and 0.72 Mean IOU for the weights provided by the authors.

This discrepancy is not entirely surprising, since reproducing results is a well-known issue of deep learning-related research, and maybe due to slight differences in implementation or training parameters. Recent research also highlighted the effect of random initialization on the estimated performance of image classification networks [14]; however, for complex deep learning architectures, such as object trackers, running multiple experiments per configuration requires substantial computational resources. In the future, we plan on exploring this issue in more detail. For the remaining experiments, we compare DenseRe^3 with the original model provided by the authors, which albeit less favourable allows an easier comparison with the previous literature.

4.2 Benchmarks Evaluation

Since we needed a state-of-the-art reference tracker to compare our results, we opted for the Recurrent Filter Learning tracker (RFL) [17]; besides its high performances, this model is based on a recurrent module thus representing an appropriate reference architecture for our experiments.

We report here the results of the One Pass Evaluation (OPE) protocol on the OTB50 and OTB100 benchmarks; while the results on OTB100 benchmark were computed at test time starting from the code as provided by the authors, those for OTB50 were already provided for multiple trackers and thus have not been re-computed. Once the baseline was defined, we executed the OPE TB100 benchmark for the RFL tracker, the original version of Re^3 and the DenseRe^3 networks; the results (Fig. 2) show that our model can outperform the original version of Re^3 in sequences characterized by low resolution, occlusions and out-of-view objects, while still performing similarly to the original architecture in other cases.

We then ran the OTB50 benchmark for the RFL, the original Re^3, the DenseRe^3 and other state-of-the-art trackers to evaluate the performances of our model with a larger pool of different architectures. The results (Fig. 3) are aligned with those obtained with OTB100; like with the other benchmark, the model performs particularly well with sequences characterized by the above attributes, reaching the top-4 positions for all attributes. In all the other sequences, it performs worse than Re^3, even though it can reach the top-5 positions.

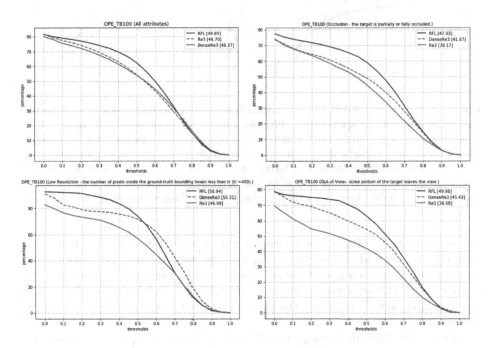

Fig. 2. Results on the OPE TB100 benchmark for RFL, the original version of Re^3, and the proposed DenseRe^3 architecture. The percentage of frames where the mean IoU is greater than a threshold (y axis) is plotted as a function of the threshold value (x axis). The success plots report the results for different sequence attributes (all attributes, occlusion, low resolution and out-of-view objects). Whilst in some cases the DenseRe^3 model scores are similar to the z version, in other cases they show a better performance of our architecture.

Finally, two example sequences from the OTB100 benchmark are reported (see Fig. 4) to depict, in a visual fashion, the achieved improvement. The first

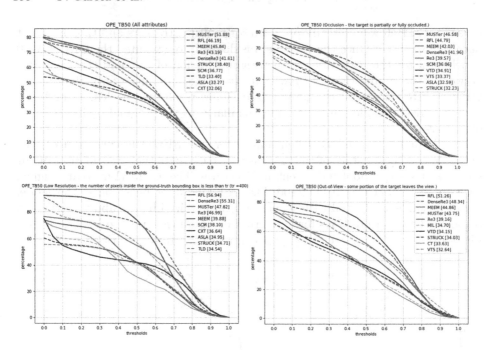

Fig. 3. Results on the OPE TB50 benchmark for the proposed DenseRe³ architecture and other state-of-the-art trackers. The percentage of frames where the mean IoU is greater than a threshold (y axis) is plotted as a function of the threshold value (x axis). The success plots reports the results for different sequence attributes (all attributes, occlusion, low resolution and out-of-view objects). In most of the subsets DenseRe³ scores are similar to those obtained by the original Re³ tracker. DenseRe³ achieves high performance in sequences with low resolution, occlusion and out-of-view target.

sequence, named "Matrix", is characterized by multiple attributes like occlusion, fast motion and illumination variation. The second sequence, named "Ironman", similarly presents multiple attributes as well, such as occlusion and out-of-view. Both the sequences have been annotated by Re³ and DenseRe³ with a red bounding-box representing the prediction of the tracker under test for each frame.

It's evident how, in the sequence "Matrix", Re³ (sub-sequence 1.a) loses the track of the object due to the fast motion of the body and the partial occlusion of the face features and consequently starts to track the hand of the second character. On the other side, DenseRe³ (sub-sequence 1.b) can keep track of the object also in presence of disturbances and it's able to progressively recover from the error caused by the occluded frames. Moreover, a robust behavior can be appreciated in the fourth frame of the sequence where the model keeps track of the object in the presence of an important variation in the illumination.

In the second example, Re³ (sub-sequence 2.a) loses the track of the object when it goes temporary out-of-the-view and it's not able to recover from its own errors thus ending up losing track of the object. On the other hand, DenseRe^3 (sub-sequence 2.b) shows again a robust behavior in case of occluded and out-of-view frames being able to keep track of the object even if with some difficulties due to the complexity of the frame.

Fig. 4. Example sequences from the OPE TB100 benchmark evaluated on DenseRe^3 and plain Re^3. The first sequence, named "Matrix", has been annotated with both Re^3 (1.a) and DenseRe^3 (1.b). Similarly the sequence named "Ironman" has been annotated by both Re^3 (2.a) and DenseRe^3 (2.b). The bounding boxes annotated by DenseRe^3 intersect with the ground-truth (green bounding box) also in case of disturbances showing thus a robust behavior of the network if compared to plain Re^3. (Color figure online)

5   Conclusions and Future Work

In this work, we explored the potential benefit of Residual and Dense LSTM in hybrid object tracking architectures. The idea of introducing residual and dense skip connections in LSTMs has been successfully explored in other applications, such as speech and text recognition.

We here investigate a case study in object tracking, in which we modified the architecture of an LSTM-based tracker, the Re^3 architecture, using both ResidualLSTM and DenseLSTM modules. Our experiments showed that both ResidualLSTM and DenseLSTM modules can be successfully used to enhance the robustness of the Re^3 tracker, as in low resolution or occlusion attributes, while keeping a parameter count comparable to the original version. In general the proposed architecture appears to be more robust to the presence of occlusions, low resolution and other disturbances. Residual and even more dense architecture allow to connect each layer not only with the previous layer, but also with previous ones. Skip connections are an essential component of deep convolutional neural networks allowing to increase the number of layers without incurring in vanishing gradients or other numerical instability. DenseRe^3 is characterized by four LSTM blocks, instead of the two blocks of the plain Re^3 tracker, thus effectively doubling the depth of the network. Nonetheless, the use of skip connections makes the information flow across the layers easier ensuring fast convergence, thus increasing performance in a comparable number of iterations. Previous works reported that increasing the number of layers in plain LSTM may lead to performance degradation, however, this phenomenon can be reduced or even reversed when residual connections are introduced [8]. We observed an even greater benefit from dense connections, but the relationship between performance and depth should be analyzed in a more systematic fashion.

We also hypothesize that in a densely connected structure, where the activations of each layer are fed to all subsequent ones, it is possible to more effectively "remember" the history of the object being tracked, thus improving the robustness in the presence of occlusions and background clutter, or when the object moves out-of-view.

We expect that similar improvements could be found on other architectures currently relying on plain LSTM modules. In the future, we plan to explore the advantages of ResidualLSTM and DenseLSTM blocks in other trackers or other visual tasks.

References

1. Gordon, D., Farhadi, A., Fox, D.: Re^3: real-time recurrent regression networks for visual tracking of generic objects. IEEE Robot. Autom. Lett. **3**(2), 788–795 (2018)
2. He, K., Zhang, X., Ren, S., Sun, J.: Deep residual learning for image recognition (2015)
3. Huang, G., Liu, Z., van der Maaten, L., Weinberger, K.Q.: Densely connected convolutional networks (2016)
4. Kim, H.I., Park, R.H.: Residual LSTM attention network for object tracking. IEEE Signal Process. Lett. **25**(7), 1029–1033 (2018)
5. Kim, J., El-Khamy, M., Lee, J.: Residual LSTM: design of a deep recurrent architecture for distant speech recognition. arXiv preprint arXiv:1701.03360 (2017)
6. Ding, Z., Xia, R., Yu, J., Li, X., Yang, J.: Densely connected bidirectional LSTM with applications to sentence classification. In: Zhang, M., Ng, V., Zhao, D., Li, S., Zan, H. (eds.) NLPCC 2018. LNCS (LNAI), vol. 11109, pp. 278–287. Springer, Cham (2018). https://doi.org/10.1007/978-3-319-99501-4_24
7. Gao, T., Du, J., Dai, L.R., Lee, C.H.: Densely connected progressive learning for LSTM-based speech enhancement. In: 2018 IEEE International Conference on Acoustics, Speech and Signal Processing (ICASSP), pp. 5054–5058. IEEE (2018)

8. Wang, J., Peng, B., Zhang, X.: Using a stacked residual LSTM model for sentiment intensity prediction. Neurocomputing **322**, 93–101 (2018)
9. Ali, A., et al.: Visual object tracking–classical and contemporary approaches. Front. Comput. Sci. **10**(1), 167–188 (2016)
10. Bertinetto, L., Valmadre, J., Henriques, J.F., Vedaldi, A., Torr, P.H.S.: Fully-convolutional siamese networks for object tracking. In: Hua, G., Jégou, H. (eds.) ECCV 2016. LNCS, vol. 9914, pp. 850–865. Springer, Cham (2016). https://doi.org/10.1007/978-3-319-48881-3_56
11. Nam, H., Han, B.: Learning multi-domain convolutional neural networks for visual tracking (2015)
12. He, K., Sun, J.: Convolutional neural networks at constrained time cost. In: Proceedings of the IEEE Conference on Computer Vision and Pattern Recognition, pp. 5353–5360 (2015)
13. He, K., Zhang, X., Ren, S., Sun, J.: Identity mappings in deep residual networks. In: Leibe, B., Matas, J., Sebe, N., Welling, M. (eds.) ECCV 2016. LNCS, vol. 9908, pp. 630–645. Springer, Cham (2016). https://doi.org/10.1007/978-3-319-46493-0_38
14. Bouthillier, X., Laurent, C., Vincent, P.: Unreproducible research is reproducible. In: International Conference on Machine Learning, pp. 725–734 (2019)
15. Marrone, S., Olivieri, S., Piantadosi, G., Sansone, C.: Reproducibility of deep CNN for biomedical image processing across frameworks and architectures. In: 2019 27th European Signal Processing Conference (EUSIPCO), pp. 1–5. IEEE (2019)
16. Wu, Y., Lim, J., Yang, M.H.: Online object tracking: a benchmark. In: IEEE Conference on Computer Vision and Pattern Recognition (CVPR) (2013)
17. Yang, T., Chan, A.B.: Recurrent filter learning for visual tracking. In: Proceedings of the IEEE International Conference on Computer Vision Workshops, pp. 2010–2019 (2017)

Theoretical Insights into the Use of Structural Similarity Index in Generative Models and Inferential Autoencoders

Benyamin Ghojogh$^{(\boxtimes)}$ (ID), Fakhri Karray (ID), and Mark Crowley (ID)

Department of Electrical and Computer Engineering,
University of Waterloo, Waterloo, ON, Canada
{bghojogh,karray,mcrowley}@uwaterloo.ca

Abstract. Generative models and inferential autoencoders mostly make use of ℓ_2 norm in their optimization objectives. In order to generate perceptually better images, this short paper theoretically discusses how to use Structural Similarity Index (SSIM) in generative models and inferential autoencoders. We first review SSIM, SSIM distance metrics, and SSIM kernel. We show that the SSIM kernel is a universal kernel and thus can be used in unconditional and conditional generated moment matching networks. Then, we explain how to use SSIM distance in variational and adversarial autoencoders and unconditional and conditional Generative Adversarial Networks (GANs). Finally, we propose to use SSIM distance rather than ℓ_2 norm in least squares GAN.

Keywords: Generative Moment Matching Network · Generative Adversarial Network · Variational Autoencoder · Adversarial autoencoder · Structural Similarity Index · SSIM kernel · Perceptual image generation

1 Introduction

Learning models can be divided into discriminative and generative [1]. Many of the generative models and inferential autoencoders produce blurry images for different reasons. Variational Autoencoder (VAE) [2] has this flaw maybe because of the lower bound approximation or restriction on the distribution. However, another reason might be the use of a non-perceptual distance in its objective [3]. Unconditional and conditional Generative Moment Matching Networks (GMMNs) [4,5] also use radial basis function kernel having ℓ_2 norm. Adversarial Autoencoder (AAE) [6] and unconditional/conditional Generative Adversarial Networks (GANs) [7,8] also use non-perceptual metrics in their objectives for comparison of real and fake data. Least Squares GAN (LSGAN) [9] uses ℓ_2 norm or Mean Square Error (MSE) in its loss function. However, MSE is shown not to be perfect for image quality assessment [3]. Structural Similarity Index (SSIM)

© Springer Nature Switzerland AG 2020
A. Campilho et al. (Eds.): ICIAR 2020, LNCS 12132, pp. 112–117, 2020.
https://doi.org/10.1007/978-3-030-50516-5_10

[10] is a perceptual measure for image quality. In this paper, we theoretically explain how SSIM can be used in different generative models and inferential autoencoders. Using SSIM can *improve the perceptual quality of the generated images* by these models. This is a poster paper and according to the expectation of the conference from a short poster paper, we suffice to the theoretical analysis and defer the empirical results to future work.

2   Structural Similarity Index, Image Structure Subspace, and SSIM Kernel

Consider two reshaped images $x_i, x_j \in \mathbb{R}^d$. The SSIM between two reshaped image blocks $\breve{x}_i = [x_i^{(1)}, \ldots, x_i^{(q)}]^\top \in \mathbb{R}^q$ and $\breve{x}_j = [x_j^{(1)}, \ldots, x_j^{(q)}]^\top \in \mathbb{R}^q$, in color intensity range $[0, l]$, is: $\mathbb{R} \ni \mathrm{SSIM}(\breve{x}_i, \breve{x}_j) := [(2\mu_{x_i}\mu_{x_j} + c_1)/(\mu_{x_i}^2 + \mu_{x_j}^2 + c_1)][(2\sigma_{x_i}\sigma_{x_j} + c_2)/(\sigma_{x_i}^2 + \sigma_{x_j}^2 + c_2)][(\sigma_{x_i,x_j} + c_3)/(\sigma_{x_i}\sigma_{x_j} + c_3)]$, where $\mu_{x_i} = (1/q)\sum_{k=1}^{q} x_i^{(k)}$, $\sigma_{x_i} = \left[(1/(q-1))\sum_{k=1}^{q}(x_i^{(k)} - \mu_{x_i})^2\right]^{0.5}$, $\sigma_{x_i,x_j} = (1/(q-1))\sum_{k=1}^{q}(x_i^{(k)} - \mu_{x_i})(x_j^{(k)} - \mu_{x_j})$, $c_1 = (0.01 \times l)^2$, $c_2 = 2c_3 = (0.03 \times l)^2$, and μ_{x_j} and σ_{x_j} are defined similarly for $\breve{x}_j$ [10]. Since $c_2 = 2c_3$, we can simplify SSIM to $\mathrm{SSIM}(\breve{x}_i, \breve{x}_j) = s_1(\breve{x}_i, \breve{x}_j) \times s_2(\breve{x}_i, \breve{x}_j)$, where $s_1(\breve{x}_i, \breve{x}_j) := (2\mu_{x_i}\mu_{x_j} + c_1)/(\mu_{x_i}^2 + \mu_{x_j}^2 + c_1)$ and $s_2(\breve{x}_i, \breve{x}_j) := (2\sigma_{x_i,x_j} + c_2)/(\sigma_{x_i}^2 + \sigma_{x_j}^2 + c_2)$. If the vectors $\breve{x}_i$ and $\breve{x}_j$ have zero mean, i.e., $\mu_{x_i} = \mu_{x_j} = 0$, the SSIM becomes $\mathbb{R} \ni \mathrm{SSIM}(\breve{x}_i, \breve{x}_j) = (2\breve{x}_i^\top \breve{x}_j + c)/(||\breve{x}_i||_2^2 + ||\breve{x}_j||_2^2 + c)$, where $c = (q - 1)c_2$ [11]. The distance based on SSIM, which we denote by $||.||_S$, is [11, 12]:

$$\mathbb{R} \ni ||\breve{x}_i - \breve{x}_j||_S := \sqrt{1 - \mathrm{SSIM}(\breve{x}_i, \breve{x}_j)} = \left[\frac{||\breve{x}_i - \breve{x}_j||_2^2}{||\breve{x}_i||_2^2 + ||\breve{x}_j||_2^2 + c}\right]^{0.5}, \quad (1)$$

where $\mu_{x_i} = \mu_{x_j} = 0$. Note that if the means of blocks $\breve{x}_i$ and $\breve{x}_j$ are not very different, $\sqrt{1 - \mathrm{SSIM}(\breve{x}_i, \breve{x}_j)}$ is still a good approximation to SSIM distance even without centering the blocks [12]. Some papers use this approximation and do not center the patches (cf. [13]).

Some works have used SSIM in machine learning for learning the image structure subspace [14] which captures the intrinsic features of an image in terms of structural similarity and distortions. In [14], a kernel, named SSIM kernel, is proposed which can be used in kernel methods in machine learning [15]. This kernel is $K = -(1/2) H D H$ where $\mathbb{R}^{n \times n} \ni H = I - (1/n)\mathbf{1}\mathbf{1}^\top$ is the centering matrix, $D \in \mathbb{R}^{n \times n}$ is the distance matrix, and n is the sample size of data. Let $D'_{i,j}$ be the distance map of two images x_i and x_j whose entry for every patch of these images is [12]:

$$\mathbb{R} \ni ||\breve{x}_i - \breve{x}_j||_S := \sqrt{2 - s_1(\breve{x}_i, \breve{x}_j) - s_2(\breve{x}_i, \breve{x}_j)}. \quad (2)$$

Note that one may use Eq. (1) for D' (and in SSIM kernel) but should center every patch while Eq. (2) does not require preprocessing but may be harder to compute. The (i, j)-th element of distance matrix is $\boldsymbol{D}(i, j) := ||\boldsymbol{D}'_{i,j}||_F$ where $||.||_F$ is the Frobenius norm. Furthermore, note that the SSIM distance is quasi-convex [12] so it is suitable for optimization [16] in different applications such as machine learning [14].

3   Generative Moment Matching Network

Maximum Mean Discrepancy (MMD), or kernel two sample test, is a measure of difference of two distributions by comparing their moments [17]. Let $\mathcal{X} := \{\boldsymbol{x}_i\}_{i=1}^{n_x}$ and $\mathcal{Y} := \{\boldsymbol{y}_i\}_{i=1}^{n_y}$ be two samples of the distributions P_x and P_y, respectively. The MMD is defined as $\mathrm{MMD}(P_x, P_y) := \sup_{f \in \mathcal{K}}(\mathbb{E}[f(\mathcal{X})] - \mathbb{E}[f(\mathcal{Y})])$ where $\mathcal{K}$ is a class of functions. If $\mathcal{K}$ is a unit ball in a universal reproducing kernel Hilbert space $\mathcal{F}$, we have: $\mathrm{MMD}^2(P_x, P_y) = ||\frac{1}{n_x}\sum_{i=1}^{n_x}\phi(\boldsymbol{x}_i) - \frac{1}{n_y}\sum_{i=1}^{n_y}\phi(\boldsymbol{y}_i)||_{\mathcal{F}}^2$

$$= \frac{1}{n_x^2}\sum_{i=1}^{n_x}\sum_{j=1}^{n_y} k(x_i, x_j) + \frac{1}{n_y^2}\sum_{i=1}^{n_y}\sum_{j=1}^{n_y} k(y_i, y_j) - \frac{2}{n_x n_y}\sum_{i=1}^{n_x}\sum_{j=1}^{n_y} k(x_i, y_j), \qquad (3)$$

where $\phi(.)$ is the pulling function and $k(.,.)$ is the kernel [15].

GMMN [4] uses Eq. (3) as the loss for training a network where the Radial Basis Function (RBF) kernel is utilized. The GMMN is a network which accepts random uniform samples in input and tries to match the moments of network's output with the batch of training data. It has two versions, i.e., in data space and code space. In the latter, the output layer of GMMN is the latent space of an autoencoder which is trained beforehand. Conditional GMMN [5] uses Conditional MMD (CMMD) where non-uniform weights are used in MMD. The CMMD is defined as $||C_{\mathcal{X}|\mathcal{Z}} - C_{\mathcal{Y}|\mathcal{Z}}||_{\mathcal{F}\otimes\mathcal{G}}^2$ where $\mathcal{Z}$ is the variable conditioned on and $\otimes$ is the tensor product (see [5] for more details).

In GMMN and conditional GMMN, the RBF kernel is used. SSIM kernel (see Sect. 2) can be used as the kernel in these two generative models. Note that only a universal kernel can be used in MMD [4,17] and CMMD [5]. Paper [18] has shown that according to the Stone-Weierstrass theorem [19], the universal kernels can be expanded in certain types of Taylor or Fourier series. The RBF kernel is an example. It is shown in [14] that the SSIM kernel can be expanded by Taylor series similar to the RBF kernel; hence, *SSIM kernel is a universal kernel* and thus can be used in GMMN and conditional GMMN.

4   Variational Autoencoder

VAE [2] can be considered as the nonlinear generalization of factor analysis [20]. Training of its encoder, with weights ψ, and decoder, with weights θ, can be seen as the E-step and M-step in expectation maximization algorithm, respectively. It maximizes the Evidence Lower Bound (ELBO) of the log likelihood of data [21]. The loss to be minimized in VAE is:

$$\mathcal{L} = -\mathrm{KL}\big(q(\boldsymbol{z}|\boldsymbol{x}, \boldsymbol{\psi}) \,||\, p(\boldsymbol{z})\big) + \mathbb{E}_{q(\boldsymbol{z}|\boldsymbol{x}, \boldsymbol{\psi})}\big[\log p(\boldsymbol{x}|\boldsymbol{z}, \boldsymbol{\theta})\big], \qquad (4)$$

where KL(.) is the KL-divergence, z is the latent variable, x is the input or re-generated data, and q and p are the conditional distributions in the encoder and decoder, respectively. The first and second terms in Eq. (4) are responsible for tuning the distribution of latent variable and better generation of data out of the latent variable, respectively. The second term, which takes care of data reconstruction, is usually replaced by the cross-entropy or ℓ_2 norm of data and generated data. However, ℓ_2 norm is not perfect for image fidelity [3]. The fact that the generated images by VAE are not perceptually satisfactory has been addressed by literature [7]. We can use SSIM distance, i.e. Eq. (1) or (2), for the second term to *measure how perceptually good the generated images are.*

5   Generative Adversarial Networks and Adversarial Autoencoder

As mentioned, GAN [7] is proposed to cover the perceptual lack of VAE. It claims that the ℓ_2 norm used in VAE is a man-made distance but a complicated distance measured by a classifier network is used in GAN. This makes GAN's generated images more perceptual. It is expected that *the generated images become much more perceptual when the power of the complicated classifier and the SSIM metric are combined.* The loss in GAN is a game-theoretical min-max problem:

$$\min_{G} \max_{D} \mathcal{L} = \mathbb{E}_{x \sim p(x)} \big[\log D(x) \big] + \mathbb{E}_{z \sim q(z)} \big[\log(1 - D(G(z))) \big], \qquad (5)$$

where p and q are the distributions of the data and the latent variable, respectively, and D and G are the discriminator (classifier) and generator, respectively. The probability of data coming from real distribution is denoted by $D(.)$. The first term in Eq. (5) is the log-likelihood of real data and the second term measures how different the generated and real data are. We can replace the second term by Eq. (1) or (2), i.e., $||x - G(z)||_S^2$, which is minimized and maximized by the generator and discriminator, respectively. This will *measure how perceptually different the generated and real data are,* resulting in perceptually better generated images because *both the generator and discriminator become more powerful in terms of perceptual differences.* Conditional GAN [8] can use the same idea in its loss to have better generated images. AAE [6], which is trained in an adversarial way like GAN, can also use the SSIM distance in its loss as was explained. Note that paper [22] has used a similar technique in GAN but with multi-scale SSIM (MS-SSIM) [23] used in Eq. (1) (see also [24]). Other ideas are used in some papers such as [25] which utilizes the middle-layer features of network for perceptually better generated images in GAN. On the other hand, LSGAN [9] uses ℓ_2 norm or MSE in its objective function to be optimized by the discriminator and generator. However, MSE is not suitable for perceptual image assessment. We propose to use SSIM distance, i.e. Eq. (1) or (2), in place of the MSE terms in the LSGAN objectives to have perceptually better generated images.

6 Conclusion

We theoretically analyzed how to use SSIM in generative models and inferential autoencoders including GMMN, VAE, AAE, GAN, and LSGAN. The use of SSIM in these models can improve the perceptual quality of the generated images as these models use non-perceptual distance metrics in their loss functions.

References

1. Ng, A.Y., Jordan, M.I.: On discriminative vs. generative classifiers: a comparison of logistic regression and Naive Bayes. In: Advances in Neural Information Processing Systems, pp. 841–848 (2002)
2. Doersch, C.: Tutorial on variational autoencoders. arXiv:1606.05908 (2016)
3. Wang, Z., Bovik, A.C.: Mean squared error: love it or leave it? A new look at signal fidelity measures. IEEE Signal Process. Mag. **26**(1), 98–117 (2009)
4. Li, Y., Swersky, K., Zemel, R.: Generative moment matching networks. In: International Conference on Machine Learning, pp. 1718–1727 (2015)
5. Ren, Y., Zhu, J., Li, J., Luo, Y.: Conditional generative moment-matching networks. In: Advances in Neural Information Processing Systems, pp. 2928–2936 (2016)
6. Makhzani, A., Shlens, J., Jaitly, N., Goodfellow, I., Frey, B.: Adversarial autoencoders. arXiv preprint arXiv:1511.05644 (2015)
7. Goodfellow, I., et al.: Generative adversarial nets. In: Advances in Neural Information Processing Systems, pp. 2672–2680 (2014)
8. Mirza, M., Osindero, S.: Conditional generative adversarial nets. arXiv preprint arXiv:1411.1784 (2014)
9. Mao, X., Li, Q., Xie, H., Lau, R.Y., Wang, Z., Paul Smolley, S.: Least squares generative adversarial networks. In: Proceedings of the IEEE International Conference on Computer Vision, pp. 2794–2802. IEEE (2017)
10. Wang, Z., Bovik, A.C., Sheikh, H.R., Simoncelli, E.P.: Image quality assessment: from error visibility to structural similarity. IEEE Trans. Image Process. **13**(4), 600–612 (2004)
11. Otero, D., Vrscay, E.R.: Unconstrained structural similarity-based optimization. In: Campilho, A., Kamel, M. (eds.) ICIAR 2014. LNCS, vol. 8814, pp. 167–176. Springer, Cham (2014). https://doi.org/10.1007/978-3-319-11758-4_19
12. Brunet, D., Vrscay, E.R., Wang, Z.: On the mathematical properties of the structural similarity index. IEEE Trans. IP **21**(4), 1488–1499 (2012)
13. Zhao, H., Gallo, O., Frosio, I., Kautz, J.: Loss functions for image restoration with neural networks. IEEE Trans. Comput. Imaging **3**(1), 47–57 (2016)
14. Ghojogh, B., Karray, F., Crowley, M.: Image structure subspace learning using structural similarity index. In: Karray, F., Campilho, A., Yu, A. (eds.) ICIAR 2019. LNCS, vol. 11662, pp. 33–44. Springer, Cham (2019). https://doi.org/10.1007/978-3-030-27202-9_3
15. Hofmann, T., Schölkopf, B., Smola, A.J.: Kernel methods in machine learning. Ann. Stat. **36**, 1171–1220 (2008)
16. Brunet, D., Channappayya, S.S., Wang, Z., Vrscay, E.R., Bovik, A.C.: Optimizing image quality. In: Monga, V. (ed.) Handbook of Convex Optimization Methods in Imaging Science, pp. 15–41. Springer, Cham (2017). https://doi.org/10.1007/978-3-319-61609-4_2

17. Gretton, A., Borgwardt, K.M., Rasch, M.J., Schölkopf, B., Smola, A.: A kernel two-sample test. J. Mach. Learn. Res. **13**(Mar), 723–773 (2012)
18. Steinwart, I.: On the influence of the kernel on the consistency of support vector machines. J. Mach. Learn. Res. **2**(Nov), 67–93 (2001)
19. De Branges, L.: The Stone-Weierstrass theorem. Proc. Am. Math. Soc. **10**(5), 822–824 (1959)
20. Harman, H.H.: Modern Factor Analysis. University of Chicago Press, Chicago (1976)
21. Kingma, D.P., Welling, M.: Auto-encoding variational Bayes. arXiv preprint arXiv:1312.6114 (2013)
22. Kancharla, P., Channappayya, S.S.: Improving the visual quality of generative adversarial network (GAN)-generated images using the multi-scale structural similarity index. In: International Conference on Image Processing, pp. 3908–3912. IEEE (2018)
23. Wang, Z., Simoncelli, E.P., Bovik, A.C.: Multiscale structural similarity for image quality assessment. In: The Thirty-Seventh Asilomar Conference on Signals, Systems & Computers, 2003, vol. 2, pp. 1398–1402. IEEE (2003)
24. Snell, J., Ridgeway, K., Liao, R., Roads, B.D., Mozer, M.C., Zemel, R.S.: Learning to generate images with perceptual similarity metrics. In: International Conference on Image Processing, pp. 4277–4281. IEEE (2017)
25. Wu, B., Duan, H., Liu, Z., Sun, G.: SRPGAN: perceptual generative adversarial network for single image super resolution. arXiv preprint arXiv:1712.05927 (2017)

Efficient Prediction of Gold Prices Using Hybrid Deep Learning

Turner Tobin[1] and Rasha Kashef[2(✉)]

[1] Computer Science Department, University of Western Ontario, London, Canada
ttobin@uwo.ca
[2] Electrical, Computer and Biomedical Engineering, Ryerson University, Toronto, Canada
rkashef@ryerson.ca

Abstract. Gold prices, in general, act counter to the market and are thus predictable to a certain degree based upon the fluctuations of other market entities. Neural networks have been applied to significant effect to difficult prediction problems and have achieved success in making predictions beyond traditional regression-based statistical models. Generally, such networks are hyperspecialized, and thus have effectiveness in a small subset of problems. This paper endeavors to take two models: RNN and CNN and hybridize them, creating a new model founded on their underlying logical theories and architectures. The primary goal in this paper is to show the effectiveness of the hybrid model in achieving a better gold price prediction that produces higher quality results at the cost of slightly increased complexity. Challenges and limitations of the proposed hybrid model are also addressed.

Keywords: Gold price prediction · Time-series · RNN · CNN · Hybrid learning

1 Introduction

While most financial assets are valued for their intrinsic properties combined with their relation to similar assets, gold tends to operate more extrinsically with no factors inherent to itself [1]. When investors lose trust in the market or centralized currencies, gold is often seen by the individual investor to be a viable option to purchase as a hedge [2]. When predicting gold prices, the data collection focuses on readily available market conditions, as opposed to having to input a company's specific financial statements and monitor its actions [3]. The performance of a predictive algorithm in forecasting gold prices comes down to the strength of the algorithm itself [4]. In this paper, we cover leading models in deep learning as applied with great success to the problem of predicting gold prices. The paper endeavors to show how the qualities that lead to success in different architectures can be combined to create new networks. Meaningful results can be accomplished through creating new networks in such a fashion, beyond creating ensembles through taking the output of one model and feeding it to another. We explore the underlying logical theories and architecture in combing both Convolutional Neural Networks (CNN) and Recurrent Neural Networks (RNN) in a hybrid model to achieve

© Springer Nature Switzerland AG 2020
A. Campilho et al. (Eds.): ICIAR 2020, LNCS 12132, pp. 118–129, 2020.
https://doi.org/10.1007/978-3-030-50516-5_11

successful prediction results. Experimental results on market data show improvement in predicting gold price two days in advance using the hybrid deep learning model. In Sect. 2, a brief discussion on CNN, RNN, and Autoregressive Integrated Moving Average (ARIMA) is given. Section 3 presents the hybrid model. Experimental results are presented and discussed in Sect. 4. Finally, derived conclusions and future directions are presented in Sect. 5.

2 Related Work and Background

Deep learning concerns itself with networks that achieve complexity through multi-layer interaction. Various models, activation functions and cost functions have been put forward and refined throughout the years [9]. It has been found that certain types of network architecture tend towards success in certain problem applications as CNN and RNN, especially for stock market prediction [17] and time-series analytics [5, 18]. In the following, a brief discussion on the convolutional neural networks, recurrent neural networks, and ARIMA predication model is presented.

2.1 Convolutional Neural Networks (CNN)

The CNN operate by reacting to input, passing that reaction forward to further neurons, and training a receptive field to interpret the response and begin to make predictions [5]. A CNN is implemented in a series of alternating layers; these layers are ordered such that convolutional layers alternating with pooling layers [6]. Convolutional layers apply an operation to the input but are not a trainable part of the network [6]. Pooling layers reduce the number of independent variables at the end of the process that is passed on to the receptive field [7]. Pooling layers can be either global or local; if the depth of the problem is shrunk between convolutional layers, it can be considered as a local pooling layer [8]. If the pooling happens at the end of the convolutions it is called global [8]. The more convolutional layers in the network the "deeper" it is considered. CNN are used to process images or visual data because of their ability to interpret spatially linked data [9]. The process begins with a 2-dimensional matrix of data that is tied together by proximity. Pooling allows for approximate answers for an area to be reached between steps to increase efficiency and to decrease the number of weights to be calculated down the line [7]. It is an effective way to take a data source with many variables, commonly pixels, and distill it into a smaller set of variables to use to predict with [5, 9].

2.2 Recurrent Neural Networks (RNN)

The RNN are networks where information is passed forward from node to node along a chain. Each node can process data from its memory. Long Short-Term Memory (LSTM) is a type of Recurrent Neural Network that can "forget" based on its predictions, using a "forget gate" [10]. A LSTM unit handles multiple operations while training. First, each unit decides how much of the factor to keep when bringing it into the unit (i.e. a weight) [11]. Then, it updates its future internal state. Finally, it determines the output to the next unit along the graph, and what to forget for later [11]. These "units" can be stacked, which creates an approximation of factors in time [12]. As CNNs are generally used in spatial data and images, RNNs are typically used in predicting the meaning behind historical data and speech [6]. Natural language processing can be applied through such RNN networks as nodes will remember things like names, genders, and plurality. Then nodes will make predictions as to what a sentence is saying using such factors because it encounters an ambiguous pronoun like a "they" or "she" [13]. It also retains the necessary information to make such decisions and forgets information as it no longer applies.

2.3 ARIMA (Autoregressive Integrated Moving Average)

ARIMA is a statistical model in which a series of coefficients are found and multiplied with some number of lagged values. This is combined with the linear combination of its margins of error over time to get a regression-based model with a combined margin of error [14]. This is covered as a standard normal statistical performance, as it is a widely accepted as a non-neural network method of forming predictions. Regression in some sense is useful to the structure of the model, as a gradient regression is applied to find the best possible weights and exponents for the series of nodes [15].

3 Hybrid CNN-RNN

The CNN have successfully modeled spatially related data, and RNN have effectively modeled temporally related data [9, 11, 13]. The ensemble approach endeavors to create a mixed model that has success modeling temporally related data, with a spatial aspect, or spatially related data with a temporal aspect. The hybrid model takes the concept of directed graphing and weighing from RNN, and pooling and convolution from CNN to form a more comprehensive model. This model is capable of solving either a problem meant for CNN or RNN or joint problems. Data is inputted into a matrix where every square in the matrix contains the root of a tree. Every tree starts with a node that contains a vector of one variable's values throughout time, a coefficient and an exponent, as well as whatever dummy nodes are required to maintain consistency in the shape of the tree as it evolves. This matrix, full of roots of trees, is then convoluted by passing a filter over the matrix. The filter takes an $n \times n$ square of the matrix and creates a new tree of these nodes, which then occupies a space in that matrix.

$$\begin{bmatrix} x_1 \ x_2 \ x_3 \\ x_4 \ x_5 \ x_6 \\ x_7 \ x_8 \ x_9 \end{bmatrix} = \begin{bmatrix} x_1 \ x_2 \\ x_3 \ x_4 \end{bmatrix}, \begin{bmatrix} x_2 \ x_3 \\ x_5 \ x_6 \end{bmatrix}, \begin{bmatrix} x_4 \ x_5 \\ x_7 \ x_8 \end{bmatrix}, \begin{bmatrix} x_5 \ x_6 \\ x_8 \ x_9 \end{bmatrix} \tag{1}$$

The sub-matrix Xsub from filtering is defined as:

$$Xsub = \begin{bmatrix} x_1 \ x_2 \\ x_3 \ x_4 \end{bmatrix} \tag{2}$$

The sub-matrix *Xsub* is represented as a tree, as shown in Fig. 1. The tree represents the values in the convolution matrix, with the final multiplication operation returning just the left side of the equation. This process continues until the matrix is 1×1, and the resulting values calculated from that node are the final predictions. This can be interpreted as a shrinking matrix, a growing tree, or an equation builder. The hybrid model uses trees to represent predictions on behalf of one section of the matrix and uses those trees to generate a future tree that could describe another scenario. The advantage of the hybrid model is that there is no need to know the best approach, as the model can find not only the optimum way to weight the factors within the nodes but also the ideal way to arrange data within the matrix to form a tree that can yield a good solution.

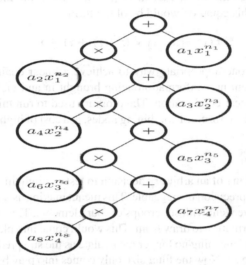

Fig. 1. Tree representation of the convolution matrix

3.1 The Hybrid Model

The input to our hybrid model is a 2D array as layers of sub-arrays, where the outside layer array is a container for sub-arrays, each contains a single variable's values through all trials and each index represents a one-time step. The output is another 2D array, where there is one sub-array which is the predicted values, with the last x being not a part of the training set when predicting. The hybrid CNN-RNN model constitutes of five main steps, the convolution step, the filtering step, the pooling step, Gradient Regression, and finally the verification step.

3.1.1 The Convolution Step

Starting with a matrix filled with trees of one useful node, a square is passed to the matrix, such that this square along with the internal nodes within this square form the new subtree. Trees are formed on the basis of the interaction between new nodes and old nodes to simulate an equation. This resolves the problem of scaling complexity in equation building. For example, if given three variables, 7 different terms are required.

$$(x) = xyz + xy + xz + zy + x + y + z \tag{3}$$

Using a tree system is an attempt to come up with close enough approximations of pre-diction answers while maintaining the directed graph-like structure of an RNN. Answers are passed upward through to make further predictions. The basic theory is such that not every combination need be checked, an answer can be made sufficiently acceptable and actionable given one possible combination and calibrated to be the best it can be. Enough combinations can be attempted and optimized to find a solution that is accept-able without having the computational cost of attempting close to every permutation. The algorithm operates the worst case of O $(r \times k \times n)$ in the model used to gather results in this study, this equation would be of the form:

$$f(x) = x * (y * (z + z) + y) + x \tag{4}$$

This example is just one appropriate way to achieve a linear chain for initial testing as it is able to represent new information being brought in and yielding a polynomial answer instead of a linear combination. The program used to run this model could also be adapted to try different forms of combining nodes, at a cost of higher time complexity.

3.1.2 Filtering Step

Filter size was something of an arbitrary decision to test through this paper, we erred on the side of the highest precision manageable. This meant making as many sub predictions as possible and thus we went with subgroups of four elements. This was thought to have the most possible information to draw from. This would come into play in a larger degree if nodes were to "lock" meaning no longer recalculate as the sub predictions would have a higher degree of impact. Now the filter size only comes into play because of the actual geometric layout of the tree. This matters as two factors that have an effect on each other may not be able to interact as they may be 3 squares from each other and not 2. If given enough attempts to form different matrix layouts this would be minimized, but we may be able to see that larger filter size has a positive effect in tree building. The other issue with filtering is the overlap. To what degree does the next square to build a tree from overlap with the last square. In this model, the overlap is maximized, as time complexity was not overly concerning with 732 trials.

3.1.3 The Pooling Step

When each tree is formed, it becomes an element of the next matrix which takes more trees as input. In the CNN sense, pooling as the matrix gets reduced to a smaller operat-ing size, data is retained throughout the process and the problem goes from being a very wide problem to very deep. Calibrating such a tree has the same complexity. The study began with the idea that in such a case the trees would at a certain point stop recalculating to save execution time, and they would become fixed positions. This would limit poten-tial solutions to problems but would save enough time that solutions could be reached given very large problems. In the model analyzed for results, all nodes remained active throughout the problem solving as it was a small sample of data.

3.1.4 Gradient Regression of Weights and Exponents

In a typical CNN, as data is added, weights would be adjusted, given enough trials weights would find their way to an optimal solution that best classifies problems [5]. In our scenario, all data is transferred to the nodes. Upon calibration, coefficients and exponents of the network are altered and then tested against the knowledge of inputs and solutions. With low size of inputs, this would lead to difficulty in reaching a meaningful solution if the precision with which we increment is too high. However, if the data sources became large, doing so in such a process would become too computationally expensive. This would require either transitioning to a model in which the values were not recalculated and instead weights were organized on an ongoing basis or potentially through taking slices. This could lead to greater error as potentially some outliers within the graph may not get captured accurately in the slices.

3.1.5 Verification Step

Upon updating nodes, we must know whether or not the solution has become closer to an optimal solution. Calculating the mean squared error was tested as a valid option, being slightly more computationally expensive than calculating the mean error and more punishing on extremely incorrect cases. When calibrating on a mean squared error basis, the hybrid model tended to find solutions that were more moderate. Instead of modeling large swings and being correct on some large variations and incorrect on others, the hybrid mode predicted results that were equally incorrect for all trials but resulted in a lower level of error overall. Calculating the number of predictions that are correct, or within a desirable range would be effective as it is accurately expressing the types of solutions. However; this type of solution leads to less accuracy when calibrating as it outputs what is essentially a step function. When calibrating a node by incrementing a coefficient by one thousandth, it may make the solution better, but not cause an increase of one more correct trial instantly, and thus the program would interpret it as no change. Therefore, further experimentation needs to go into a verification method that outputs a metric by the number of successes, possibly could yield stronger results.

3.2 Challenges in the Hybrid Model

In the hybrid model for this problem instance there are three main challenges: (1) the occurrence of a prime numbered quantity of free variables, (2) data gathered in business days, and (3) the inability to use growth vectors due to the shortcomings of the data.

3.2.1 Prime Numbered Variables

In the case of odd-numbered variables, data needs to be filled into the matrix such that proper filtering can occur. This means that for every subsection of the matrix, that area of the matrix is populated. Originally, the solution was that when filling a matrix if the data had run out but there were still areas to populate, those areas should be filled in with variable number one. This variable acts as an arbitrary number that would be automatically factored out if irrelevant. Later in testing, it was seen that while filling the matrix, a simple solution would be to take the index to the mod of the number of variables, so it was not always the origin being used as a filler.

3.2.2 Data Challenges

Inherent limits to financial data are that there is only data for weekdays. The markets are only open Monday to Friday. The data used was additionally weak as it was collected once per day. This means per week there is only 5 data points per variable. For the model to be able to produce results with significance for the future, the model uses data two days out to make predictions. A prediction on Monday using Thursday's data would follow a different decision-making pattern than a prediction for Wednesday using Monday's data. Thus, the decision was made to keep all decisions within the week. This limits the effective data, as per week now there are only three predictions and thus there is 3/5 of an already small data pool. This means to get enough data to train a model, it must be trained over many years. Previous studies have shown financial models train better on a tighter timescale and stretching past a number of years leads to decreased accuracy in results [16]. Future work on using data across different time zones to reduce the gap between Friday and Monday is recommended.

3.2.3 Growth Vectors

Instead of holding absolute values in each node, it is recommended to use a growth vectors. The model can find the best combination of all growth vectors, then the estimated growth of the output variable from one trial to another would increase the ability to bring unlike data sources together. The growing of two factors, and the shrinking of another factor may lead the output to grow at an exponential rate, and the model then would not generate an absolute number. For financial data, multiplying the trading volume by the share price to get an answer in dollars of gold price. This does not make inherent sense and would not be in appropriate units. The model will still output a good prediction, but the prediction would make more sense as how much the gold price would grow, given the growth rate of the trading volume and share price of another asset. This would potentially solve problems in data sets such as this one where outputs vary from in the range of 80 to the range of 20. The model needs to only predict the drop and then predict it stagnates, as opposed to predicting its exact position. The issue is that to get growth numbers, numbers from -1 days are needed. However, we have daily data, and 2-day lead time, thus we only have predictions for Thursday and Friday, and we would have 2/5 of the sample size. The main purpose of switching this model specifically to a growth-based algorithm would be the ability to include things like sentiment analysis of news sources and Google Trends data into predictions. This will not give an output of trading volume only but increase in gold searches and a decrease in the gold price, thus this could have a meaning in terms of an increase in trading volume.

4   Experimental Results

In this section, we test the performance of the hybrid model to that of ARIMA, CNN and LSTM as benchmark [4]. The dataset used Yahoo Finance data, future datasets will be considered. The given time period used in the contrasting study was used. Data was gathered on the performance of the S&P 500 index and the Barrick Gold price, which included their daily highs and lows, trading volume, open, close and adjusted close. The data, other than the output variable of daily high for gold price, was shifted back 2 days so the model gave actionable information. The variables were not duplicated and shifted back since there is only data on weekdays, which makes shifting it a challenge. The output predictions are for the same entity (GOLD), and the data gathered is also at intervals of one day [4]. Figures 2, 3 and 4 show the prediction results of the ARIMA, CNN, and LSTM, respectively. We can observe that ARIMA has the worst performance and the LSTM achieves significant forecasting results as compared to the ARIMA and CNN. To increase comparability, we will show results of the hybrid model up to the same point in time as the other models. The analysis of the Mean Squared Errors (MSE) in Table 1 shows that the hybrid model outperforming the CNN, LSTM and ARIMA.

Table 1. Mean squared errors (MSE) comparison

Model	Mean squared error
ARIMA	1.90E+07
CNN	8.65E+00
LSTM	7.12E+00
Hybrid CNN-RNN	6.00E+00

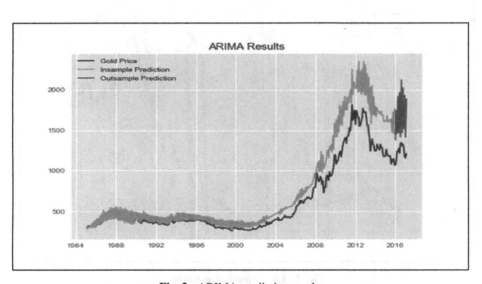

Fig. 2. ARIMA prediction results

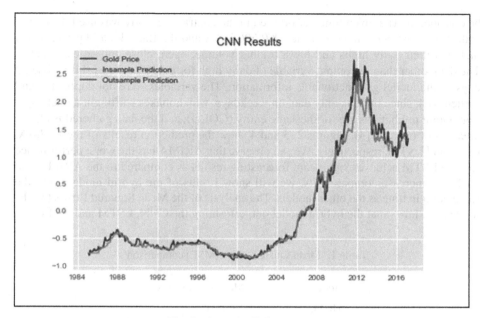

Fig. 3. CNN prediction results

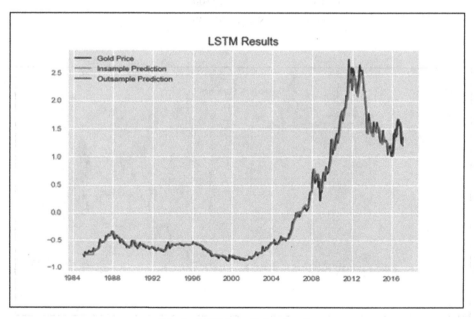

Fig. 4. LSTM prediction results

Figures 5.a and 5.b show the performance of the Hybrid Model, with one series representing predictions and the other represents the actual prices up until the end of the comparable data for data collected in 2017 and 2016, respectively.

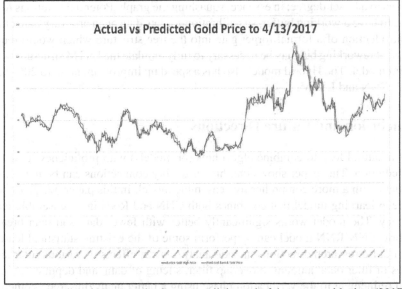

(a) Predicted Vs. Actual prices up until the end of the comparable data (2017)

(b) Predicted Vs. Actual prices up until the end of the comparable data (2016)

Fig. 5. Hybrid CNN-RNN prediction results

Looking at the graphed results in Figs. 4 and 5, we can see that the LSTM and hybrid model perform to similar degrees of accuracy, and these mean squared errors represent deviations away from the standard pattern. Two large drops in the price cause a series of large mispredictions (Fig. 5.b). This was covered in the verification section, as switching to verifying solution quality by mean squared error ended up making the same predictions to a lesser degree: in essence, squashing the graph. Potential solutions to this would be to have a working buffer, so calibration forgets data at an arbitrary date cutoff, or the introduction of a formal forget gate into the tree structure which would require some logical reworking but may be necessary to fully emulate the LSTM structure within the hybrid model. The Hybrid model also has a speed up improvement up to 20% faster than both CNN and LSTM.

5 Conclusion and Future Directions

It seems a natural leap to combine algorithms (or models) with proficiency to achieve better prediction. This paper shows that higher quality connections can be made using limited data with a more comprehensive ensemble model. In this paper, we proposed a hybrid deep learning model that combines both CNN and RNN in an ensemble learning strategy. The model works significantly better with fewer data and data breadth. The Hybrid CNN-RNN model can outperform some of the existing statistical learning models. Future research directions include analyzing the effects of various filter sizes, quantities of filler data, amount to overlap filter, slicing of data, and depth cutoffs. An extended refinement to the verification phase using a better analyzing tree geometry is also of future investigation. Lastly, whether the calibration should occur while building the tree or at the end from the top down is a future research direction. The model needs further optimization by adding data from more market variables as well as adding sentiment analysis or Google Trends data. In addition, obtaining better quality data with more sources to bypass the one entry per day restraints and experimenting with growth vectors would be an extension to the work in this paper.

References

1. Cai, J., Cheung, Y.-L., Wong, M.C.S.: What moves the gold market? J. Futures Mark. **21**(3), 257–278 (2001)
2. Baur, D.G., Lucey, B.M.: Is gold a hedge or a safe haven? An analysis of stocks, bonds and gold. Fin. Rev. **45**(2), 217–229 (2010)
3. Aggarwal, R., Soenen, L.A.: The Nature and efficency of the gold market. J. Portfolio Manag. **14**, 201–213 (1988)
4. Srivastava, S.: Deep Learning in Finance. Towards Data Science (2017)
5. Razavian, A.S., Azizpour, H., Sullivan, J., Carlsson, S.: CNN features off-the-shelf: an astounding baseline for recognition. In: The IEEE Conference on Computer Vision and Pattern Recognition (CVPR) Workshops (2014)
6. Abdel-Hamid, O., Deng, L., Yu, D.: Exploring convolutional neural network structures and optimization techniques for speech recognition. In: INTERSPEECH, Lyon (2013)
7. Simonyan, K., Zisserman, A.: Very deep convolutional networks for large-scale image recognition. In: ICLR, San Diego (2015)
8. Lin, M., Chen, Q., Yan, S.: Network in network. arXiv (2014)
9. Fischer, M.M.: Computational neural networks—tools for spatial data analysis. In: Spatial Analysis and Geo Computation, pp. 79–102 (2006). https://doi.org/10.1007/3-540-35730-0_6
10. Gers, F.A., Schmidhuber, J., Cummins, F.: Learning to forget: continual prediction with LSTM. In: 9th International Conference on Artificial Neural Networks: ICANN 1999, Edinburgh (1999)

11. Hochreiter, S., Schmidhuber, J.: Long short-term memory. Neural Comput. **9**(8), 1735–1780 (1997)
12. Hansen, J.V., McDonald, J.B., Nelson, R.D.: Time series prediction with genetic-algorithm designed neural networks: an empirical comparison with modern statistical models. Comput. Intell. **15**, 171–184 (2002)
13. Mikolov, T., Karafiát, M., Burget, L., Černocký, J., Khudanpur, S.: Recurrent neural network based language model. In: INTERSPEECH (2010)
14. Sowell, F.: Modeling long-run behaviour with the fractional ARIMA model. J. Monetary Econ. **29**, 277–302 (1992)
15. Adebiyi, A., Adewumi, A.O., Ayo, C.K.: Comparison of ARIMA and artificial neural networks models for stock price prediction. J. Appl. Math. **2014**, 614342:1–614342:7 (2014)
16. Sami, I., Nazir, K.: Predicting future gold rates using machine learning approach. Int. J. Adv. Comput Sci. Appl. **8**(12), 92–99 (2018)
17. Kim, T., Kim, H.Y.: Forecasting stock prices with a feature fusion LSTM-CNN model using different representations of the same data. PLoS ONE **14**(2), e0212320 (2019)
18. Tan, X., Kashef, R.: Predicting the closing price of cryptocurrencies: a comparative study. In: Proceedings of the Second International Conference on Data Science, E-Learning and Information Systems (DATA 2019). Association for Computing Machinery, New York, NY, USA (2019). Article 37, 1–5. https://doi.org/10.1145/3368691.3368728

Exploring Information Theory
and Gaussian Markov Random Fields
for Color Texture Classification

Cédrick Bamba Nsimba[✉] and Alexandre L. M. Levada

Department of Computer Science, Federal University of São Carlos,
São Carlos, SP, Brazil
cedrick.bamba@ifsp.edu.br, alexandre@dc.ufscar.br

Abstract. This paper proposes a novel approach to compute information theory measures from Gaussian Markov Random Field (GMRF) for color texture classification task. We firstly transform the three color channels of the input image into three set of sub-bands of the form LL, HH, HL and LH using three Discret Wavelet Transforms. We then visualize each sub-band as a GMRF from which we generate features by computing Fisher information matrix and Shannon's entropy to encode the local spatial dependency. The concatenation of the computed features are then used as the texture descriptor, which in turn is used as input for the classifiers referred to in this work. Experiments were performed with color texture images from public databases widely used in the literature that demonstrate the efficiency of the proposed method.

Keywords: Color texture classification · Gaussian Markov random field · Information theory · Fisher information · Texture descriptors

1 Introduction

In recent years, researches in computer vision have increased considerably and demonstrated significant successes in real-world challenges. Color and texture are two main aspects of any natural images and have proved to be an important element in a lot of computer vision applications such as object recognition [1,2], flower recognition [3,4], texture classification [5], material recognition [6,7], remote sensing [10], tissue recognition in medical images [11,21] and many more. As a consequence of this fact, many researches have been carried out over the last years in order to create new powerful tools for feature extraction in texture images. Some of these works are mentioned in the next paragraphs, specially those ones related to this proposal.

Some recent comprehensive surveys on texture representation for texture classification can be found in [22]. Among state-of-the-art propositions, a great number of texture analysis methods have been developed over the years, each one exploring a novel approach to extract the image's texture information. For instance, we have classical methods based on second-order statistics such as

© Springer Nature Switzerland AG 2020
A. Campilho et al. (Eds.): ICIAR 2020, LNCS 12132, pp. 130–143, 2020.
https://doi.org/10.1007/978-3-030-50516-5_12

Co-occurrence Matrices (GLCM) [25], Haralick [26], Histogram of Oriented Gradients (HoG) [27] and Local Binary Patterns (LBP) [23]. Over many years, these four descriptors have gained a large amount of interest in many computer vision researching groups. The features extracted using the aforementioned descriptors have proven to be discriminative in classifying texture patterns. However, due to the fact that they are applied to gray-scale images, their performance on natural images is limited without exploiting color information. To overcome this issue, an extension of the classical LBP and several other recent strategies have been proposed to incorporate these local patterns with color features. In [28], a novel approach is proposed to encode cross-channel texture correlation for color texture classification task. The authors quantitatively studied the correlation between different planes using LBP and Shannon's information theory to measure the correlation between channel pairs.

The another family of methods that has drawn the attention of many researchers and has provided quite effective image classification performance is CNN-based framework. Despite its successes in image classification, CNNs alone are not very suitable for texture classification. It is known that directly applying CNNs on texture classification problems results in only moderate accuracy [29,30]. At the same time, a key discovery in learning CNNs was that its features pretrained on very large datasets were found to transfer well to many other problems, including texture analysis, with a relatively modest adaptation effort [22]. Some CNN-based texture models that can be stated here are the model of Cimpoi et al. in [31] which combined CNN features with Fisher Vector encoder (FV-CNN). The authors in [33] proposed texture CNN (T-CNN) which is a specialized version of CNN for texture classification.

In this work, we would like to develop a powerful color texture image classification strategy strongly based on the capacity of information theory measures to capture significant textural information from the input color image. Our motivation here is to represent and characterize an input image by a set of local descriptors generated only from a whole interaction of a given pixel with its neighbours inside a wavelet sub-band of a given channel (R, G or B). The idea of using measurements strongly based on information theory, namely Fisher information matrix and Shannon's entropy is to explore color texture as a set of R, G and B wavelet sub-bands where the coefficients of each sub-band are modeled by GMRF. So each sub-band can be considered as a complex and unique system, from which we can estimate parameters and compute the Fisher information matrix (metric tensor components) and Shannon entropy, given an output of the random field model. All these measures are functions of the model parameters, more precisely, of the variance and the inverse temperature ($\beta = \frac{1}{T}$), which is used to control the global spatial dependence structure of the system. By embedding both color and theoretic information, by mean of GMRF and wavelet tools, we propose a novel approach for color texture image description.

2   Overview

Over the last decades, wavelets [18–20] have been widely applied on texture analysis due to the possibility of analyzing a variability of features at different scales. Several works have been carried out on the statistical modeling of wavelet distribution for texture characterization. Due to the variability of the wavelet histogram shapes, it is important to have flexible statistical models that are capable of capturing each of them. In this context, it has been shown that the distribution of the wavelet coefficients within a sub-band can be modeled by a Generalized Gaussian Distribution (GGD) with zero mean [12,13] and its variants [14,15].

On the other hand, since Gaussian Markov random field (GMRF) has proved to offer a powerful framework for texture analysis [16,17] due to its local conditional probability distribution which encapsulates spatial dependencies between a pixel and its neighbours [34], many authors have proposed to conduct texture classification using texture features derived from model based GMRF method in combination with different wavelet transforms [38–40].

However, the above conjunction approach involving GMRF with wavelets just transforms the input image to the wavelet domain and then, applies GMRF to extract features from the resulting wavelet coefficients of the whole image, and ignores to model the whole interaction between coefficients of the same sub-bands which carries texture correlation information. In fact, there exists texture patterns that may appear to be locally irrelevant (inside a sub-band) and may become extremely informative in the more global perspective. This is an exact consequence of any system in witch pieces of data along different locations and scales are related by the intricate non-linear relationship [41]. Encoding these intricate non-linear relationships effectively can greatly boost the discriminative ability of the feature vector.

Within this context, information theoretic measures play a fundamental role for revealing what is crucially informative or not since they represent statistical knowledge in a systematic manner. Entropy and Fisher information are two of the most popular measures in information theory. Shannon's entropy [35] measures a degree of uncertainty about any source of information while Fisher information [36,37] measures the amount of information a random sample conveys about an unknown parameter, besides having a geometric interpretation [42].

In this paper, we propose a novel method to compute information theory measures from a wavelet sub-band modeled by a GMRF in order to conduct texture classification task. In practice, we transform an input image into n wavelet sub-bands and visualize each sub-band as a GMRF. We then compute Fisher information matrix and Shannon's entropy from such sub-bands. These measurements of n sub-bands are finally concatenated into a final feature vector. The novel intricate non-linear relationship encoding strategy of the proposed method can guarantee discriminative texture descriptors, which then allow higher accuracy in classification task.

Roughly speaking, the contributions of this work are three-fold: (a) Information-theoretic descriptor for color texture classification using wavelets

and Gaussian Markov random fields; (b) Incorporation of non-Euclidean geometry by means of the Fisher information matrix, which defined the metric tensor of
the underlying parametric model and (c) Combination of color, wavelet features
and statistical properties defining a novel approach based on complex systems
analysis.

3   Proposed Framework for Color Texture Classification

The proposed algorithm consists of two primary stages: the modeling of wavelet
coefficients using an isotropic pairwise GMRF and color texture descriptors to
characterize each input image. Each of them is now described in detail. Then,
the complete framework is presented. First of all, we introduce, in few lines, the
notion of Fisher information in GMRF in order to facilitate the understanding
of the proposal.

Gaussian random fields are important models in studying non-linear interactions between elements of a stochastic complex system along time. In the context
of the current work, these models allow us to derive exact closed-form expressions
for two relevant quantities: (1) estimators for the coupling parameter (β); and
(2) the expected Fisher information matrix of each wavelet sub-band. With the
equivalence between Gibbs random fields (global models) and Markov random
fields (local models) stated by Hammersley-Clifford theorem [24] it is possible to
characterize an isotropic pairwise Gaussian random field by a set of local conditional density functions (LCDFs), avoiding computations with the joint Gibbs
distribution (due to the partition function).

An isotropic pairwise Gaussian Markov random field related to a local neighborhood system η_i defined on a lattice $S = \{s_1, s_2, ..., s_n\}$ is completely characterized by a set of n local conditional density functions
$p(x_i|\eta_i, \theta)$, given by

$$p(x_i|\eta_i, \theta) = \frac{1}{\sqrt{2\pi\sigma^2}} \exp\left\{-\frac{1}{2\sigma^2}\left[x_i - \mu - \beta\sum_{j\in\eta_i}(x_j - \mu)\right]^2\right\} \tag{1}$$

where $\theta = (\mu, \sigma^2, \beta)$ is the parameters vector composed of the expected value
(mean), the variance of the random variables in the field and the inverse temperature or coupling parameter, which is used to control the global spatial dependence structure of the system. In Eq. (1), the intensity value of the pixel at the
position i is given by x_i, and η_i is its local neighboring pixels.

Under the information theory perspective, it is known that Fisher information matrix of a statistical model carries important information about the
intrinsic geometric properties of its parametric space. The objective here is to
extract Fisher information matrix components from GMRF model defined on
each wavelet sub-band. In this regard, lets assume here that $p(x_i|\eta_i, \theta)$ is the local
conditional probability density function of a GMRF as defined in the Eq. (1).

Then the Fisher information matrix, in turn, is defined as

$$
\begin{aligned}
\{I(\theta)\}_{ij} &= E\left[\left(\frac{\partial}{\partial\theta_i}\log p(x_i|\eta_i,\theta)\right)\left(\frac{\partial}{\partial\theta_j}\log p(x_i|\eta_i,\theta)\right)\right] \\
&= -E\left[\frac{\partial^2}{\partial\theta_i\partial\theta_j}\log p(x_i|\eta_i,\theta)\right] \; for \; i,j = 1,...,n.
\end{aligned}
\tag{2}
$$

Shannon's entropy, in turn, is defined as the expected value of a self-information according to the Eq. (3).

$$
\begin{aligned}
H_\beta &= -E\left[\log L(\theta; X)\right] \\
&= -E\left[\log \prod_{i=1}^{n} p(x_i|\eta_i,\theta)\right]
\end{aligned}
\tag{3}
$$

Replacing Eq. (1) into (3), we have an expression for the entropy of a Gaussian-Markov random field:

$$
H_\beta = \frac{1}{2}\left[log\left(2\pi\sigma^2\right)+1\right] - \frac{1}{\sigma^2}\left[\beta\sum_{j\in\eta_i}\sigma_{ij} - \frac{\beta^2}{2}\sum_{j\in\eta_i}\sum_{k\in\eta_i}\sigma_{jk}\right]
\tag{4}
$$

where σ_{ij} denotes the covariance between the central variable x_i and one of its neighbors $x_j \in \eta_i$ and σ_{jk} denotes the covariance between two variables x_j and x_k in the neighborhood η_i.

A fundamental step in our method is the computation of the Fisher information matrix (metric tensor components) and entropy, given an output of the random field model. All these measures are function of the model parameters, more precisely, of the variance and the inverse temperature β. In all the experiments conducted in this paper, the Gaussian random field parameters μ and σ^2 are both estimated by the sample mean and variance, respectively, using the maximum likelihood estimatives. However, maximum likelihood estimation is intractable for the inverse temperature parameter estimation (β), due to the existence of the partition function in the joint Gibbs distribution. An alternative, proposed by Besag [32], is to perform maximum pseudo-likelihood estimation, which is based on the conditional independence principle. The basic idea with this proposal is to replace the independence assumption by a more flexible conditional independence hypothesis, allowing us to use the local conditional density functions of the random field model in the definition of a likelihood function, called pseudo-likelihood.

Let an isotropic pairwise Markov random field model be defined on a rectangular lattice $S = \{s_1, s_2, \ldots, s_n\}$ with a neighborhood system η_i. Assuming that $\mathbf{X} = \{x_1, x_2, \ldots, x_n\}$ denotes the set corresponding to the observations, the pseudo-likelihood function of the model is defined by:

$$
L\left(\vec{\theta}; \mathbf{X}\right) = \prod_{i=1}^{n} p(x_i|\eta_i,\vec{\theta}).
\tag{5}
$$

Plugging Eq. (1) into Eq. (5) leads to:

$$log\, L\left(\vec{\theta}; \mathbf{X}\right) = -\frac{n}{2}log\left(2\pi\sigma^2\right) - \frac{1}{2\sigma^2}\sum_{i=1}^{n}\left[x_i - \mu - \beta\sum_{j\in\eta_i}(x_j - \mu)\right]^2. \qquad (6)$$

We can also compute this estimative from the covariance matrix of the configuration patterns. Let Σ_p be the covariance matrix of the random vectors $\vec{p}_i, i = 1, 2, \ldots, n$, obtained by lexicographic ordering the local configuration patterns $x_i \cup \eta_i$. In this work, we choose a second-order neighborhood system, making each local configuration pattern a 3×3 patch. Thus, since each vector $\vec{p}_i$ has 9 elements, the resulting covariance matrix Σ_p is 9×9. Let Σ_p^- be the sub-matrix of dimensions 8×8 obtained by removing the central row and central column of Σ_p (these elements are the covariances between the central variable x_i and each one of its neighbors $x_j \in \eta_i$). Also, let $\vec{\rho}$ be the vector of dimensions 8×1 formed by all the elements of the central row of Σ_p, excluding the middle one (which denotes the variance of x_i actually).

In other words, given $\mathbf{X}$, all the measures we need to estimate β are based solely in the covariance matrix of the patches, Σ_p. The maximum pseudo-likelihood of β can be computed by:

$$\hat{\beta}_{MPL} = \frac{\|\vec{\rho}\|_+}{\|\Sigma_p^-\|_+} \qquad (7)$$

where $\|A\|_+$ denotes the sum of the elements of the matrix A (not to be confused with the matrix norm).

3.1 Modeling Wavelet Coefficients Using an Isotropic Pairwise GMRF

In what follows below, we adopt the Discrete Wavelet Transform (DWT) to decompose the input texture image through a 3-level decomposition scheme, which means that each color channel will be transformed into 10 sub-bands, which we denote here by:

$$\mathcal{S_R} = \{LL_3^{\mathcal{R}}, LH_3^{\mathcal{R}}, HL_3^{\mathcal{R}}, HH_3^{\mathcal{R}}, LH_2^{\mathcal{R}}, HL_2^{\mathcal{R}}, HH_2^{\mathcal{R}}, LH_1^{\mathcal{R}}, HL_1^{\mathcal{R}}, HH_1^{\mathcal{R}}\} \quad (8)$$

$$\mathcal{S_G} = \{LL_3^{\mathcal{G}}, LH_3^{\mathcal{G}}, HL_3^{\mathcal{G}}, HH_3^{\mathcal{G}}, LH_2^{\mathcal{G}}, HL_2^{\mathcal{G}}, HH_2^{\mathcal{G}}, LH_1^{\mathcal{G}}, HL_1^{\mathcal{G}}, HH_1^{\mathcal{G}}\} \quad (9)$$

$$\mathcal{S_B} = \{LL_3^{\mathcal{B}}, LH_3^{\mathcal{B}}, HL_3^{\mathcal{B}}, HH_3^{\mathcal{B}}, LH_2^{\mathcal{B}}, HL_2^{\mathcal{B}}, HH_2^{\mathcal{B}}, LH_1^{\mathcal{B}}, HL_1^{\mathcal{B}}, HH_1^{\mathcal{B}}\} \quad (10)$$

The Eqs. (8), (9) and (10) represent the set of wavelet sub-bands associated to $\{\mathcal{R}, \mathcal{G}, \mathcal{B}\}$ color channels.

The GMRF model for each decomposition level can be defined by considering a 3×3 local configuration pattern $x_i \bigcup \eta_i$ because we chose a second-order neighborhood system (i.e., of size $\Delta = 8$, usual choices of Δ are $4, 8, 12, 20, 24$, etc.). In each sub-band, $\vec{p}_i$ represents a random variable that associates a pixel x_i with its neighbors (η_i). We assume that the wavelet coefficients from a given sub-band can be modeled by a Gaussian Markov Random Field which is specified through the local characteristics $p(x_i | \eta_i, \theta)$ as defined in Eq. (1). The motivation behind switching to the wavelet domain is that by exploiting the multi-resolution property of the DWT a neighborhood of some fixed order is assumed to already capture texture characteristics at different resolutions.

3.2   Feature Extraction

The objective of this section is to compute Fisher information matrix and Shannon's entropy from GMRF models as stated in the Sect. 3.1. From each wavelet sub-band (which can belong either to $\mathcal{S}_R$, $\mathcal{S}_G$ or to the $\mathcal{S}_B$) modeled by GMRF, we first generate a dataset with patches of $n \times n$ (where $n \times n$ is equal to the value of the largest positive integer $\leq$ neighborhood-order $+1$). Each row of the generated dataset will then be composed by nine elements (due to the use of 3×3-patch dimension). We then compute the Fisher information matrix components and Shannon's entropy from this dataset. For purposes of notation, we define the Fisher information matrices as:

$$g^{(1)}(\vec{\theta}) = \begin{pmatrix} I_{\mu\mu}^{(1)}(\vec{\theta}) & I_{\mu\sigma^2}^{(1)}(\vec{\theta}) & I_{\mu\beta}^{(1)}(\vec{\theta}) \\ I_{\sigma^2\mu}^{(1)}(\vec{\theta}) & I_{\sigma^2\sigma^2}^{(1)}(\vec{\theta}) & I_{\sigma^2\beta}^{(1)}(\vec{\theta}) \\ I_{\beta\mu}^{(1)}(\vec{\theta}) & I_{\beta\sigma^2}^{(1)}(\vec{\theta}) & I_{\beta\beta}^{(1)}(\vec{\theta}) \end{pmatrix} \tag{11}$$

and

$$g^{(2)}(\vec{\theta}) = \begin{pmatrix} I_{\mu\mu}^{(2)}(\vec{\theta}) & I_{\mu\sigma^2}^{(2)}(\vec{\theta}) & I_{\mu\beta}^{(2)}(\vec{\theta}) \\ I_{\sigma^2\mu}^{(2)}(\vec{\theta}) & I_{\sigma^2\sigma^2}^{(2)}(\vec{\theta}) & I_{\sigma^2\beta}^{(2)}(\vec{\theta}) \\ I_{\beta\mu}^{(2)}(\vec{\theta}) & I_{\beta\sigma^2}^{(2)}(\vec{\theta}) & I_{\beta\beta}^{(2)}(\vec{\theta}) \end{pmatrix} \tag{12}$$

where $g^{(1)}(\vec{\theta})$ is the type-I Fisher information matrix and $g^{(2)}(\vec{\theta})$ is the type-II Fisher information matrix. Since we have three channels, this procedure will result in three feature vectors, each one representing a specific wavelet sub-band of a specific channel. These feature vectors are of the form :

$$\mathcal{V}_\mathcal{R} = \left[I_{\mu\mu}^{(1)}(\theta)^\mathcal{R}, I_{\mu\mu}^{(2)}(\theta)^\mathcal{R}, I_{\sigma^2\sigma^2}^{(1)}(\theta)^\mathcal{R}, I_{\sigma^2\sigma^2}^{(2)}(\theta)^\mathcal{R}, I_{\sigma^2\beta}^{(1)}(\theta)^\mathcal{R}, I_{\sigma^2\beta}^{(2)}(\theta)^\mathcal{R}, I_{\beta\beta}^{(1)}(\theta)^\mathcal{R}, I_{\beta\beta}^{(2)}(\theta)^\mathcal{R}, H_\beta^\mathcal{R} \right] \tag{13}$$

$$\mathcal{V}_\mathcal{G} = \left[I_{\mu\mu}^{(1)}(\theta)^\mathcal{G}, I_{\mu\mu}^{(2)}(\theta)^\mathcal{G}, I_{\sigma^2\sigma^2}^{(1)}(\theta)^\mathcal{G}, I_{\sigma^2\sigma^2}^{(2)}(\theta)^\mathcal{G}, I_{\sigma^2\beta}^{(1)}(\theta)^\mathcal{G}, I_{\sigma^2\beta}^{(2)}(\theta)^\mathcal{G}, I_{\beta\beta}^{(1)}(\theta)^\mathcal{G}, I_{\beta\beta}^{(2)}(\theta)^\mathcal{G}, H_\beta^\mathcal{G} \right] \tag{14}$$

$$\mathcal{V}_\mathcal{B} = \left[I_{\mu\mu}^{(1)}(\theta)^\mathcal{B}, I_{\mu\mu}^{(2)}(\theta)^\mathcal{B}, I_{\sigma^2\sigma^2}^{(1)}(\theta)^\mathcal{B}, I_{\sigma^2\sigma^2}^{(2)}(\theta)^\mathcal{B}, I_{\sigma^2\beta}^{(1)}(\theta)^\mathcal{B}, I_{\sigma^2\beta}^{(2)}(\theta)^\mathcal{B}, I_{\beta\beta}^{(1)}(\theta)^\mathcal{B}, I_{\beta\beta}^{(2)}(\theta)^\mathcal{B}, H_\beta^\mathcal{B} \right] \tag{15}$$

where H_β denotes the Shannon's entropy. Since we have ten wavelet sub-bands for each channel, the feature vectors of all the three channels will be of the form:

$$\mathcal{F}_{\mathcal{R}} = [\mathcal{V}'_{R1} \circ \mathcal{V}'_{R2} \circ \mathcal{V}'_{R3} \circ \mathcal{V}'_{R4} \circ \mathcal{V}'_{R5} \circ \mathcal{V}'_{R6} \circ \mathcal{V}'_{R7} \circ \mathcal{V}'_{R8} \circ \mathcal{V}'_{R9} \circ \mathcal{V}'_{R10}] \tag{16}$$

$$\mathcal{F}_{\mathcal{G}} = [\mathcal{V}'_{G1} \circ \mathcal{V}'_{G2} \circ \mathcal{V}'_{G3} \circ \mathcal{V}'_{G4} \circ \mathcal{V}'_{G5} \circ \mathcal{V}'_{G6} \circ \mathcal{V}'_{G7} \circ \mathcal{V}'_{G8} \circ \mathcal{V}'_{G9} \circ \mathcal{V}'_{G10}] \tag{17}$$

$$\mathcal{F}_{\mathcal{B}} = [\mathcal{V}'_{B1} \circ \mathcal{V}'_{B2} \circ \mathcal{V}'_{B3} \circ \mathcal{V}'_{B4} \circ \mathcal{V}'_{B5} \circ \mathcal{V}'_{B6} \circ \mathcal{V}'_{B7} \circ \mathcal{V}'_{B8} \circ \mathcal{V}'_{B9} \circ \mathcal{V}'_{B10}] \tag{18}$$

where $\mathcal{V}'_{ij} = \mathcal{V}_{ij} max(\mathcal{V}_{ij})$ is the normalized version of $\mathcal{V}_{ij}$, which in turn, denotes the feature vector for the j-th wavelet sub-band of the i-th image channel and $\circ$ denotes the vector concatenation. It follows that i and j assume the values $\{\mathcal{R}, \mathcal{G}, \mathcal{B}\}$ and $\{1, 2, 3, 4, 5, 6, 7, 8, 9, 10\}$ respectively. Finally, the feature vector that will represent the input color texture image will be as shown below

$$\mathcal{F} = [\mathcal{F}_{\mathcal{R}} \circ \mathcal{F}_{\mathcal{G}} \circ \mathcal{F}_{\mathcal{B}}] \tag{19}$$

Algorithm 1 summarizes the whole process of the proposed color texture classification scheme in the form of a sequence of logical and objective steps. At the same time Fig. 1 provides a graphical representation of the proposed scheme in order to facilitate the understanding of the proposal. According to Fig. 1, the first stage consists in transforming the input color image into R, G and B channel images. From each channel image, a 3-level DWT is achieved which results in 30 sub-bands (10 sub-bands for each channel). We then visualize each sub-band as a GMRF from which we compute the respective Fisher information matrix components and Shannon's entropy. This results in $270 = n_s \times (n_f + e_s) \times n_c$ features where n_s, n_f, e_s and n_c denote the number of sub-bands for each color channel, the number of Fisher information matrix components for each channel, the entropy of the related sub-band and the number of color channels respectively. Here are the values that we took into account in this work, $n_s = 10$, $n_f = 8$, $e_s = 1$ and $n_c = 3$.

Algorithm 1. Proposed Color Texture Feature Extraction Scheme

---

1: **function** COLORTEXTTUREFEATUREEXTRACTION(X)
2:     Split the input image $X_{m \times n}$ into R, G and B images.
3:     Apply the DWT to decompose each of the three images.
4:     **for** each subband of the decomposition **do**
5:         **a.** Extract 3×3 patches
6:         **b.** Compute the covariance matrix of
7:         the patches
8:         **c.** Estimate the parameter beta with
9:         the equation (7)
10:         **d.** Compute Fisher information with
11:         the equations (20) to (27)
12:         **e.** Compute entropy with the equation (3)
13:     **end for**
14:     Normalize all the values.
15:     Concatenate the values from the three channels into a single vector $\mathcal{F}$.
16:     **return** $\mathcal{F}$
17: **end function**

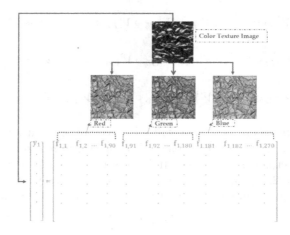

Fig. 1. Schematic diagram showing the proposed texture feature extraction approach in RGB space.

The non-zero components of the metric tensor $g^{(1)}(\vec{\theta})$ (type-I Fisher information matrix) can be expressed as:

$$I_{\mu\mu}^{(1)}(\vec{\theta}) = \frac{1}{\sigma^2}\left(1 - \beta\Delta\right)^2 \left[1 - \frac{1}{\sigma^2}\left(2\beta\left\|\vec{\rho}\right\|_+ - \beta^2\left\|\Sigma_p^-\right\|_+\right)\right] \qquad (20)$$

$$\begin{aligned}I_{\sigma^2\sigma^2}^{(1)}(\vec{\theta}) &= \frac{1}{2\sigma^4} - \frac{1}{\sigma^6}\left[2\beta\left\|\vec{\rho}\right\|_+ - \beta^2\left\|\Sigma_p^-\right\|_+\right] \\ &\quad + \frac{1}{\sigma^8}\left[3\beta^2\left\|\vec{\rho}\otimes\vec{\rho}\right\|_+ - 3\beta^3\left\|\vec{\rho}\otimes\Sigma_p^-\right\|_+ + 3\beta^4\left\|\Sigma_p^-\otimes\Sigma_p^-\right\|_+\right]\end{aligned} \qquad (21)$$

$$\begin{aligned}I_{\sigma^2\beta}^{(1)}(\vec{\theta}) = I_{\beta\sigma^2}^{(1)}(\vec{\theta}) &= \frac{1}{\sigma^4}\left[\left\|\vec{\rho}\right\|_+ - \beta\left\|\Sigma_p^-\right\|\right] \\ &\quad - \frac{1}{2\sigma^6}\left[6\beta\left\|\vec{\rho}\otimes\vec{\rho}\right\|_+ - 9\beta^2\left\|\vec{\rho}\otimes\Sigma_p^-\right\|_+ + 3\beta^3\left\|\Sigma_p^-\otimes\Sigma_p^-\right\|_+\right]\end{aligned} \qquad (22)$$

$$\begin{aligned}I_{\beta\beta}^{(1)}(\vec{\theta}) &= \frac{1}{\sigma^2}\left\|\Sigma_p^-\right\|_+ \\ &\quad + \frac{1}{\sigma^4}\left[2\left\|\vec{\rho}\otimes\vec{\rho}\right\|_+ - 6\beta\left\|\vec{\rho}\otimes\Sigma_p^-\right\|_+ + 3\beta^2\left\|\Sigma_p^-\otimes\Sigma_p^-\right\|_+\right]\end{aligned} \qquad (23)$$

where $\left\|A\right\|_+$ denotes the summation of all the entries of the vector/matrix A and $\otimes$ denotes the Kronecker (tensor) product.

Similarly, the non-zero components of the metric tensor $g^{(2)}(\vec{\theta})$ (type-II Fisher information matrix) can be expressed as:

$$I_{\mu\mu}^{(2)}(\vec{\theta}) = \frac{1}{\sigma^2}\left(1 - \beta\Delta\right)^2 \qquad (24)$$

$$I^{(2)}_{\sigma^2\sigma^2}(\vec{\theta}) = \frac{1}{2\sigma^4} - \frac{1}{\sigma^6}\left[2\beta\,\|\vec{\rho}\|_+ - \beta^2\,\|\Sigma^-_p\|_+\right] \tag{25}$$

$$I^{(2)}_{\sigma^2\beta}(\vec{\theta}) = I^{(1)}_{\beta\sigma^2}(\vec{\theta}) = \frac{1}{\sigma^4}\left[\|\vec{\rho}\|_+ - \beta\,\|\Sigma^-_p\|\right] \tag{26}$$

$$I^{(2)}_{\beta\beta}(\vec{\theta}) = \frac{1}{\sigma^2}\,\|\Sigma^-_p\|_+ \tag{27}$$

4 Experimental Study

4.1 Image Databases

Two texture image datasets were used in the experiments:

- **KTH-TIPS2b** [8] is a collection of 11 classes of images. Each class consists of 432 texture images. All images are captured varying scales, viewing angles and lighting conditions in order to study the generalization ability of material recognition methods to the new material instance. We used 80% of the images for training the SVM classifier and the remaining ones (20%) for testing.
- **Salzburg** [9] contains a collection of 476 color texture images that have been captured around Salzburg (Austria). For each texture class (from total of 10 classes used in this paper), there were 128×128 source images, of which 80% was used for training the classifier, while the other 20% was used for testing.

4.2 Evaluation of the Proposed Method

Table 1 shows the performance of our method on two publicaly used datasets. Using Sazlburg dataset, our approach significantly out performs most of conventional and state-of-the-art methods for color texture classification, specially those ones that use the combination of wavelet, information theory and GMRF. Similarly, using the KTH-TIPS2b, our method greatly outperforms most of the traditional ones. Meanwhile, our method performed very well on Salzburg (with 100% of accuracy) than on KTH-TIPS2b (having 95.17% of accuracy) datasets.

The variance of 3.3% in accuracy between Salzburg and KTH-TIPS2b may be related to three items: (a) the increasing in the size of the used datasets, (b) the difference in the DWT level decomposition and (c) the used color space. These points are very interesting and worthy of our attention and they deserve to be tackled in the future work (as referred in Sect. 5).

With the recent increase in popularity of deep learning methods (deep learning-based feature extractor), such as convolutional neural network, used in many applications since they are end-to-end with good performance for image classification, we then performed a comparison with CNN. However, applying CNNs directly (from scratch) on texture datasets is known to achieve only moderate accuracy in texture classification. In this experiment we evaluated several CNN-based texture models and locally transferred FV-CNN (LFV-CNN) achieved state of the art results on KTHTIPS2b (82.6%) and on Salzburg (93.22%), as shown in Table 1.

Table 1. Classification rate on Salzburg and KTH_TIPS2b databases.

Database	Success rate (%)	
	Proposed approach	CNN
Salzburg	100.00	93.22
KTH_TIPS2b	95.17	82.6

5 Conclusions

In this work, a novel approach for texture classification has been carried out, putting emphasis on color texture feature extraction. The main idea consists in splitting the input texture image into R, G and B images and decompose each channel into 10 wavelet sub-bands using a 3-level-based DWT. Each sub-band is then visualized as a GMRF from which Fisher information matrix components and Shannon's entropy are computed. The resulting features of all sub-bands of the three channels are then concatenated to form a vector of features that will represent the input texture image. This vector of features is used as input to SVM classifier.

A set of experiments have been carried out using two benchmark datasets (KTH-TIPS2b and Salzburg datasets) of the state-of-the-art to evaluate the proposed method. The results show that the proposed algorithm reaches a better accuracy compared with the state-of-the-art algorithms, showing that the current proposal has a potential to become a powerful tool for texture analysis tasks.

With regard to future work, we will work very hard on the following three items to reach an higher performance (very close to 100% of accuracy for most of datasets): (a) increase the size of the used datasets so the need for dimensionality reduction techniques will be fundamental. In this way, we will adopt manifold learning techniques that proved to be suitable for texture features space [43], (b) study the different DWT level decomposition in order to find out which level is adequate to a given dataset size to boost the classification performance and (c) study another color spaces than RGB, for instance HSV in order to find out which specific color space is suitable for a given dataset.

References

1. Khan, F.S., van de Weijer, J., Vanrell, M.: Top-down color attention for object recognition. In: 2009 IEEE 12th International Conference on Computer Vision, pp. 979–986 (2009)
2. van de Sande, K., Gevers, T., Snoek, C.: Evaluating color descriptors for object and scene recognition. IEEE Trans. Pattern Anal. Mach. Intell. **32**(9), 1582–1596 (2010)
3. Nilsback, M., Zisserman, A.: Automated flower classification over a large number of classes. In: 2008 Sixth Indian Conference on Computer Vision, Graphics Image Processing, pp. 722–729, December 2008

4. Qi, X., Xiao, R., Li, C., Qiao, Y., Guo, J., Tang, X.: Pairwise rotation invariant co-occurrence local binary pattern. IEEE Trans. Pattern Anal. Mach. Intell. **36**(11), 2199–2213 (2014)
5. Pietikainen, M., Maenpaa, T., Viertola, J.: Color texture classification with color histograms and local binary patterns. In: Workshop on Texture Analysis in Machine Vision, January 2002
6. Li, W., Fritz, M.: Recognizing materials from virtual examples. In: Fitzgibbon, A., Lazebnik, S., Perona, P., Sato, Y., Schmid, C. (eds.) ECCV 2012. LNCS, vol. 7575, pp. 345–358. Springer, Heidelberg (2012). https://doi.org/10.1007/978-3-642-33765-9_25
7. Sharan, L., Liu, C., Rosenholtz, R., Adelson, E.H.: Recognizing materials using perceptually inspired features. Int. J. Comput. Vision **103**(3), 348–371 (2013)
8. Hayman, E., Caputo, B., Fritz, M., Eklundh, J.-O.: On the significance of real-world conditions for material classification. In: Pajdla, T., Matas, J. (eds.) ECCV 2004. LNCS, vol. 3024, pp. 253–266. Springer, Heidelberg (2004). https://doi.org/10.1007/978-3-540-24673-2_21
9. Kwitt, R., Meerwald, P.: Salzburg texture image database. http://www.wavelab.at/sources/STex/. Accessed Feb 2018
10. Jiang, L., Rich, W., Buhl-Brown, D.: Texture analysis of remote sensing imagery with clustering and Bayesian inference. Int. J. Image Graph. Sig. Proces. **7**, 1–10 (2015)
11. Lerski, R.A., Straughan, K., Schad, L.R., Boyce, D.V.M., Bluml, S., Zuna, I.: MR image texture analysis-an approach to tissue characterization. Magn. Reson. Imaging **11**(6), 873–87 (1993)
12. Westerink, P.H., Biemond, J., Boekee, D.E.: Sub-band Image Coding, Kluwer Academic (1991). chapter Sub-band coding of color images
13. Mallat, S.G.: A theory of multiresolution image decomposition: the wavelet representation. IEEE Trans. Pattern Anal. Mach. Intell. **11**(7), 647–693 (1989)
14. Do, M.N., Vetterli, M.: Wavelet-based texture retrieval using generalized Gaussian density and Kullback-Leibler distance. IEEE Trans. Image Process. **11**(2), 146–158 (2002). https://doi.org/10.1109/83.982822
15. Allili, M.S.: Wavelet modeling using finite mixtures of generalized Gaussian distributions: application to texture discrimination and retrieval. IEEE Trans. Image Process. **21**(4), 1452–1464 (2012). https://doi.org/10.1109/TIP.2011.2170701
16. Li, S.Z.: Markov Random Field Modeling in Image Analysis. Springer, Berlin (2001). https://doi.org/10.1007/978-4-431-67044-5
17. Petrou, M., Sevilla, P.G.: Image Processing. Texture: Dealing with Texture, 1st edn. Wiley John and Sons, West Sussex (2006)
18. Van de Wouwer, G., Scheunders, P., Dyck, D.: Statistical texture characterization from discrete wavelet representation. IEEE Trans. Image Process. **8**, 592–598 (1999). https://doi.org/10.1109/83.753747
19. Vetterli, M., Kovacevic, J.: Wavelets and Subband Coding. Prentice-Hall, Englewood Cliffs (1995)
20. Raju, U.S.N., Vijaya Kumar, V., et al.: Texture classification based on extraction of skeleton primitives using wavelets. Inf. Technol. J. **7**(6), 883–889 (2008)
21. Ong, S., Jin, X., Jayasooriah, Sinniah, R.: Image analysis of tissue sections. Comput. Biol. Med. **26**(3), 269–279 (1996). Information Retrieval and Genomics
22. Liu, L., Chen, J., Fieguth, P., Zhao, G., Chellappa, R., Pietikainen, M.: From bow to CNN: two decades of texture representation for texture classification. Int. J. Comput. Vision **127**(1), 74–109 (2019)

23. Pietikainen, M., Hadid, A., Zhao, G., Ahonen, T.: Computer Vision Using Local Binary Patterns. Computational Imaging and Vision. Springer, London (2011). https://doi.org/10.1007/978-0-85729-748-8. https://books.google.com.br/books?id=wBrZz9FiERsC

24. Hammersley, J.M., Clifford, P.: Markov field on finite graphs and lattices, preprint (1971). www.statslab.cam.ac.uk/grg/books/hammfest/hamm-cliff.pdf

25. Haralick, R., Shanmugam, K., Dinstein, I.: Texture features for image classification. IEEE Trans. Syst. Man Cybern. **3**, 610–621 (1973)

26. Haralick, R.M.: Statistical and structural approaches to texture. Proc. IEEE **67**, 786–804 (1979). https://doi.org/10.1109/proc.1979.11328

27. Dalal, N., Triggs, B.:. Histograms of oriented gradients for human detection. In: IEEE Computer Society Conference on Computer Vision and Pattern Recognition, CVPR 2005, vol. 1, pp. 886–893 (2005). http://ieeexplore.ieee.org/xpls/abs_all.jsp?arnumber=1467360

28. Qi, X., Qiao, Y., Li, C.-G., Guo, J.: Exploring Cross-Channel Texture Correlation for Color Texture Classification (2013). https://doi.org/10.5244/C.27.97

29. Fujieda, S., Takayama, K., Hachisuka, T.: Wavelet convolutional neural networks for texture classification. arXive-prints, arXiv:1707.07394, July 2017

30. Hafemann, L.G., Oliveira, L.S., Cavalin, P.: Forest species recognition using deep convolutional neural net-works. In: 2014 22nd International Conference on Pattern Recognition, pp. 1103–1107, August 2014

31. Cimpoi, M., Maji, S., Kokkinos, I., Vedaldi, A.: Deep filter banks for texture recognition, description, and segmentation. Int. J. Comput. Vision **118**(1), 65–94 (2016)

32. Besag, J.: Spatial interaction and the statistical analysis of lattice systems. J. Roy. Stat. Soc. Ser. B. **36**, 192–236 (1974)

33. Andrearczyk, V., Whelan, P.: Using filter banks in convolutional neural networks for texture classification. Pattern Recogn. Lett. **84**, 63–69 (2016)

34. Zhao, Y., Zhang, L., Li, P., Huang, B.: Classification of high spatial resolution imagery using improved Gaussian Markov random-field-based texture features. IEEE Trans. Geosci. Remote Sens. **45**(5), 1458–1468 (2007)

35. Shannon, C., Weaver, W.: The Mathematical Theory of Communication. University of Illinois Press, Urbana (1949)

36. Frieden, B.R.: Science from Fisher Information: A Unification. Cambridge University Press, Cambridge (2004)

37. Frieden, B.R., Gatenby, R.A.: Exploratory Data Analysis Using Fisher Information. Springer, London (2006). https://doi.org/10.1007/978-1-84628-777-0

38. Hafner, G.M., Liedlgruber, A., Uhl, M., Vécsei, A., Wrba, F.: Combining Gaussian Markov random fields with the discrete-wavelet transform for endoscopic image classification. In: DSP 2009: 16th International Conference on Digital Signal Processing, Proceedings, pp. 1–6 (2009). https://doi.org/10.1109/ICDSP.2009.5201226

39. Mani, M.R., Subbaiah, K.V.: Texture Classification Method using Wavelet Transforms Based on Gaussian Markov Random Field (2010)

40. Porter, R., Canagarajah, N.: Robust rotation-invariant texture classification: wavelet, Gabor filter and GMRF based schemes. IEE Proc. Vision Image Sig. Process. **144**(3), 180–188 (1997). https://doi.org/10.1049/ip-vis:19971182

41. Levada, A.L.M.: learning from complex systems: on the roles of entropy and fisher information in pairwise isotropic Gaussian Markov random fields. Entropy, Special Issue Inf. Geometry. **16**, 1002–1036 (2014)

42. Levada, A.L.M.: Information geometry, simulation and complexity in Gaussian random fields. Monte Carlo Methods Appl. **22**(2), 81–107 (2016)
43. Nsimba, C.B., Levada, A.L.M.: Nonlinear dimensionality reduction in texture classification: is manifold learning better than PCA? In: Rodrigues, J.M.F., et al. (eds.) ICCS 2019. LNCS, vol. 11540, pp. 191–206. Springer, Cham (2019). https://doi.org/10.1007/978-3-030-22750-0_15

Anomaly Detection for Images Using Auto-encoder Based Sparse Representation

Qiang Zhao[✉] and Fakhri Karray[✉]

Center for Pattern Analysis and Machine Intelligence, Department of Electrical and Computer Engineering, University of Waterloo, Ontario, Canada
{qiang.zhao,karray}@uwaterloo.ca

Abstract. Anomaly detection is a pattern recognition task that aims at distinguishing abnormal patterns from normal ones. In this paper, we propose a convolutional auto-encoder based model to detect anomaly images by producing a sparse representation in the latent space. The proposed approach is able to represent the normal images using sparse encoding and the encoding can be well reconstructed by the decoder. However, the learned convolutional filters are not able to represent the abnormal images in a sparse way. Therefore, the decoder can not reconstruct the abnormal images with high quality. By assessing the reconstruction performance, we can distinguish the abnormal images from the normal ones. The experimental results show the superiority of our proposed model over other variants of auto-encoder based anomaly detection models in terms of AUC. In addition, the results show that the sparse representation based anomaly detection method could apply to different scenarios.

Keywords: Anomaly detection · Auto-encoder · Sparse representation

1 Introduction

Anomaly detection is a pattern recognition task with the goal of distinguishing abnormal patterns from normal ones. However, some density based anomaly detection techniques such as clustering and nearest neighbor are based on the concept of local density that fail to apply to data with high local fluctuations such as images. Based on this fact, researchers have focused on exploring various generative models for pattern recognition purpose. As examples, the variational auto-encoder [1] and the Generative Adversarial Networks (GANs) [2] provide attractive alternatives for other maximum likelihood techniques.

Recently, anomaly detection has become an area of active research. Using deep learning methods for anomaly detection task has been extensively studied across a range of domains in [3]. The lack of generality to unseen anomalies is a major obstacle for anomaly detection task in images. Among these models,

© Springer Nature Switzerland AG 2020
A. Campilho et al. (Eds.): ICIAR 2020, LNCS 12132, pp. 144–153, 2020.
https://doi.org/10.1007/978-3-030-50516-5_13

the auto-encoder combine the feature extraction concept with dimensionality reduction which make it becoming an expected candidate for image anomaly detection task. In addition, what makes the auto-encoder a favorable model for image anomaly detection is that it only depends on normal data (which is the common situation in anomaly detection tasks) during the training process. The proposed model exploring the potential of using auto-encoder based method to detect anomalies in images. The proposed method is able to find the low-dimensional data representations for which normal images and abnormal images are expected to have significantly different expressions. And these representations can be leveraged to quantitative measurements.

Over the last decade, deep convolution neural networks have achieved remarkable success on various computer vision tasks such as image classification and object detection. On the one hand, the characteristics of the parameter sharing and local connectivity of the convolutional layers make it sensitive to unpredictable changes in local area of images. On the other hand, theoretical research [4] demonstrates that the hierarchical architectures achieves good performance on universal visual pattern recognition tasks. Therefore, a hierarchical convolution network is developed as a pattern encoder to obtain the latent representations for normal images.

Especially, in the case of auto-encoder, the encoding layers of the auto-encoder are developed to preserve quantity of information rather than quality of information. Thus, auto-encoder may lose critical information that is most relevant to the image anomaly detection task. To address this problem, the related work [5] suggest to corrupt the particular feature and divide the input data into the effective reconstructed part and the noise part. This related work is inspired by robust principal component analysis that fail to meet our expectation on generating the sparse feature representation in the latent space.

As state in [5], another common challenge of using auto-encoder for feature representation learning in high dimensional data is to guarantee a robust reusable representation of the images in the latent space. Even further, the auto-encoder is trained to restore as much information as possible and is not able to determine what kind of information is relevant to the specific problem we are trying to solve during training process. In this case, another obstacle of using the auto-encoder method for image anomaly detection task is to avoid producing a perfect copy of input as output. Based on this fact, the problem of image anomaly detection is transferred to explore a criteria that could train the auto-encoder to obtain a latent representation that can effectively distinguish abnormal images from normal ones. The related stacking denoise auto-encoder [6] addresses this problem by forcing the auto-encoder to denoise the corrupted input images and [7] further improve the robustness of the feature representations by performing a layer-wise initialization while partially corrupting the input. However, both work take the risks of destroying the original normal patterns of images and increase the computational complexity for training the model.

Consider that the normal images are supposed to follow the consistent sparse representation in the latent space, that is, the feature representation of normal images always satisfy the sparsity condition. The related work [8] based on dictionary learning and learns dictionaries from the latent space of the variational auto-encoder, but this work needs to make an assumption on the probability distribution of the data. We introduce the sparse coding to train the encoder learning a sparse representation in the latent space. Previous work like [9] applied sparse coding to detect the unusual event of the videos in a real time basis by continuously updating the learned dictionary. Also, the abnormality of the image is determined by proposed sparse reconstruction cost as shown in [10]. However, both works only use sparse coding as a measurement of evaluating the normality of event rather than a standard criteria for representation learning.

Our work bridges the gap between the success of convolution neural network for image feature learning and the consistency of the auto-encoder in learning a sparse representation in the latent space. By evaluating the reconstruction performance of each image, we can effectively distinguish the abnormal image from the normal one.

The rest of the paper is organized as follows: Sect. 2. introduces the proposed approach. The details of the experiment setup and results are described in Sect. 3. Finally, the conclusion is presented in Sect. 4.

2   Proposed Approach

Anomaly detection in images by sparse representation is based on the assumption that the abnormal images will show inconsistent embedded behavior such as located separately from the normal samples in the latent space. Based on that assumption, we assume that the abnormal images can not be reconstructed from the sparse representation with high quality. Thus, the reconstruction loss of normal images and abnormal images will have large differences and it is reasonable to treat the reconstruction loss as anomaly score for detecting the abnormal images.

2.1   Sparse Representation

We make use of the advantage of auto-encoder based model that it is not sensitive to the specific texture of the anomaly images. The proposed model benefits from the work proposed in [11] and extends the convolution auto-encoder based model to learn the sparse representations that are used to reconstruct the normal images.

The traditional training process of auto-encoder is to minimize the pixel-wise independent mean square error with respect to encoder parameter set Θ and the decoder parameter set Φ, given by:

$$L_{res}_{\Theta,\Phi} = \frac{1}{N} \sum_{i=1}^{N} ||x_i - g_\Phi(f_\Theta(x_i))||_2^2 \tag{1}$$

where N is the number of training samples, x_i is each normal image, Θ is the parameter set learned during the convolutional mapping in encoder $f_\Theta(x)$ and the Φ is the parameter set learned during the deterministic reconstruction process in decoder $g_\Phi(x)$.

An explicit criteria is proposed to force the model to learn the identity sparse representations of the normal images in the latent space. The formulated sparse representation in the latent space is obtained by adding a constrained penalty term, that is, instead of using mean square error as loss function to optimize the reconstruction performance, the L1 norm constrain is added on the intermediate representations that train the model to produce a sparse feature representation in the latent space. The novelty of the proposed model is based on the belief that the sparsity of the latent representation is the key to improve the accuracy of image anomaly detection task. The trade-off here is to force the model maintain the sparse representation required to reconstruct the input without holding redundant noise features. Also, the encoding convolutional filters are trained to produce a much divergent set of solutions that sensitive to abnormal patterns. The very first intuition of adding L1 norm as a regularisation term for training is to prevent overfitting and improve the generalisation ability of the model. But in this case, we introduce the idea of sparse coding as a "regularizer" to punish the complexity of the intermediate representations in the latent space. Therefore, by adding a parameterized L1 norm constrain on images' latent representations, the loss function is formulated that aims to train the proposed model both sensitive to the reconstruction performance and the representation sparsity, the proposed sparsity based loss function given by:

$$L_{sparse}_{\Theta,\Phi} = \frac{1}{N} \sum_{i=1}^{N} ||x_i - g_{\Phi}(f_{\Theta}(x_i))||_2^2 + \lambda ||f_{\Theta}(x_i)||_1 \qquad (2)$$

where N is the number of training samples, Θ is the encoder parameter set learned in encoder and Φ is the decoder parameter set, λ is the tuned parameter that controls the degree of sparsity of the latent representation.

2.2 Anomaly Score

From anomaly detection perspective, the anomaly score evaluates how well a given test image follows the normal patterns. In the evaluating stage, we assume that the concatenated convolutional filters trained with normal images are not able to represent the anomaly images in a sparse way. Therefore, the decoder can not reconstruct the abnormal images with high quality. By comparing the reconstruction performance, we can distinguish the abnormal images from the normal ones. Based on the above, the reconstruction error (1) is adopted as the anomaly score to assess the abnormality of the test images.

3   Experiment

We test the effectiveness of our model on two well-known image anomaly detection scenarios respectively. Also, to emphasize the effectiveness of the proposed model, we compare it with several variants of auto-encoder models in terms of AUC score in both scenarios.

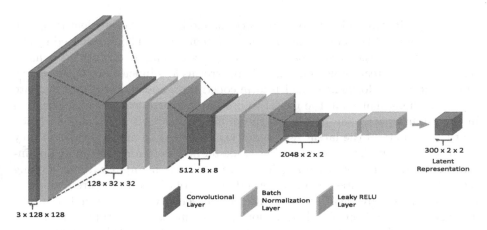

Fig. 1. The architecture of the encoder of the proposed convolutional auto-encoder. The encoder consists of four convolution blocks, the first block only consist of convolutional layer and RELU layer, for the next three blocks, there is a batch normalization layer between the convolutional layer and RELU layer.

As shown in Fig. 1 that the encoder has four successive convolutional modules. The proposed model's architectural is based on the several improvements developed in recent years. [12] suggests that replacing the down-sampling layer with consequent strided convolutional layer will not compromise accuracy for feature selection. Each convolutional layer is followed by the batch normalization layer that helps stabilizing the distribution of each hidden unit during training. In particular, the batch normalization proposed in [13] addresses the problem of decaying learning rate in a deep neural network and therefore accelerates the training process with less efforts made during the initialization process. However, to prevent the model oscillation caused by applying batch normalization to every convolutional layer, we omit the batch normalization at the input layer of the encoder and the output layer of the decoder as suggested in the work of [14]. The last convolutional layer in encoder and decoder is sent to the Sigmoid function and the Tanh function respectively. Furthermore, [15] suggests that a non-zero slope for negative part in rectified linear unit could preserve the information when data is transferred through the deep layers. Therefore, we adopt the leaky RELU activation layer after each convolutional section and we found out that it improves the performance of the proposed model when training cycle is relatively short.

3.1   Data Set

The HAM10000 Data Set: The HAM10000 data set is a open-source dermatoscopic images that consists several pigmented skin lesions [16]. The relevant disease information of the data set is shown in Table 1, the data set contains: melanocytic nevi (NV), actinic keratoses and intraepithelial carcinoma/bowen's disease (AKIEC), basal cell carcinoma (BCC), benign keratosis-like lesions (BKL), dermatofibroma (DF), melanoma (MEL) and vascular lesions (VASC). In our experiments, we treat NV images as normal skin images and separate them into 100 test set and 6605 training set. During the test stage, for each run, we randomly select 100 images from each class of abnormal skin diseases to calculate the anomaly score using (1) and combine it with the anomaly score of the NV images from test set to compute the AUC score.

Table 1. The HAM10000 data set:

Disease	NV	AKIEC	BCC	BKL	DF	MEL	VASC
Number	6705	327	514	1099	115	1113	142

Daytime Driving Distraction Data Set: The daytime distraction driving image data set [17] contains three different distraction driving behaviors (talking, texting and focusing on the GPS near the gear stick of the simulation while driving) as well as normal driving behavior. The relevant information of the data set shown in Table 2. The images are collected based on the integral view of 25 drivers' upper body movements while they are driving in a simulated environment during the daytime. The 25 participants consist of 16 males and 9 females and they are from different countries and in different ages. The images are extracted from a video stream in 5 frames per second basis. Because of the low frame rate, images from adjacent time stamps present different body poses. Note that some images are deleted due to the unexpected disturbances occurred during the image collecting process. For training the model in each run, we randomly pick 16 drivers' normal driving images and testing on the three different distraction driving images from the remaining 9 drivers respectively.

Table 2. Daytime Distraction Driving data set:

Behavior	Normal	Talking	Texting	GPS
Number	4993	4921	4991	4926

3.2   Experiment Setting

As shown in Fig. 1, the encoder part of the proposed model have four convolution blocks. Similarly, the decoder has four symmetrical deconvolution blocks. During the training stage, we feed the model with normal samples $\{x_1, x_2, ..., x_n\}$ for each $x \in \mathbb{R}^{d \times d}$ that we considered to be i.i.d. sample from the unknown prior distribution.

To show the superior representation learning ability of the proposed model, we compare its performance with several variants of auto-encoder based models, including the baseline auto-encoder model trained by binary cross entropy loss, the VAE [1] and the proposed model without sparse representation. In order to maintain the reliability of the results, the architecture of each model is constructed to have similar convolution blocks as our proposed model. We set the prior to be a Gaussian distribution during training process of the VAE in both data sets.

- The same data pre-processing is performed for both data sets, that is to transform each image to the size 128 × 128 and normalize each channel of images to the range of −1 to 1.
- The dimension of the latent space is set to 300 and the sparse representation parameter λ is set to 0.002 for HAM10000 data and 4e−4 for driving distraction data.
- The batch size for training both data sets is set to 64 and the model is trained for 70 epoch in each run, optimized with the Adam optimizer [18], the learning rate is set to 1e−4 with the weight decay of 1e−5.

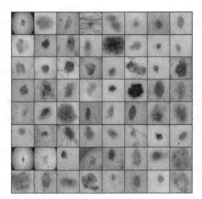 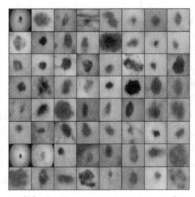

(a) the original images from the test set   (b) the images reconstructed from their sparse representations

Fig. 2. The HAM10000 images and its corresponding images reconstructed from sparse representations

3.3   Experiment Results

The original images and their corresponding reconstructed ones are shown in Fig. 2 and Fig. 3 respectively. As we can see, the images reconstructed from the sparse representations are blurred but also highlight the objects and in a sense clear out the interference from their backgrounds. To reduce the impact of the random initialization of deep layered architecture on experiment results, the experiment results of comparative models are obtained based on the average of 10 run for each model. As shown in Table 3 and Table 4, our proposed model achieves overall better results compared to other variants of auto-encoder based models in both data sets.

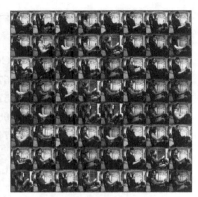

(a) the original images from the
test set

(b) the images reconstructed
from their sparse representations

Fig. 3. The driving distraction images and its corresponding images reconstructed from sparse representations

Table 3 shows the proposed model with sparse encoding improves the anomaly detection performance on skin disease data, especially for those types of anomaly diseases that perform relatively poor in the baseline model. For example, the proposed model greatly improves the detection of AKIEC disease from 0.59 AUC to 0.80 AUC. Same for VASC disease, the proposed model improves the detection performance from 0.48 AUC to 0.66 AUC. This also illustrates that the proposed model could effectively represent the normal images in a sparse way. As a result, the proposed model trained with normal images can not represent the abnormal images in a sparse way. Therefore, the anomaly images can not be well reconstructed from the latent sparse representation and result in large reconstruction error. In this way, the proposed model performs well on detecting the anomaly cases. Same with distraction driving data set, as shown in Table 4, the proposed model greatly improves the detection of GPS distraction behavior from 0.63 AUC to 0.83 AUC. However, the proposed model only slightly improves the detection of texting distraction behavior from 0.67 AUC to 0.68 AUC and improves the detection of talking on the phone distraction behavior from 0.56 AUC to 0.64 AUC. This fact also indicate that the proposed model could detect the anomaly images that may seem similar to the normal images in human sight, which shows that the encoder trained by normal images extracted useful normal patterns by sparse coding. Therefore, the abnormal images can not reconstructed well from the sparse representations and result in relatively large reconstruction errors.

Table 3. The AUC results on HAM10000 data set.

| | AUC | | | | | | |
	MEL	BCC	AKI	BKL	DF	VAS	ALL
Baseline	0.60	0.57	0.59	0.71	0.57	0.48	0.59
VAE	0.78	0.57	0.68	0.60	0.56	0.59	0.63
CAE	0.80	0.65	0.76	0.69	0.60	0.60	0.68
CAE + sparse	0.79	0.74	0.78	0.70	0.65	0.66	0.72

Table 4. The AUC results on driving distraction data set.

| | AUC | | | |
	TALK	TEXT	GPS	ALL
Baseline	0.56	0.67	0.63	0.62
VAE	0.61	0.64	0.75	0.67
CAE	0.61	0.66	0.82	0.70
CAE + sparse	0.64	0.69	0.83	0.72

4   Conclusion

In this paper, we attempt to answer the question: what makes a good representation for anomaly detection task? We explore the potential of combining sparse coding with the auto-encoder for image anomaly detection task. The proposed model aims at detecting anomaly images by representing the images in a sparse way in the latent space. The experiment results show that our work have a better overall performance than other variants of the auto-encoder based models on image anomaly detection task. Take the advantage of auto-encoder based model that only depend on normal images for training, the proposed model can apply to a quite common situation when only few labeled normal images are available. Also, the proposed model can potentially be extended to other anomaly detection tasks such as the abnormal event detection in video streams and the text misspelling detection.

Acknowledgements. The authors would like to thank the Natural Sciences and Engineering Research Council of Canada (NSERC) for its generous support of the project under the Strategic Partnership Grants (NSERC SPG-G).

References

1. Kingma, D.P., Welling, M.: Auto-encoding variational Bayes. In: The 2nd International Conference on Learning Representations (ICLR) (2014)
2. Goodfellow, I.J., et al.: Generative adversarial nets. In: Conference on Neural Information Processing Systems (NIPS) (2014)

3. Chalapathy, R., Chawla, S.: Deep learning for anomaly detection: a survey. arXiv:1901.03407 (2019)
4. Fukushima, K.: Neocognitron: a hierarchical neural network capable of visual pattern recognition. Neural Networks 1, 119–130 (1988)
5. Zhou, C., Paffenroth, R.C.: Anomaly detection with robust deep autoencoders. In: Proceedings of the 23rd ACM SIGKDD International Conference on Knowledge Discovery and Data Mining, pp. 665–674 (2017)
6. Vincent, P., Larochelle, H., Lajoie, I., Bengio, Y., Manzagol, P.-A.: Stacked denoising autoencoders: learning useful representations in a deep network with a local denoising criterion. J. Mach. Learn. Res. 11, 3371–3408 (2010)
7. Vincent, P., Larochelle, H., Bengio, Y., Manzagol, P.-A.: Extracting and composing robust features with denoising autoencoders. In: The 25th International Conference on Machine Learning, pp. 1096–1103 (2008)
8. Sun, J.Y., Wang, X.Z., Xiong, N.X., Shao, J.: Learning sparse representation with variational auto-encoder for anomaly detection. IEEE Access 6, 33353–33361 (2018)
9. Zhao, B., Fei-Fei, L., Xing, E.P.: Online detection of unusual events in videos via dynamic sparse coding. In: Computer Vision and Pattern Recognition (CVPR), pp. 3313–3320 (2011)
10. Cong, Y., Yuan, J., Liu, J.: Sparse reconstruction cost for abnormal event detection. In: Computer vision and pattern recognition (CVPR), pp. 3449–3456 (2011)
11. Lee, H., Battle, A., Raina, R., Ng, A.Y.: Efficient sparse coding algorithms. In: Advances in neural information processing systems (NIPS), pp. 801–808 (2007)
12. Springenberg, J.T., Dosovitskiy, A., Brox, T., Riedmiller, M.: Striving for simplicity: the all convolutional net. CoRR, abs/1412.6806 (2014)
13. Ioffe, s., Szegedy, C.: Batch normalization: accelerating deep network training by reducing internal covariate shift. In: ICML (2015)
14. Radford, A., Metz, L., Chintala, S.: Unsupervised representation learning with deep convolutional generative adversarial networks. In: ICLR (2015)
15. Xu, B., Wang, N., Chen, T., Li, M.: Empirical evaluation of rectified activations in convolution network. arXiv preprint arXiv:1505.00853 (2015)
16. Philipp, T.: The HAM10000 dataset, a large collection of multi-source dermatoscopic images of common pigmented skin lesions. Online (2018)
17. Ou, C., Zhao, Q., Karray, F., Khatib, A.E.: Design of an end-to-end dual mode driver distraction detection system. In: Karray, F., Campilho, A., Yu, A. (eds.) ICIAR 2019. LNCS, vol. 11663, pp. 199–207. Springer, Cham (2019). https://doi.org/10.1007/978-3-030-27272-2_17
18. Kingma, D.P., Ba, J.L.: Adam: a method for stochastic optimization (2015)

Medical Image Analysis

A Framework for Fusion of T1-Weighted and Dynamic MRI Sequences

João F. Teixeira[1,2](✉) (iD), Sílvia Bessa[2] (iD), Pedro F. Gouveia[3,4],
and Hélder P. Oliveira[2,5] (iD)

[1] Faculty of Engineering, University of Porto, Porto, Portugal
jpfteixeira.eng@gmail.com
[2] INESC TEC, Porto, Portugal
{silvia.n.bessa,helder.f.oliveira}@inesctec.pt
[3] Breast Unit, Champalimaud Clinical Centre,
Champalimaud Foundation, Lisbon, Portugal
pedro.gouveia@fundacaochampalimaud.pt
[4] Faculty of Medical, Lisbon University, Lisbon, Portugal
[5] Faculty of Sciences, University of Porto, Porto, Portugal

Abstract. Breast cancer imaging research has seen continuous progress throughout the years. Innovative visualization tools and easier planning techniques are being developed. Image segmentation methodologies generally have best results when applied to specific types of exams or sequences, as their features enhance and expedite those approaches. Particular methods have more purchase with the segmentation of particular structures. This is the case with diverse breast structures and the respective lesions on MRI sequences, over *T1w* and *Dyn*.

The present study presents a methodology to tackle an unapproached task. We aim to facilitate the volumetric alignment of data retrieved from *T1w* and *Dyn* sequences, leveraging breast surface segmentation and registration. The proposed method revolves around Canny edge detection and mending potential holes on the surface, in order to accurately reproduce the breast shape. The contour is refined with a Level-set approach and the surfaces are aligned together using a restriction of the Iterative Closest Point (ICP) method. This could easily be applied to other paired same-time, volumetric sequences.

The process seems to have promising results as average two-dimensional contour distances are at sub-voxel resolution and visual results seem well within range for the valid transference of other segmented or annotated structures.

Keywords: Rigid registration · Segmentation · Space transfer · MRI · Breast

This work was funded by the ERDF - European Regional Development Fund through the Norte Portugal Regional Operational Programme (NORTE 2020), under the POR-TUGAL 2020 Partnership Agreement and through the Portuguese National Innovation Agency (ANI) as a part of project BCCT.Plan–NORTE-01-0247-FEDER-01768 and by Fundação para a Ciência e a Tecnologia (FCT) within PhD grants number SFRH/BD/135834/2018 and SFRH/BD/115616/2016.

© Springer Nature Switzerland AG 2020
A. Campilho et al. (Eds.): ICIAR 2020, LNCS 12132, pp. 157–169, 2020.
https://doi.org/10.1007/978-3-030-50516-5_14

1   Introduction

Breast cancer is the most common cancer in women worldwide. With an expected 10-years survival rate of 80% in western countries, patients will live long lives with the full consequences of their treatment. Recently, oncoplastic techniques have improved aesthetical outcomes up to having a clear positive impact quality of life. Computer vision developments in breast cancer imaging has emerged with innovative visualization tools that can further improve surgical planning and their outcomes.

Surgical oncoplastic treatment planning would benefit from three dimensional (3D) representations of the internal tissues and anatomical structures of the breast and torso. These representations can enhance clinical planning discussions during pre-surgical multidisciplinary meetings and even facilitate the development of educational platforms.

Magnetic Resonance Imaging (MRI) is often performed on breast cancer patients and allows 3D image reconstruction of the breast as it extracts regular interval image slices from the patient's torso. Image processing methods are often used to analyse these challenging MRI sequences, namely focusing on T1-weighted ($T1w$) and MRI Dynamic Contrast-Enhanced (MRI-DCE, Dyn henceforth). Some anatomical torso structures are less complex to annotate, or automatically segment, in either sequences and, as such, it stands to reason to focus the development efforts towards dealing with the clearer options first.

The present study aims to automatically obtain the breast anterior surface on $T1w$ or Dyn at instant zero ($Sd0$). For this, we propose a 3D segmentation method that focuses on collecting the relevant edges, mend contour gaps and get a full volume, in order to reach a verisimilar smooth surface. In particular, we employ the Canny detector [2], Convex Hull [1] and Active Contours [3] approaches.

Additionally, it is our intention to cross the information from both sequences, specifically unifying lesion annotations, acquired from the Dyn sequence, along with the remaining anatomy reference points, obtained from $T1w$. Consequently, this work also contemplates a rigid registration task between the two, in which the Iterative Closest Point algorithm [16] was used, with some transformation restrictions enabled by the known data setup.

1.1   Related Work

Across multiple cancer specificities, and breast cancer research in particular, there is a significant amount of data and methods associated with segmentation and multi-modal registration or fusion.

Segmentation approaches range from Maximum a Posteriori Estimation approaches [14], Expectation Maximization–Markov Random Field techniques [5] and Atlas-based approaches [12] to U-Net methodologies [6]. Nonetheless, research does not focus as much on $T1w$ or $T2w$ sequences as it does on Dyn, due to the lower difficulty of lesion segmentation on that sequence.

Furthermore, those segmentation procedures are largely directed only towards the lesion and generally customarily the breast surface.

Breast multi-modal registration tasks generally involve fusing the *Dyn* sequences to other entirely different modalities, such as Positron emission tomography (PET) and computer tomography (CT) [4], or between the 3D three-dimensional (3D) MRI and 2D data such as X-ray Mammography [9,15]. There are also human biology based works that focus on intra-modality alignment, commonly associated with the monitoring of some disease's progression, as usually found applied to brain CT scans [10]. Nevertheless, some similar work is also done with breast *Dyn* sequences [8,11,14], among others.

However, to the best of our knowledge, this seems to be the first paper concerning *MRI* intra-patient registration between *T1w* and *Dyn* or their derived sub-sequences (*i.e.* T1w *to* Sd0). Hence, we start experimenting with an approach using low complexity methods such as edge detection [2], contour refinement [3] and registration using ICP [16].

2 A Framework for Fusing Different Sequences of Breast Volume Data

2.1 Dataset

The dataset used was provided by the BCCT.plan project[1] and consists of *T1-weighted* thoracic MRI exams (*T1w*) from 27 breast cancer patients, obtained with a Philips Ingenia 3.0T MRI scanner. Each exam comprises 60 gray-scale axial images, with the approximate dimensions of 3 mm thickness and resolution of 720×720 pixels (0.3–0.5 mm/pixel). Additionally, each *T1w* acquisition has a corresponding dynamic contrast study (*Dyn*) that includes the sequence data at the instant zero (*Sd0*). In turn, this sequence comprises 300 gray-scale sagittal images, with 1 mm thickness and a resolution of 300×300 pixels (0.5–0.6 mm/pixel), in a narrower field of view.

Furthermore, the axial exams have corresponding annotations consisting on binary masks of the breasts, as it is easier to segment it in this sequence. On the other hand, the sagittal exams have analogous annotations, encompassing only the Lesions, as the contrast sequence provides a better means for the segmentation of the structure. Examples are shown if Fig. 2.

All annotations were manually performed by experts with more than 5 years of experience. The combination of the two annotations would greatly enhance the automatic production of 3D atlas for multiple clinical reasons.

2.2 Methodology

The tasks intended to be tackled include *T1w* and *Sd0* sequences' segmentation and subsequent registration of both segmented breast surfaces. For the segmentation task, both sequence volumes (*T1w* and *Sd0*) are processed using the same

---

[1] BCCT.plan - 3D tool for planning breast cancer conservative treatment - NORTE-01-0247-FEDER-017688.

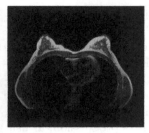

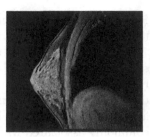

(a) *T1w* central slice

(b) *Sd0* central slice

Fig. 1. MRI images

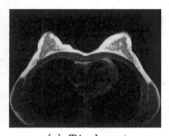

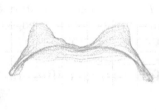

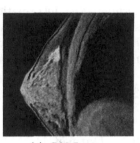

(a) *T1w* breast

(b) *T1w* breast perimeter

(c) *Sd0* Lesion

Fig. 2. Annotation images

pipeline, despite their difference in field of view, voxel resolution and extent. To adjust to the difference in the prevalent slice perspective of each sequence, the sagittally acquired *Sd0* volume undergoes a coarse pre-processing transformation (two 90° rotations on different axes) so that both anatomical volumes face the same direction, on the axial view.

Breast Surface Segmentation: The followed approach builds towards completely enclosing the breast surface in a solid region, ignoring internal structures as much as possible, so as to set the conditions to enable the accurate generation of the breast surface. An overview of the processing pipeline for the segmentation task is presented in Fig. 3.

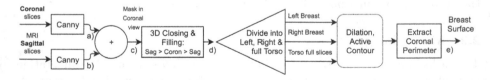

Fig. 3. Segmentation pipeline for one MRI sequence

Specifically, the initial step is to employ the Canny edge detector [2] to obtain the salient boundaries of the MRI volume in question (weak threshold 0.025, strong 0.4, $\sigma = \sqrt{2}$). This detection is performed in both coronal and sagittal directions, as seen in Figs. 4a and 4b, to compensate potential gaps due to perspective, namely in respect to hidden information pertaining to the inframammary folds. The edge maps are joined together (Fig. 4c) and are closed with a 3D sphere kernel (3 pixel radius). Flood-filling operations ensue, first along the sagittal perspective, then along the coronal and, finally, along the sagittal view again (Fig. 4d).

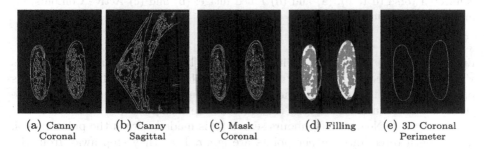

(a) Canny   (b) Canny   (c) Mask   (d) Filling   (e) 3D Coronal
   Coronal   Sagittal   Coronal   Perimeter

Fig. 4. Pipeline intermediate results for $Sd0$

The following step required the application of morphological dilation (5 radius disk) and a Chan-Vese [3] level-set block (100 iterations, Smooth factor 11 and Contraction bias set to 0), over the coronal perspective, as shown in Figs. 5b and 5e. However, slices that had not reached the torso yet, and thusly had two main objects (left and right breasts), could inadvertently have both breasts connected by the active contours approach. Consequently, we separated them in left and right images for individual processing. Figures 5a to 5c exemplify the processing of a slice only using the left side, while Figs. 5d to 5f show the progression of the last full torso slice, across the same final methods of the pipeline. The transition between left/right processing and the first torso slice corresponds to that when the two breast objects become connected, after the Filling step. The advancement throughout the slices halts when the segmented slice area stops increasing, roughly reaching the axilla regions or the point where the field of view starts to crop the lateral torso limits. Lastly, the final surface perimeter is subsequently extracted from those slice outcomes, and examples are shown in Fig. 4e, 5a and 5f.

Registration: The registration is preceded by converting the coordinates of the point clouds to real world values (i.e., the respective voxel resolution is applied to the point list). As previously reported, the point clouds suffer a 90° rotation across 2 axes, such that the objects appear on the same orientation. The point clouds are subsequently processed by a variant of the ICP algorithm [16], that imposes the restriction of no rotation, as both sequences of data are captured

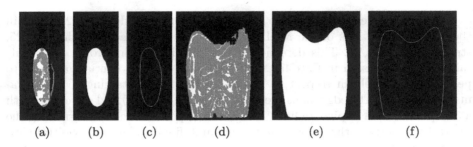

(a)   (b)   (c)   (d)   (e)   (f)

Fig. 5. Further intermediate processing results for *Sd0* left breast (a to c) and full connected torso (d to e). (a) and (d) Filled image; (b) and (e) Active Contours; (c) and (f) Final Perimeter

during the same session, and with the patient lying trying not to move. This will enable to accurately convert the special volumetric data, here the lesion's opintsm acquired on the *Sd0*, and place it on the respective location among the structure data on the base view (in this case the *T1w*).

Briefly, a K-Nearest Neighbours search [7] is made between the point clouds, across both directions, outlier points are ignored, as they step away from the standard deviation (10%). Using the average position calculated from each point cloud match list, the displacement is calculated, and the two point clouds are moved closer to each other with every iteration. The transformation model learning stops when the iteration update falls below a given threshold (10^{-6}).

2.3   Metrics

This work comprises two main procedures, the 3D segmentation of *T1w* and *Sd0* sequences and the registration between those surfaces' outputs. However, the Ground Truth data available only includes the *T1w* breast region annotation mask (*T1wGT*). Specifically, this encompasses the full area between the anterior and posterior limits of the breast, and is laterally confined by the *Latissimus Dorsi* Muscle. This poses an evaluation restriction as, for instance, area metrics against the *T1wGT* would be severely prejudicial on assessing the segmentation, as the surface result will not be closed, and thus cannot be filled in.

Following this rationale, the main measurements employed are of surface distances, firstly, 2D slice-by-slice, as it relates with the domain in which the *T1w* segmentation is commonly considered, and second, across full surfaces (global 3D errors), which relate to the outcome presentation space. Naturally, this requires that the slice-wise perimeters of the *T1wGT* are obtained and be considered as the *de facto* ground truth, as revealed in Sect. 2.1.

This presents yet another challenge, as this perimeter includes anterior and posterior breast contours. Again, this only enables to provide a fair measurement in the direction towards the *T1wGT*: the average of the minimum euclidean distance between the points of the first contour towards the *T1wGT* [13]. Furthermore, logic dictates that the registration task, and respective comparison should not be performed against the *T1wGT* for two reasons. First, the posterior part of the annotation perimeter may inadvertently influence the registration process, as the segmented objects do not acquire such detail. Second, it stands to reason that, the employment of the same method should give close results for two objects, which theoretically present the same anatomy. As such, the *Sd0* segmentation result can only be evaluated fairly when towards the *T1w* segmentation, rather than a manual annotation.

Standing on equal footing, the resulting, aligned surfaces can enjoy the legitimacy of metric calculations in both directions and even a combined average distance, weighted with the number of points from each surface (AD) [13].

Finally, the 3D results also show the maximum and minimum values of the distances across all patients.

3   Results and Discussion

Starting with 2D distances, a visual example of the segmentation results is shown in Figs. 6. At first glance, it can be gathered from the images that the method produces segmentation results for the *T1w* which are close to the *T1wGT*. Although, in general, the segmentation seems to follow an internal contour of the breast, instead of following the skin. On the other hand, the annotators themselves did not seem particularly consistent on delineating the breast contour in regards of including the skin. In fact, the cases in Figs. 6b and 6c (in green) present a GT line that mostly follows a centerline of the skin, instead of its outer limits. On the right sides of Fig. 6a the annotation closely follows the internal posterior skin limit.

Considering a more wide perspective, Fig. 7 presents a boxplot of the average slice distance towards the *T1wGT*, and describes how the error varies across the anatomical vertical regions. It is apparent that the central portion of the torso, which includes the breasts' larger portions has a more consistent and well-defined result, since distinctly shows the lower distance values (roughly 2.5 mm) and variations.

This was to be expected as the segmentation method operates on the sagittal and coronal perspectives, with focus on the last one, with approaches that enlarge a segmentation in order to get a smooth, gap-less surface. This effect, in turn, tends to expand when the torso vertical extremities are reached, less homogeneous intensities are found and the results of initial steps need further correction from level-set extrapolation.

The registration process of the segmented surfaces, depicted in Fig. 8, presents visually acceptable alignments, where larger mismatches may be attributable to segmentation errors at the vertical extremities, in particular of the *T1w* surface.

We were able to plot the 3D distance per patient and, overall, it was larger on the *Sd0* to *T1w* direction (graphic not shown). This lead us to Fig. 9, that shows a profile similar to that of Fig. 7 in the sense that vertical extremities are the main location of larger distances. The central portions of the slice sequence

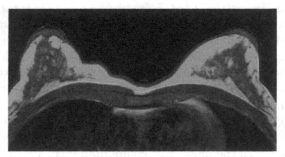

(a) Contours overview

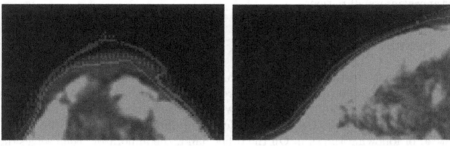

(b) Details on the Left nipple area   (c) Details on the Right breast

Fig. 6. Visual results of the segmentation and registration methods, and *T1wGT*. Segmentation: *T1w* (red), Registered *Sd0* (blue); and GT perimeter (green) (Color figure online)

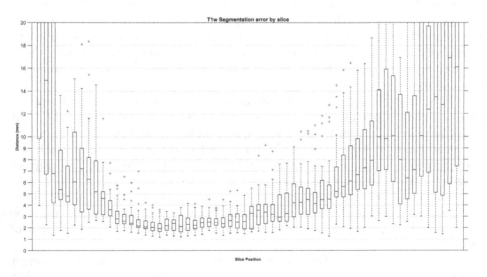

Fig. 7. Distribution of patients' segmented contour distances from *T1w* to *T1wGT*, per slice. Largest box: median 12.8, 75th percentile 83.3, max 284.2.

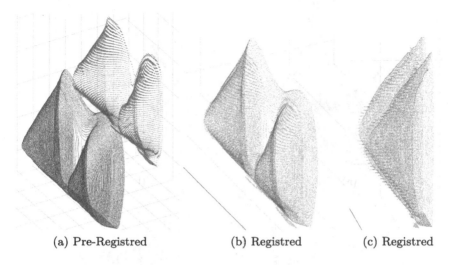

(a) Pre-Registred   (b) Registred   (c) Registred

Fig. 8. Illustration of the registration step

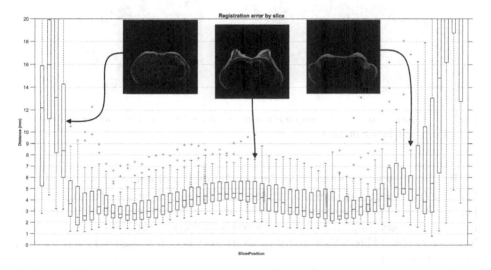

Fig. 9. Distribution of patients' registered contours' distances per slice. Illustration consist of slices 4, 31 and 53 of a total of 60; corresponding, respectively, to the upper abdominal part, breast cross-section and rough upper termination of Manubrium. Largest box: median 34.7, 75 percentile 76.3, max 180.4.

present the smallest errors, although there seems to be a slight raise at the very center. Roughly, these slices match those of the nipple region and often large fibro-glandular openings, which tend to be poorly segmented by the method, specifically on the *T1w* domain. Despite that detail, the breast shape acquired from both sequences seems to match enough to conduct reliable registration, producing an average error around the central slices of approximately the 4 mm.

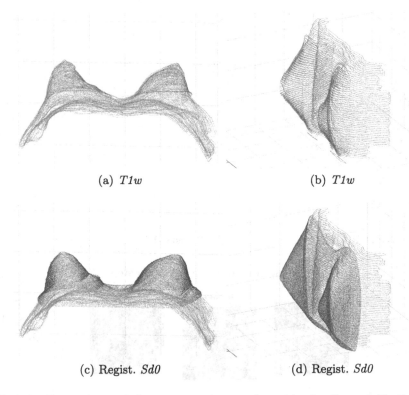

(a) *T1w*   (b) *T1w*

(c) Regist. *Sd0*   (d) Regist. *Sd0*

Fig. 10. Comparison of the segmentation results with the Ground Truth contour (Green) (Color figure online)

It is also relevant to note that, for the 2D registration error to be produced, the *Sd0* 3D points that landed between the slices of *T1w*, had to be *flattened*, i.e., their value was rounded to match the closest slice and pixel in the *T1w* resolution space. This, naturally, imposed additional errors to the comparison, however enabled this rather useful analysis, to where the segmentation method can be improved.

Analysing in more detail the 2D contours, something that is visible in both Fig. 6 and 10 is how closely the *Sd0* follows the outer skin interface (Figs. 6b, 6c, 10c and 10d) while the *T1w* segmentation follows the inner one (Figs. 6c, 10a and 10b), in many cases leaving the manual annotation averaging between both.

Table 1 presents further numerical evidence of the 3D results. The 3D *T1w* segmentation results corroborate the boxplot's central average error of about 4 to 7 pixels, which is a good positioning result, considering some of the method's limitations and the reliability of the dataset.

Table 1. Segmentation and registration 3D errors

	Min	Avg. dist.	AD	Max
T1w to *T1wGT*	1.23	2.20 (0.47)	*n.a.*	3.05
Sd0 to *T1w*	1.75	2.91 (1.30)	2.57 (1.01)	6.17
T1w to *Sd0*	1.37	1.77 (0.47)		3.76

All metrics in mm (min is best). Avg. Dist. and AD present values averaged across all patients, and respective standard deviations in parenthesis.

On the registration side, the trend of larger *Sd0* to *T1w* error, can also be immediately corroborated with the results across all patients. This was anticipated as the *Sd0* has roughly 3 times more slices than *T1w*, for the same vertical extent. Consequently, the individual distances of the *Sd0* slices between the *T1w*'s slices contribute, and further add to the global surface error. Nonetheless, general errors are quite low considering that *T1w* has a slice thickness of 3 mm. In particular, the Avg. Dist. values are not too different, which confirms that both segmentation results have fairly similar shapes and extents, which in turn testifies to the even performance across both sequences. Additionally, this also argues for the capacity of the proposed ICP setup for the particular objective, the accurate placement of the lesions onto the correct space of the breast across sequences.

Finally, the weighted metric, AD, naturally has a value between the Average Distance towards each sequence, leaning more to the *Sd0* to *T1w* direction, as *Sd0* consistently has more points on the respective clouds than *T1w*.

4 Conclusion

An early pipeline for the untried fusion between the breast outer contour of *T1w* and *Sd0 MRI* sequences has been proposed. The main objective was to unify both sequences under the same orientation and 3D space, to enable to indistinctly use annotations made on one of the sequences together with the annotations of the other, while on the same space reference. This is particularly helpful as specific structures are easier to segment on a particular sequence and vice-versa.

For this, the Canny edge detection approach was used along with contour filling and refining, employing Convex Hull and Level-Set methods. The registration final step applied an ICP process to bring the segmented surfaces together.

Both visual and quantitative outcomes show encouraging results for the main objective of the study, managing average contour distances below *T1w*'s slice thickness value. On the other hand, for detailed breast segmentation purposes, this work presents some limitations. For instance there is an inability to take in the lower intensity transitions around the skin and consequently not including the nipple. Additionally, the upper and lower contour of the breast tend to present considerably larger error than those at the center, which includes the main portion of the breast. We believe these are trade-offs that have to be adjusted with the Canny method, or the approach requires further inclusion criteria for those regions, as the skin may not be perfectly delimited or particularly visible on the $T1w$ segmented surface. In sum, the segmentation results might require further adjustments or post-processing for atlas and other goals, however, the proposed approach seems perfectly suited to fulfill the purpose of joining the annotations of these two sequences.

References

1. Barber, C.B., Dobkin, D.P., Dobkin, D.P., Huhdanpaa, H.: The Quickhull algorithm for convex hulls. ACM Trans. Math. Softw. **22**(4), 469–483 (1996). https://doi.org/10.1145/235815.235821
2. Canny, J.: A computational approach to edge detection. IEEE Trans. Pattern Anal. Mach. Intell. **PAMI–8**(6), 679–698 (1986). https://doi.org/10.1109/TPAMI.1986.4767851
3. Chan, T.F., Vese, L.A.: Active contours without edges. IEEE Trans. Image Process. **10**(2), 266–277 (2001). https://doi.org/10.1109/83.902291
4. Dmitriev, I.D., Loo, C.E., Vogel, W.V., Pengel, K.E., Gilhuijs, K.G.A.: Fully automated deformable registration of breast DCE-MRI and PET/CT. Phys. Med. Biol. **58**(4), 1221–1233 (2013). https://doi.org/10.1088/0031-9155/58/4/1221
5. Doran, S.J., et al.: Breast MRI segmentation for density estimation: do different methods give the same results and how much do differences matter? Med. Phys. **44**(9), 4573–4592 (2017). https://doi.org/10.1002/mp.12320
6. Fashandi, H., Kuling, G., Lu, Y., Wu, H., Martel, A.L.: An investigation of the effect of fat suppression and dimensionality on the accuracy of breast MRI segmentation using U-nets. Med. Phys. **46**(3), 1230–1244 (2019). https://doi.org/10.1002/mp.13375
7. Friedman, J.H., Bentley, J.L., Finkel, R.A.: An algorithm for finding best matches in logarithmic expected time. ACM Trans. Math. Softw. **3**(3), 209–226 (1977). https://doi.org/10.1145/355744.355745
8. Gong, Y.C., Brady, M.: Texture-based simultaneous registration and segmentation of breast DCE-MRI. In: Krupinski, E.A. (ed.) IWDM 2008. LNCS, vol. 5116, pp. 174–180. Springer, Heidelberg (2008). https://doi.org/10.1007/978-3-540-70538-3_25
9. Hopp, T., Baltzer, P., Dietzel, M., Kaiser, W.A., Ruiter, N.V.: 2D/3D image fusion of X-ray mammograms with breast MRI: visualizing dynamic contrast enhancement in mammograms. Int. J. Comput. Assist. Radiol. Surg. **7**(3), 339–348 (2012). https://doi.org/10.1007/s11548-011-0623-z
10. Klein, A., et al.: Evaluation of 14 nonlinear deformation algorithms applied to human brain MRI registration. NeuroImage **46**(3), 786–802 (2009). https://doi.org/10.1016/j.neuroimage.2008.12.037
11. Kumar, R., et al.: Application of 3D registration for detecting lesions in magnetic resonance breast scans. In: Loew, M.H., Hanson, K.M. (eds.) Medical Imaging 1996: Image Processing, vol. 2710, pp. 646–656. International Society for Optics and Photonics, SPIE (1996). https://doi.org/10.1117/12.237968

12. Ortiz, C.G., Martel, A.L.: Automatic atlas-based segmentation of the breast in MRI for 3D breast volume computation. Med. Phys. **39**(10), 5835–5848 (2012). https://doi.org/10.1118/1.4748504
13. Song, E., et al.: Hybrid segmentation of mass in mammograms using template matching and dynamic programming. Acad. Radiol. **17**(11), 1414–1424 (2010). https://doi.org/10.1016/j.acra.2010.07.008
14. Xiaohua, C., Brady, M., Lo, J.L.-C., Moore, N.: Simultaneous segmentation and registration of contrast-enhanced breast MRI. In: Christensen, G.E., Sonka, M. (eds.) IPMI 2005. LNCS, vol. 3565, pp. 126–137. Springer, Heidelberg (2005). https://doi.org/10.1007/11505730_11
15. Zhang, Y., Qiu, Y., Goldgof, D.B., Sarkar, S., Li, L.: 3D finite element modeling of nonrigid breast deformation for feature registration in X-ray and MR images. In: 2007 IEEE Workshop on Applications of Computer Vision (WACV 2007), pp. 38–38, February 2007. https://doi.org/10.1109/WACV.2007.2
16. Zhang, Z.: Iterative point matching for registration of free-form curves and surfaces. Int. J. Comput. Vis. **13**(2), 119–152 (1994). https://doi.org/10.1007/bf01427149

Contributions to a Quantitative Unsupervised Processing and Analysis of Tongue in Ultrasound Images

Fábio Barros[1,2(✉)], Ana Rita Valente[1,2], Luciana Albuquerque[1,3],
Samuel Silva[1,2], António Teixeira[1,2], and Catarina Oliveira[1,4]

[1] Institute of Electronics and Informatics Engineering of Aveiro,
University of Aveiro, Aveiro, Portugal
{fabiodaniel,rita.valente,lucianapereira,sss,ajst,
coliveira}@ua.pt

[2] Department of Electronics Telecommunications and Informatics,
University of Aveiro, Aveiro, Portugal

[3] Center for Health Technology and Services Research,
University of Aveiro, Aveiro, Portugal

[4] School of Health Science, University of Aveiro, Aveiro, Portugal

Abstract. Speech production studies and the knowledge they bring forward are of paramount importance to advance a wide range of areas including Phonetics, speech therapy, synthesis and interaction. Several technologies have been considered to study static and dynamic features of the articulators and speech motor control, such as electromagnetic articulography (EMA), real-time magnetic resonance (RTMRI) and ultrasound (US) imaging. While the latest advances in RTMRI provide a wealth of data of the full vocal tract, it is an expensive resource that requires specialized facilities. In this sense, US is a more affordable alternative for several contexts, enabling the acquisition of larger datasets, but demands adequate computational approaches for processing and analysis. While the literature is prolific in proposing methods for tongue segmentation from US, the noisy nature of the images and the specificities of the equipment often dictate a poor performance on novel datasets, a matter that needs to be assessed, before large data acquisition, to devise suitable acquisition and processing methods. In the scope of a line of research studying speech changes with age, this work describes the first results of an automatic tongue segmentation method from US, along with a characterization of the main challenges posed by the image data. Even though improvements are still needed, particularly to ensure temporal coherence, at its current stage, this method can already provide the required data for an automatic analysis of maximum tongue height, a relevant parameter to assess speech changes on vowel production.

Keywords: Ultrasound tongue imaging · Automatic tongue's landmarks extraction · Image segmentation · Speech production · European Portuguese vowels

© Springer Nature Switzerland AG 2020
A. Campilho et al. (Eds.): ICIAR 2020, LNCS 12132, pp. 170–181, 2020.
https://doi.org/10.1007/978-3-030-50516-5_15

1   Introduction

Ultrasound (US) imaging is a non-invasive, safe, portable and fast technology which can provide important information for different areas within speech research, including aging speech, linguistic, disordered speech or speech processing (e.g., [1,24,27,39]). It is commonly used to demonstrate the midsagittal surface contour of the tongue during production [20]. The increased availability of ultrasonographic equipment enables its consideration as a reasonably priced alternative to study dynamic aspects of speech production [5].

One of the major challenges to be addressed refers to the processing, visualization and also the effective and easy use of US images in speech production studies with a large amount of data. Without an appropriate approach to these aspects (automatic edge extraction and fast accurate measurements of tongue contours), the results tend to be below the potential of the technology. Manual edge identification is tedious, time consuming and subjective hindering accuracy and measurement reliability, leading to non-reproducible results (e.g, [1,22,32]). To avoid manual segmentation of the tongue in US images, several methods have been developed.

The development of automatic tongue contour methods helps overcome the disadvantages found in manual analysis. Nevertheless, US artifacts, corrupting noise, and the presence of spurious edges are also major challenges for the automatic processing of US images. Acoustic shadow, mirror and refracting artifacts produce high intensity areas unrelated with the contour as well as discontinuities in tongue surface (e.g. [1,29,32,34]). Anatomical structures near the tongue can also be a problem for automatic detection as they can function as US reflectors, blocking the US wave passage before reaching the tongue body. Although several algorithms have been proposed to simplify US tongue contour extraction, whether they are semi-automatic or automatic, they persist to be strongly susceptible to errors, requiring time consuming manual tracking [6,22]. This is also a first barrier to consider machine learning methods, at this stage, given the lack of correctly annotated sets of data for training and testing.

Our team is engaged in studying the impact of aging in European Portuguese (EP) speech production using ultrasonography of the vocal tract. In this regard, and before initiating the desirable large number of data acquisitions, also involving older adults, we needed to devise how the resulting data could be tackled to extract the tongue contours and, in the process, potentially further tune the acquisition procedure (e.g., imaging parameters, US probe positioning, and participant selection based on anatomical criteria).

In this context, we propose a first iteration of an automatic segmentation method to extract the tongue contour, from US images, and provide a characterization of the imaging features and anatomical regions posing the hardest challenges. The performance assessment was accomplished through the comparison of the outcomes of the proposed method with manually corrected segmentations.

Paper Structure: Sect. 2 provides an overview of representative works dealing with US segmentation in the context of speech production studies; Sect. 3 presents the main aspects of the tongue contour segmentation; Sect. 4 reveal the performance of the proposed method; finally, Sect. 5 presents some conclusions and highlights the main research directions for future improvements of the present work.

2   Background and Related Work

Since the tongue shape changes during speech production, understanding the mechanism of the tongue movement can increase our knowledge of normal speech and speech disorders [35]. US has been used in articulatory studies of vowels and, mainly, the tongue height (vertical position of the highest point of the tongue) and the tongue advancement (location of the constriction in the front-back dimension) have been considered [7,12,17,30], however it is also important to explore the raising of the whole tongue body [7] to capture the overall modifications in tongue shape and location [12].

In these speech studies, one of the steps required to enable going beyond a qualitative analysis based on the visual inspection of the images is the segmentation of the tongue contours. Several methods have been developed in the last decades for semi-automatic or automatic tongue contour detection. Concerning the semi-automatic approaches, active contour based methods (snakes) [22,37] and graph based methods [33] are different examples that require manual processing to initialize the tongue contour determination. EdgeTrak [24] is one of the most used tools for US tongue contour segmentation based on active contours. It requires the identification of a few points by the user, in the first frame of the sequence, and the software optimizes the snake for that frame and considers it to be used in the next frame, imposing some temporal consistency. Although this software fails in rapid tongue curvature changes, producing uncharacteristic tongue contours [22]. Laporte and Ménard [21,22] implement a semi-automatic tongue contour algorithm based on the snake method proposed by Li et al. [24] but using a particle filter to overcome limitations in the method, mainly the limited capture range and the difficult to recovery from errors.

Methods based on neural networks typically provide a fully automatic segmentation, avoiding manual initialization (e.g., [26,36,38]). Autotrace [10] is an example of a software used to US tongue identification based on deep neural networks that use manual labeled data for the training phase and that is highly dependent on similarities between training and data [22]. Other authors [15] implement the same algorithm developed by Fasel and Berry [10] but using automatic labeled training data, although with weak temporal consistency constraints. Tang et al. [32] introduce a method based on a machine learning approach combined with a high-order Markov Random Field energy minimizing frame that performs spatial-temporal segmentation of tongue within a single optimization framework. This method is implemented in a software known as TongueTrack, which considers temporal and spatial constraints in a flexible manner and does not require training data. However, it is not adapted for rapid tongue movements [22]. Xu et al. [37] show that the robustness of EdgeTrak and TongueTrack could be improved by a periodical reinitialization of the tracker, which improves recovery from previous errors. Other notable algorithms were proposed based on deep neural networks (e.g., [8,9]) and convolutional neural networks (e.g., [26,39]). The algorithms described use different data for accuracy determination (e.g., syllable repetition, sentences, reading data, words or non-sense words). Research considering corpora with vowels mostly studies corner vowels (e.g., [a], [i] and/or [u]) in isolated productions (e.g.,[35]), syllables (e.g., [20]), sentences (e.g.,[24]) or vowel-consonant-vowel segments of nonsense words (e.g.,[13,33]).

The aforementioned methods, mainly those based on machine learning present different advantages, such as: are fully automatic (i.e., do not require manual initialization or tuning during the extraction process [10,26]) and the accuracy levels are higher in comparison with e.g., snake-based methods [26]. Additionally, several disadvantages are also reported: the need of different amounts of segmented images for training (e.g., [10]); non adaptation to rapid tongue movement [22]; computational intensiveness (e.g., TongueTrack [32]); to be highly dependent of a similarity between training and test data (e.g., [8,10]). Csapó and Lulich [6] compared three different automatic algorithms to extract tongue contour tracing – EdgeTrak, TongueTrack and AutoTrace – with a manual tracing of 1145 images. The authors conclude that manual tracing presents good agreement and can be considered the "gold standard"; also, auto-matic tracing presented good agreement under appropriate conditions, mainly the AutoTrace 3.5. Convolutional neural networks [26] have been shown to pro-vide more accurate outcomes than deep belief neural networks or active contours. Recently, Karimi et al. [16] presented a method which automatically generates contour points for initialization (i.e., does not require training or manual inter-vention) and their method shows similar results of those obtained by Laporte and Ménard [22], whose results are superior to the state of art algorithms, namely EdgeTrak [24], TongueTrack [32] and Autotrace [10].

Concluding, there are several difficulties observed when attempting to apply previous methods to novel datasets, whether due to the intrinsic (and noisy) nature of the images, the anatomy of the imaged subjects, or, such as in Tongue-Track, the effort required to understand and tweak several parameters.

3 Methods

The overall steps required to perform speech production studies considering US data of the tongue are presented in Fig. 1. In what follows, we describe the steps followed to perform segmentation of the tongue contour.

3.1 US Data Acquisition

The work carried out considers a set of pilot data acquisitions to test the protocol and imaging apparatus. Synchronous acquisition of US images and speech sounds

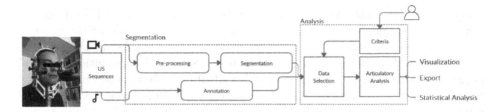

Fig. 1. Pipeline depicting the main stages of the proposed US processing and analysis method.

through Articulate Assistant Advanced software (AAA) [2] took place in a quiet room, using an endocavitary probe (65EC10EA) with 90° field of view positioned under the participants' chin using a stabilization helmet [3]. US was collected with a Mindray DP6900 at a frame rate of 60 Hz. Audio was collected with a Philips SBC ME400 microphone connected to an external sound system (UA-25 EX USB).

The corpus consisted of 9 repetitions for each of 9 EP oral vowels ([i], [e], [ɛ], [a], [o], [ɔ], [u], [ɐ] and [ɨ]), acquired for three different speakers. Corpus acquisition began with the production of the sequence /tatatata/ to assess sound and image synchronization and with swallowing saliva for hard palate delineation. The recorded data was collect as video and audio synchronized with a SyncBrightUp unit [4], and the audio is automatically segmented, at phoneme level, using Web-MAUS [18]. These data were acquired with approval by the Ethics Committee of Escola Superior de Enfermagem de Coimbra (ref. 639/12-2019).

3.2   US Pre-processing

The noisy nature of the US images (Fig. 2) sometimes renders the tongue contour quite difficult to identity, particularly for tongue configurations that result in attenuated contours, e.g., due to a higher position of the tongue. To reduce the impact of this issue, pre-processing for noise reduction is applied to the US images. As a first step, the US image is cropped with the aim of removing irrelevant information for our studies, roughly selecting only the region of interest (ROI) where the tongue will appear (Fig. 2). Subsequently, to enhance the outline corresponding to the tongue surface [16], a phase symmetry filter [19] is applied to the cropped region. In particular, we adopt a phase symmetry using oriented filters [28] with the number of wavelet scales $n = 4$ and the number of filter orientations $r = 14$, shown to be adequate for US images [16]. After filtering, we produce a final image (I_b) considering the cropped image (I_c) and filtered image (I_f). To do that, we first calculate a threshold level based on the median value of intensity of (I_f) (excluding the zero values). The resulting thresholded binary image (I_t), has pixels assigned 1 if their intensity, in (I_f), is greater than the threshold and 0, otherwise. Finally, we produce I_b, similar to (I_t), but containing the pixel intensity from the cropped image, I_c. Equation 1 show how this can be mathematically expressed.

$$I_b(i,j) = 1 - I_t(i,j) + I_t(i,j)I_c(i,j) \tag{1}$$

3.3 Tongue Segmentation

The overall rationale for the proposed segmentation approach derives from how US images are produced. US waves are emitted by a probe, located under the speaker's chin, and these propagate radially with a field of view and depth depending on the probe characteristics and configuration. In the US sequences, the focal point (probe origin) is represented by a semi-circular corona, at the bottom of the image. Therefore, our method analyses the image by travelling radially from the focal point outwards. Given an orientation, the intensity values for all pixels along it are collected and ordered in descending order of intensity and the one presenting the highest intensity (located inside the ROI) is selected as candidate of the tongue contour. After a first pass over all radials, a second pass is performed to detect points that may be misplaced. To this effect, the average radius for the currently selected candidates is computed, $\overline{Radials}$, and those detected as going beyond 2σ of that value are recomputed by selecting the next pixel from the initial ordering that meets the criterion.

Figure 2 shows the points-of-interest extracted from the tongue surface obtained from the original US images.

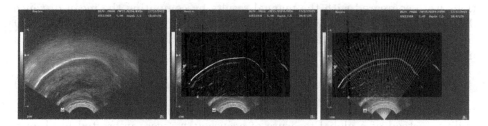

Fig. 2. Different steps along the segmentation of the tongue surface from US. From left to right, the initial US frame, the image after pre-processing, and after the identification of the points along the tongue.

4 Performance Assessment

At this stage, and to obtain an objective assessment of how the proposed method performs, to inform further improvements, our goal was to compare the obtained segmentation with those revised by two annotators with experience in speech production studies.

4.1   Segmentation Revision and Difference Computation

Manual Segmentation Revision: To ensure a wide variety of tongue configurations, representative of the data the method needs to tackle, based on the audio annotation, one frame from the beginning, two from the middle and one from the end of each vowel were selected, for each speaker. The frames were selected randomly from the repetitions, yielding a total of 108 images (4 images per vowel of each speaker. The images were segmented using the proposed method and the outcomes was provided to the speech scientists for revision. Each of the annotators received a unique set of 108 images to revise. Additionally, they revised a same set of 54 images, covering all vowels and speakers, to enable gathering data on inter-annotator variability.

While a full segmentation of the tongue might had been performed by the annotators, as we were more interested in identifying the issues of the proposed method, we opted for a revision procedure. In the revision process, the annotators could freely displace the points proposed by the method in two ways: 1) click on the left mouse button if they were confident about their re-positioning or 2) click on the right mouse button if they were uncertainty of the tongue contour. This latter case might happen when, given their knowledge about the tongue's shape (and spatial coherence), they would have an intuition of were the point should be, but no tongue contour was clearly visible.

Evaluation: To evaluate the segmentation accuracy, the automatic segmentations were compared with the manual results by computing the mean distance difference between them ($Seg_{diff}.$), for each radial. This can be expressed by Eq. 2, where R_{seg} means the radial length of automatic segmentation and R_{rev} means the radial length of manual revision.

$$Seg_{diff}.(\%) = 100 \times (R_{seg} - R_{rev})/R_{rev} \qquad (2)$$

A similar process were also implemented to assess the inter-annotator variability. Instead of a global measure of accuracy, we were interested in observing distance differences for the different regions of the tongue. Additionally, we intended to understand how distance differences patterns varied, for each vowel.

4.2   Results

In this section, we evaluate the accuracy of the proposed automatic method through the comparison of their results with the manual segmentation revision of the tongue contour.

Figure 3 shows representative examples of the segmentations provided by the proposed method and depicts the outcomes of comparing them for a central low ([a], left column) and height back vowel ([u], right column) with those performed by the annotators. The contours superimposed over the US images (Figs. 3a and 3b) show a more demanding segmentation setting posed by back vowels, due to a less discernible tongue contour, with notable segmentation errors at the tip and back regions of the tongue.

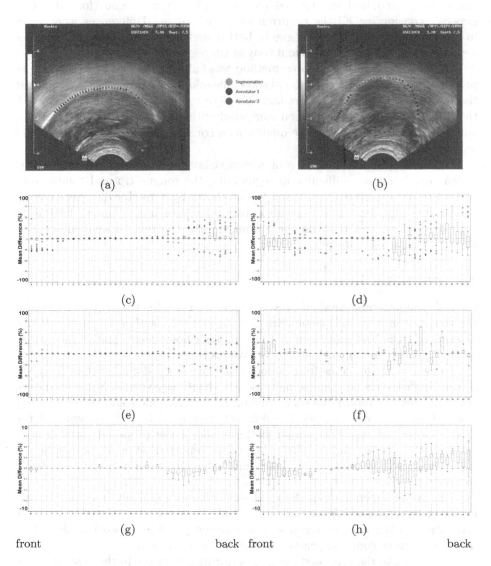

front   back front   back

Fig. 3. Overall illustrative results for the proposed segmentation method and its comparison with human revised segmentations for vowels [a] (left column) and [u] (right column). First row, illustrative US images and corresponding segmented and revised tongue contours; second row, average differences (in %), along the tongue, between the automatic segmentations and their revised versions for all occurrences of the respective vowel; third row, same as row two, but only considering points marked as confidently revised by the annotators; fourth row, differences between the two annotators.

Figures 3c and 3d show the mean distance differences (%) between the automatic segmented and the revised contours, along the tongue (for all radial sweeps), considering all the occurrences of [a] and [u]. Differences were lower in the central region of the tongue in both vowels. However, higher differences were found in the tongue root and tongue tip region for vowel [u].

To understand if the proposed method was failing in regions were the images present good quality, Figs. 3e and 3f shows mean distance differences (%) only at the points where the annotators had complete confidence. As can be observed, the differences are reduced, when compared with the previous results (Figs. 3c and 3d), hinting on poor image quality as a considerable reason for disparities among segmentations.

Finally, Figs. 3g and 3h present a comparison between the two annotators which corroborates a difficulty in segmenting the tongue tip and tongue root regions (more prominently in vowel [u]) as shown by a greater disagreement.

Additionally, for vowel [i] we also observed a poor image quality leading to high differences even between the automatic and the revised segmentation as between the two annotators.

4.3   Discussion

Overall, the proposed method presents high accuracy for the central regions of the tongue contour, and particularly for the central vowels ([a], [ɐ], [ɨ]). Thus, as expected, the best images come from sounds whose tongue surfaces are fairly flat and gently curved, such as the low vowels (e.g., [a]).

The worst images come from sounds that have steep slopes, such as [i] or [u], so that high vowels may image more poorly than lower vowels. Some studies of vowel production report the same difficulties, and exclude some vowels (mostly vowel [i] or back vowels), and/or some tokens due to bad US imaging of the final analysis [11, 30]. This poorer performance of the proposed method is in agreement with, for example, Stone [31] which reports that edges perpendicular to the beam will image best and edges more than 50° from perpendicular begin to image poorly.

This vowel dependence of the results constitutes a difficult challenge for our intended research – implying full coverage of the vowel inventory of Portuguese – pointing to the need to reconsider acquisition procedure, possibly departing from the current common practice described in the literature.

At this stage, the proposed method performs accurately in the central region of the tongue, where the highest tongue point is typically located, which is an important result. The highest point of the tongue is one measure often considered in US studies to determine tongue high and tongue advancement, two articulatory measures highly associated with the first and second formant frequencies of the vowels, respectively (e.g., [14, 23, 25]). So, the automatic method studied may enable an unsupervised extraction and analysis of this tongue measures.

The proposed method tackles, albeit to a limited degree, the issue of keeping a reasonable spatial coherence, among the points extracted for a single frame. One aspect that might improve the method is the consideration of both spatial and temporal (i.e., along the different frames of the sequences) coherence, due to the fact that adjacent frames might show slightly better defined segments of the tongue contour, which would enable solving some of the missing data.

5 Conclusions

The work described presents a tongue segmentation approach to serve the processing and analysis of a novel US dataset to study the effects of age on EP vowel production. Through the use of the proposed method, the tongue height and the tongue advancement can be extracted to investigate the speech production with age.

Even though our results present some limitations, mainly in the tongue tip and root regions and require further testing, at its current stage, the developed system is already valuable to make possible the otherwise time prohibitive study of the tongue in Phonetic research [16]. Conditions are created for the continuation of the research of age effect in vowel articulation, now employing direct automatic measures of tongue height and advancement/retraction.

Due to the difficulty to keep the probe in the same position and the tongue scanned on the same angle for all participants, it is challenging to compare contours between different speakers or different recording sessions of the same speaker [11]. For that, in the future, work is needed to obtain speaker independent measures. Also, positioning of the US probe needs to be optimized, considering: (1) angle optimization and/or (2) vowel dependent positioning of the probe.

Acknowledgements. This research was financially supported by the projects VoxSenes (POCI-01-0145-FEDER-03082) and MEMNON (POCI-01-0145-FEDER-028976) – COMPETE2020 under POCI and FEDER, and by national funds (OE), through FCT/MCTES, SOCA – Smart Open Campus CENTRO-01-0145-FEDER-000010 (Portugal 2020 under POCI and FEDER) and by IEETA Research Unit funding (UIDB/00127/2020). Luciana Albuquerque's work is funded by the FCT through grant SFRH/BD/115381/2016.

References

1. Akgul, Y.S., Stone, C., Maureen, K.: Automatic extraction and tracking of contours. Trans. Med. Imaging **18**(10), 1035–1045 (1999)
2. Articulate Assistant Ltd.: Articulate Assistant Advanced Ultrasound Module User Manual (2014)
3. Articulate Instruments Ltd.: Ultrasound Stabilisation Headset Users Manual (2008)
4. Articulate Instruments Ltd.: SyncBrightUp Users Manual (2010)
5. Chen, Y., Lin, H.: Analysing tongue shape and movement in vowel production Using SS ANOVA in ultrasound imaging. In: ICPhS, pp. 124–127 (2011)
6. Csapó, T.G., Lulich, S.M.: Error analysis of extracted tongue contours from 2D ultrasound images. In: INTERSPEECH, pp. 2157–2161. ISCA, Dresden (2015)

7. Dokovova, M., Sabev, M., Scobbie, J.M., Lickley, R., Cowen, S.: Bulgarian vowel reduction in unstressed position: an ultrasound and acoustic investigation. In: 19th ICPhS, pp. 2720–2724 (2019)
8. Fabre, D., Hueber, T., Bocquelet, F., Badin, P.: Tongue tracking in ultrasound images using EigenTongue decomposition and artificial neural networks. In: INTERSPEECH, pp. 2410–2414. ISCA, Dresden (2015)
9. Fabre, D., Hueber, T., Girin, L., Alameda-Pineda, X., Badin, P.: Automatic animation of an articulatory tongue model from ultrasound images of the vocal tract. Speech Commun. **93**, 63–75 (2017). https://doi.org/10.1016/j.specom.2017.08.002
10. Fasel, I., Berry, J.: Deep belief networks for real-time extraction of tongue contours from ultrasound during speech. In: International Conference on Pattern Recognition, pp. 1493–1496 (2010). https://doi.org/10.1109/ICPR.2010.369
11. Georgeton, L., Antolík, T.K., Fougeron, C.: Effect of domain initial strengthening on vowel height and backness contrasts in French: acoustic and ultrasound data. JSLHR **59**(6), S1575–S1586 (2016)
12. Georgeton, L., Kocjančič Antolík, T., Fougeron, C.: Domain initial strengthening and height contrast in French: acoustic and ultrasound data. In: 10th ISSP, Cologne, pp. 142–145 (2014). https://halshs.archives-ouvertes.fr/halshs-01401388
13. Hall, K.C., Allen, C., Mcmullin, K., Letawsky, V., Turner, A.: Measuring magnitude of tongue movement for vowel height and backness. In: ICPhS (2015)
14. Hillenbrand, J., Getty, L.A., Clark, M., Wheeler, K.: Acoustic characteristics of American English vowels. J. Acoust. Soc. Am. **97**(5), 3099–3111 (1995). http://ukpmc.ac.uk/abstract/MED/7759650
15. Jaumard-Hakoun, A., Xu, K., Roussel-ragot, P., Stone, M.L.: Tongue contour extraction from ultrasound images. In: 18th International Congress of Phonetic Sciences (ICPhS) (2015)
16. Karimi, E., Ménard, L., Laporte, C.: Fully-automated tongue detection in ultrasound images. Comput. Biol. Med. **111**(103335), 1–13 (2019). https://doi.org/10.1016/j.compbiomed.2019.103335
17. Kirkham, S., Nance, C.: An acoustic-articulatory study of bilingual vowel production: advanced tongue root vowels in Twi and tense/lax vowels in Ghanaian English. J. Phon. **62**, 65–81 (2017)
18. Kisler, T., Reichel, U., Schiel, F.: Multilingual processing of speech via web services. Comput. Speech Lang. **45**, 326–347 (2017). https://doi.org/10.1016/j.csl.2017.01.005
19. Kovesi, P., et al.: Symmetry and asymmetry from local phase. In: Tenth Australian Joint Conference on Artificial Intelligence, vol. 190, pp. 2–4. Citeseer (1997)
20. Lancia, L., Rausch, P., Morris, J.S.: Automatic quantitative analysis of ultrasound tongue contours via wavelet-based functional mixed models. J. Acoust. Soc. Am. **137**(2), EL178–EL183 (2015). https://doi.org/10.1121/1.4905881
21. Laporte, C., Ménard, L.: Robust tongue tracking in ultrasound images: a multi-hypothesis approach. In: Proceedings of the Annual Conference of the International Speech Communication Association, INTERSPEECH, pp. 633–637 (2015)
22. Laporte, C., Ménard, L.: Multi-hypothesis tracking of the tongue surface in ultrasound video recordings of normal and impaired speech. Med. Image Anal. **44**, 98–114 (2018). https://doi.org/10.1016/j.media.2017.12.003
23. Lee, S.H., Yu, J.F., Hsieh, Y.H., Lee, G.S.: Relationships between formant frequencies of sustained vowels and tongue contours measured by ultrasonography. Am. J. Speech Lang. Pathol. **24**(4), 739–749 (2015)
24. Li, M., Kambhamettu, C., Stone, M.: Automatic contour tracking in ultrasound images. Clin. Linguist. Phon. **19**(6–7), 545–554 (2005)

25. Morrison, G.S., Assmann, P.F.: Vowel Inherent Spectral Change: Modern Acoustics and Signal Processing. Springer, Heidelberg (2013). https://doi.org/10.1007/978-3-642-14209-3

26. Mozaffari, M.H., Lee, W.S.: Domain adaptation for ultrasound tongue contour extraction using transfer learning: a deep learning approach. J. Acoust. Soc. Am. **146**(5), EL431–EL437 (2019). https://doi.org/10.1121/1.5133665

27. Mozaffari, M.H., Wen, S., Wang, N., Lee, W.: Real-time automatic tongue contour tracking in ultrasound video for guided pronunciation training. In: 14th International Joint Conference on Computer Vision, Imaging and Computer Graphics Theory and Applications (VISIGRAPP 2019), vol. 1, pp. 302–309 (2019). https://doi.org/10.5220/0007523503020309

28. Muldal, A.: Python-phasepack (2016). https://github.com/alimuldal/phasepack

29. Noble, A., et al.: Ultrasound image segmentation : a survey. IEEE Trans. Med. Imaging **25**, 987–1010 (2006)

30. Song, J.Y.: The use of ultrasound in the study of articulatory properties of vowels in clear speech. Clin. Linguist. Phon. **31**(5), 351–374 (2017). https://doi.org/10.1080/02699206.2016.1268207

31. Stone, M.: A guide to analysing tongue motion from ultrasound images. Clin. Linguist. Phon. **19**(6–7), 455–501 (2005). https://doi.org/10.1080/02699200500113558

32. Tang, L., Bressmann, T., Hamarneh, G.: Tongue contour tracking in dynamic ultrasound via higher-order MRFs and efficient fusion moves. Med. Image Anal. **16**(8), 1503–1520 (2012). https://doi.org/10.1016/j.media.2012.07.001

33. Tang, L., Hamarneh, G.: Graph-based tracking of the tongue contour in ultrasound sequences with adaptive temporal regularization. In: Computer Society Conference on Computer Vision and Pattern Recognition - Workshops (CVPRW 2010), pp. 154–161. IEEE (2010). https://doi.org/10.1109/CVPRW.2010.5543597

34. Unser, M., Stone, M.: Automated detection of the tongue surface in sequences of ultrasound images. J. Acoust. Soc. Am. **91**(5), 3001–3007 (1992). https://doi.org/10.1121/1.402934

35. Wang, H., Wang, S., Denby, B., Dang, J.: Automatic tongue contour tracking in ultrasound sequences without manual initialization. In: Asia-Pacific Signal and Information Processing Association Annual Summit and Conference (APSIPA), pp. 200–203. IEEE (2015). https://doi.org/10.1109/APSIPA.2015.7415503

36. Wen, S.: Automatic tongue contour segmentation using deep learning. Master of Applied Science in Electrical and Computer Engineering, University of Otawa (2018)

37. Xu, K., et al.: Robust contour tracking in ultrasound tongue image sequences. Clin. Linguist. Phon. **30**(3–5), 313–327 (2016). https://doi.org/10.3109/02699206.2015.1110714

38. Zhu, J., Styler, W., Calloway, L: Automatic tongue contour extraction in ultrasound images with convolutional neural networks. J. Acoust. Soc. Am. **143**(3), 1966 (2018). https://doi.org/10.1121/1.5036466

39. Zhu, J., Styler, W., Calloway, I.: A CNN-based tool for automatic tongue contour tracking in ultrasound images. eprint arXiv:1907.10210, pp. 1–6 (2019)

Improving Multiple Sclerosis Lesion Boundaries Segmentation by Convolutional Neural Networks with Focal Learning

Gustavo Ulloa[1]([✉]), Alejandro Veloz[2][ID], Héctor Allende-Cid[3][ID],
and Héctor Allende[1]

[1] Universidad Técnica Federico Santa María, Av. España 1680, Valparaíso, Chile
gustavo.ulloa@gmail.com
[2] Universidad de Valparaíso, Blanco 951, Valparaíso, Chile
[3] Pontificia Universidad Católica de Valparaíso, Av. Brasil 2950, Valparaíso, Chile

Abstract. Multiple sclerosis lesions segmentation is an important step in the diagnosis and tracking in the evolution of the disease. Convolutional Neural Networks (CNN) have been obtaining successful results in the task of lesion segmentation in recent years, but still present problem segmenting boundaries of the lesions. In this work we focus the learning process on hard voxels close to the boundaries of the lesions by means of a stratified sampling and the use of focal loss function that dynamically increase the penalization on this kind of voxels. This approach was applied on the 2015 Longitudinal MS Lesion Segmentation Challenge dataset (ISBI2015 (https://smart-stats-tools.org/lesion-challenge)), obtaining better results than approaches using binary cross entropy loss and focal loss functions with uniform sampling.

Keywords: Convolutional Neural Networks · Image segmentation · Multiple sclerosis lesions · Magnetic Resonance Imaging · Stratified sampling · Focal loss

1 Introduction

Multiple sclerosis (MS) is a chronic autoimmune, inflammatory neurological disease affecting the Central Nervous System (CNS) in which the autoimmune system attacks the myelin sheath, myelin producing cells and the axons present in the white matter brain tissue [6]. This can cause progressive loss of motor, sensor, visual or cognitive brain functions in people suffering this condition. Magnetic Resonance Imaging (MRI) is used to detect MS white matter lesions, in particular, employing T1-weighted (T1-w), T2-weighted (T2-w), PD-weighted (PD-w), and fluid attenuated inversion recovery T2 (T2-FLAIR) sequences.

Supported by Fondecyt Grant 1170123.

MS lesions are expressed in MRI as small hyperintensities regions in T2-w, PD-w and T2-FLAIR MRI sequences, and small hypointensities regions in T1-w MRI sequences [4].

From the radiologist point of view, extracting quantitative information about the volume and the amount of MS-related lesions in white matter is important for assisting the diagnosis, for assessing the disease progression, and for testing new drugs [5,14]. In this scenario, image processing techniques such as image segmentation helps to radiologists to detect and delineate more accurately MS-related lesions. The automatic segmentation of MS-related lesions also decreases considerably the time needed by the radiologist to analyze the progression of this condition, also decreasing the intra and inter-expert variability in assessing it.

In the last years, using deep learning architectures based on Convolutional Neural Networks (CNNs), state-of-the-art results have been obtained in most of visual recognition tasks; in particular in image classification, object detection and semantic segmentation. The usual way to perform the segmentation task is by using as input a patch surrounding the interest voxel to classify in one of the defined classes. Typically, the patches correspond to a set of intensities determined by uniform sampling in a square neighborhood centered in the voxel of interest.

The problem of MS-related lesion segmentation is difficult mainly due to the imbalance between voxels that belong to the MS class versus voxels of normal tissues. e.g., the ISBI2015 has a ratio of about $\sim 1 : 130$ MS: normal voxels. Although there are techniques designed for dealing with this imbalanced problem, e.g. weighted cross entropy or focal loss-based methods [9], using these approach is complicated from the computational point of view as they require a huge amount of $n \times n$ patches for each available image modality. For this reason, there is some level of consensus in using a set of candidate lesion voxels with an average imbalance of $\sim 1 : 50$, which are further balanced by randomly under-sampling the non MS-related lesion class [1,11].

Additionally, for segmenting MS-related lesions, some image features, such as partial volume effect, variability of MS lesions intensities and the overlapping between MS lesions and normal brain tissues intensities distributions complicated even more the problem of MS lesion segmentation [4]. Due to this, an accurate segmentation of the MS lesions borders is difficult to determine at glance by radiology practitioners or by using machine learning models. In fact, in these MS lesion domains, i.e. borders, the higher proportion of false positive and false negative errors are concentrated. Although CNNs have obtained prominent segmentation results, they still are not comparable in terms of accuracy to the segmentation that radiologist can produce manually.

This work focuses in focalizing the CNN learning to voxels that are difficult to classify, i.e. the voxels that belong to the MS-related lesions borders. For doing this, we use the focal loss function in the problem of MS-related lesions segmentation. We combine this approach with a stratified strategy for sampling normal tissue voxels with the purpose of increasing the amount of MS-related lesion voxels that are close to the lesion borders in the training set. Although the focal loss function was introduced to deal with the class imbalance in object detection problems by assuming that the instances that are difficult to classify correspond to the minority class [9]. In this work, we use focal loss for its ability to assign dynamically higher losses to hard-to-classify voxels. The underlying hypothesis of our proposal is that hard-to-classify voxels are located in image locations close to MS lesions, that potentially correspond to borders of such lesions.

The paper is organized as follows: In Sect. 2 we presents the background related to focal loss function, the proposed methodology is presented in Sect. 3. Section 4 presents discussions about the results obtained on the 2015 Longitudinal MS Lesion Segmentation Challenge dataset. Finally, in Sect. 5 conclusions and future work are presented.

2   Background: Focal Loss Function

In [9] the problem of class imbalance is addressed in the task of object detection, specifically when this task is performed by CNNs of a one-stage. This loss function is based on the binary cross entropy loss function, which is given by the expression:

$$BCE(y, p) = -y \log p - (1 - y) \log(1 - p) \tag{1}$$

where $y \in \{0, 1\}$ is the ground truth where $y = 0$ and $y = 1$ correspond to the non-lesion and lesion classes respectively and $p \in [0, 1]$ is the probability for the class with label $y = 1$ estimated by the model. For notational convenience, p_t is defined as:

$$p_t = \begin{cases} p & \text{if } y = 1, \\ 1 - p & \text{otherwise,} \end{cases} \tag{2}$$

Thus, for the t-th input instance the BCE can be expressed as $BCE(y, p) = BCE(p_t) = -\log(p_t)$.

In the focal loss function, a weighting term given by $(1 - p_t)^\gamma$ is added to the binary cross entropy loss function. According to this, the focal loss function can be expressed as:

$$FL(p_t) = -(1 - p_t)^\gamma \log(p_t), \tag{3}$$

where γ is a parameter that modulates the weighting effect provided to this loss function. In this way, the focal loss weights dynamically the training examples losses, down-weighting easy-to-classify training examples and assigning higher losses to examples with low prediction accuracy, i.e. hard-to-classify training examples.

In [9] also was presented an α-balanced variant of focal loss, in which a new weighting α parameter is added to perform an additional class balancing to the one implicitly performed by focal loss. Although the authors report a slight increase in accuracy, this was not used because we have already adopted the procedure of subsampling the majority non-lesion class as a balancing method.

3   Methods

The CNN architecture used with different loss functions and with or without augmented sampling on the hard examples voxels close of the boundaries lesions were evaluated on the data set of the 2015 Longitudinal MS Lesion Segmentation Challenge [2]. The data set consist of longitudinal multi-channel MR images from 19 different patients, where the training set is conformed by the MR images from five patients (segmented data set) and the test set from fourteen patients. Since in the testing set we do not have the segmentation done by experts we used only the training data set. 4.4 average time points were acquired by patient (4, 5 or 6 time points by patient) in the training and testing data sets on a 3.0 Tesla MRI scanner, where four channels were available for each patient, corresponding to T1-weighted, T2-weighted, PD-weighted and T2-FLAIR. Each MR image has a 1 [mm] isotropic cubic voxel resolution. The training data includes binary masks with manual lesion segmentations.

3.1   Preprocessing

Prior to the extraction of the MR image patches that are used to train the CNN, two preprocessing tasks were applied. Each one consisting in several subtasks that are explained below. The first preprocessing task consists in the standard MRI processing typically made within the medical imaging community and that was also applied for the 2015 Longitudinal MS Lesion Segmentation Challenge dataset [2], i.e. correction of the MR intensity inhomogeneities, skull-stripping, dura mater stripping, and rigid-body registration to a 1 [mm] isotropic MNI template. The second preprocessing task has the following steps [1]: Intensities truncation to the percentiles within the range $[0.01, 0.99]$, intensities scaling to the range $[0, 1]$, rigid-body registration of the T2-FLAIR images to the ICBM452 probabilistic atlas [7], and, finally, extraction of a subset of supra-threshold voxels. The voxels that belong to this subset are conformed by intersecting the set of voxels that exceeds one threshold on the T2-FLAIR image and another threshold in the probability map of the white matter. This is because MS-related lesions appear as hyper-intense regions on T2-FLAIR images [4] and also are mostly located in the white matter.

3.2   CNN Architecture

We choose a V-Net CNN architecture in which each convolutional block includes a convolutional layer with different number of filters, filters sizes, and a 2×2 max pooling layer [1]. All convolutional blocks are fully connected layers that use Leaky ReLU activation functions.

The V-Net CNN architecture receives as input four patches of size 33×33. These patches are extracted from the axial plane of T1-weighted, T2-weighted, and T2-FLAIR channels. It must be noted that PD-weighted images were excluded from the input data, despite they are typically used in MS-related lesion segmentation. This is due to the fact that including this image modality the segmentation results became worse. This was verified empirically according to the Dice coefficient. Figure 1 schematizes our CNN architecture.

The 2D patch size of 33×33 was chosen due to the fact that it was the size that obtained the best results reported in [1] using the ISBI2015 dataset. After the third convolutional block there is two fully connected layers of sixteen nodes and an output layer of one node with sigmoid activation function, which yield output probability for lesion classes by the center voxel of the input patch.

Fig. 1. CNN architecture used, based on the V-Net network [1]

3.3   Training Details

The weights were updated with the stochastic gradient descent and the loss functions used were the binary cross entropy and focal loss. As regularization method we used dropout with probability of 0.1 after each layer empirically selected. The negative slope coefficient of the Leaky ReLU activation function was empirically setting to $\beta = 0.3$. The learning rate selected was 0.01 with decay learning rate of 1×10^{-6}, momentum parameter of 0.9 with Nesterov momentum and batch size of 64. The data set for the training was labeled into two classes, non-lesion tissue and lesion tissue.

The voxels belonging to the class without injury were sampled in two different ways from the set of voxels that were injury candidates (Step 2 (d)) that were not labeled as an injury by the expert. The first corresponds to a random sampling through a uniform distribution and the second type of sampling was stratified, that is, this set is stratified into two strata. A p proportion of voxels of the final sample was extracted from the first stratum corresponding to a neighborhood of voxels adjacent to the 2 mm wide lesions and the remaining $1 - p$ ratio was extracted from the second stratum corresponding to the rest of candidate voxels.

As is common in medical imaging problems, the proportion of voxels belonging to the pathological classes, in this case lesion tissue are the minority classes, for that, it is a imbalanced classification problem. To handle this, we generated augmented data from the lesion class which balanced the data set and obtained even better results than the approach of only down-sampling the majority class. It consisted of the artificial generation of new patches centered on voxels of the lesion class, by applying linear transformations to a percentage of randomly selected patches. These transformations consisted of rotated images from angles drawn from uniform distribution $U[0, 360]$ and horizontal and vertical reflections.

To compare the segmentation performance, we implemented a cross validation group where five identical models were trained, corresponding to the 5 available patients in the dataset, that is, each group corresponds to a patient along with the MR images that were taken over time. Thus, 5 folds were made, whose details are found in Table 1.

Table 1. Group cross validation sets.

Train	Validation	Test
2, 3, 4	5	1
1, 4, 5	3	2
2, 4, 5	1	3
1, 3, 5	2	4
1, 2, 3	4	5

A training of 100 epochs with early stopping (blocks of 20 epochs) without diminishing the loss of the validation set was arranged.

The implementation and training of the CNN was carried out using Keras Deep Neural Network Library [3] with TensorFlow as backend numerical engine. The model was trained using a NVIDIA GeForce GTX 1080TI Graphic Card.

3.4 Evaluation Metrics

To compare the proposal, we used three metrics: the True Positive Rate (TPR), Positive Predictive Value (PPV) and Dice coefficient. TPR indicates the rate of voxels correctly segmented as lesions (Eq. (4)), PPV indicates the rate of segmented voxels from ones estimated as lesions (Eq. (5)) and the Dice coefficient (Eq. (6)). Dice coefficient is an overlap measure commonly used as overlap measure between two binary label masks, the expert and the automatic segmentation for imbalanced binary classification and segmentation problems, which value are in the range $[0, 1]$, where 0 indicates no agreement and 1 means that the two masks are identical. The $\mathcal{M_R}$ mask corresponds to the lesion segmentation performed by the human and $\mathcal{M_A}$ is the generated mask by the algorithm, TP, FN and FP correspond to the amount of true positive values, false negative values and false positive values.

$$\text{TPR}(\mathcal{M_R}, \mathcal{M_A}) = \frac{|\mathcal{M_R} \cap \mathcal{M_A}|}{|\mathcal{M_R} \cap \mathcal{M_A}| + |\mathcal{M_R} \cap \mathcal{M_A^C}|} = \frac{TP}{TP + FN}, \quad (4)$$

$$\text{PPV}(\mathcal{M_R}, \mathcal{M_A}) = \frac{|\mathcal{M_R} \cap \mathcal{M_A}|}{|\mathcal{M_R} \cap \mathcal{M_A}| + |\mathcal{M_R^C} \cap \mathcal{M_A}|} = \frac{TP}{TP + FP}, \quad (5)$$

$$\text{Dice}(\mathcal{M_R}, \mathcal{M_A}) = \frac{1}{\frac{1}{\text{TPR}(\mathcal{M_R},\mathcal{M_A})} + \frac{1}{\text{PPV}(\mathcal{M_R},\mathcal{M_A})}} = \frac{2TP}{2TP + FN + FP}. \quad (6)$$

4 Experimental Design and Results

Table 2 shows the performance of the V-Net convolutional network architecture (Fig. 1) when equipped with the binary cross entropy (BCE) loss functions or the loss function focal loss (FL), in addition to the type of sampling used, uniform or stratified. The column LF of Table 2 corresponds to the loss function used, p is the proportion of voxels near the edges of the lesions in stratified sampling, $-$ indicates uniform sampling, γ it is the adjustable parameter of the focal loss function (see Eq. (3)) and u is the threshold that maximizes the Dice coefficient to discriminate the CNN output as a lesion class.

Table 2. Results.

LF	p	γ	u	TPR(sd)	PPV(sd)	Dice(sd)	p-value
BCE	–	–	0.96	0.7051(0.0229)	0.6788(0.0217)	0.6786(0.0082)	7.3×10^{-05}
FL	–	0.5	0.91	0.6961(0.0268)	0.6855(0.0221)	0.6756(0.0142)	1.3×10^{-04}
FL	–	1.0	0.86	0.6818(0.0253)	0.7030(0.0173)	0.6790(0.0156)	5.9×10^{-03}
FL	–	2.0	0.77	0.6849(0.0245)	0.7008(0.0209)	0.6779(0.0165)	3.2×10^{-03}
FL	–	5.0	0.63	0.6835(0.0316)	0.6791(0.0235)	0.6633(0.0172)	9.2×10^{-09}
BCE	0.1	–	0.92	0.6891(0.0223)	0.6977(0.0250)	0.6793(0.0104)	7.8×10^{-04}
FL	0.1	0.5	0.85	0.6914(0.0203)	0.7000(0.0210)	0.6822(0.0117)	2.4×10^{-02}
FL	0.1	1.0	0.80	0.6851(0.0154)	0.7127(0.0198)	0.6877(0.0091)	–
FL	0.1	2.0	0.71	0.6811(0.0204)	0.7035(0.0233)	0.6791(0.0100)	5.0×10^{-04}
FL	0.1	5.0	0.60	0.6687(0.0220)	0.6951(0.0274)	0.6668(0.0124)	4.2×10^{-10}
BCE	0.2	–	0.89	0.6855(0.0244)	0.6842(0.0276)	0.6690(0.0141)	7.4×10^{-08}
FL	0.2	0.5	0.83	0.6856(0.0277)	0.6882(0.0318)	0.6712(0.0093)	1.8×10^{-09}
FL	0.2	1.0	0.77	0.6790(0.0195)	0.6901(0.0218)	0.6701(0.0088)	1.3×10^{-10}
FL	0.2	2.0	0.70	0.6484(0.0223)	0.7175(0.0302)	0.6652(0.0114)	8.1×10^{-12}
FL	0.2	5.0	0.59	0.6489(0.0171)	0.6905(0.0254)	0.6464(0.0191)	7.9×10^{-14}

In Table 2 the average performance and standard deviation of TPR, PPV and Dice coefficient can be observed after performing $n = 30$ repetitions for each combination of parameters represented by columns LF, p and γ. The best result for TPR occurs when BCE is used with uniform sampling without stratification, which contrasts with the worst result for PPV. This situation also reflects the worst FP performance, which is due to the fact that in this type of sampling there is the smallest proportion of difficult voxels of the non-injury class, that is, near the edges of the lesions.

The best result for PPV happens when a stratified sampling with a proportion of $p = 0.2$ is performed and Focal Loss is used with $\gamma = 2.0$, which indicates that one way to reduce false positives is to pay more attention to the difficult examples of the non-lesion class close to the edges of the lesions. In the latter case, on the other hand, the highest proportion of false negatives and therefore the worst TPR performance was obtained.

As is known, Dice coefficient is the most widely used metric in the literature because it is an overlap measure between expert and model masks by imbalanced binary tasks, which quantifies the relationship or trade off between false positives and false negatives when combining TPR metrics (Eq. 4) and PPV (Eq. 5) using the harmonic mean. In Table 2 it can be seen that the best segmentation result using Dice coefficient corresponds to the use of focal loss with parameter $\gamma = 1.0$ and when the proportion of voxels near the edge of the lesions is $p = 0.1$. This result was contrasted with all other combinations of parameters by means of a t-test with the assumption of different and unknown variances (Welch's test), where the null hypothesis corresponds to the assumption that there are no differences of means Dice with the different combinations and the alternative hypothesis says that this result has a higher mean Dice. It is observed that all p-values were less than 0.03.

The use of paired t-test of differences of means as in other papers was rejected because they are not paired or dependent samples and if so assumed, there is a high variability in the statistic because it depends on the combination used in the matching of the couples.

We can also observe that the use of focal loss drastically influenced the decrease of the threshold $u = 0.80$, especially when stratified sampling was carried out that increases the proportion of non-lesion voxels close to the edges in the training set, which reinforces us the hypothesis that in these areas are the most difficult voxels to classify.

Considering the variance in CNN weights inherent in the training mechanism, it seems appropriate to graph the expected behavior of the segmentation algorithms by using the simultaneous truth and performance level estimation (STAPLE) algorithm [13]. STAPLE is an expectation-maximization algorithm for the statistical fusion of binary segmentations, thus generating a consensus segmentation of the $N = 30$ binary segmentations for each combination of parameters (LF, p and γ).

Figure 2 shows the consensus of the automatic segmentation obtained using different combination of loss functions and parameters. Figure 2(a) shows a T2-FLAIR axial slice of Patient number 2. Figure 2(b) presents the ground truth. Figures 2(c) and (d) present the automatic segmentation obtained by using the binary cross entropy loss with both uniform sampling and stratified sampling strategies ($p = 0.1$), respectively. Similarly, Figs. 2(e) and (f) correspond to the automatic segmentation obtained by the proposed method using focal loss ($\gamma = 1.0$) with both uniform sampling and stratified sampling ($p = 0.1$), respectively. True positive voxels are presented in red, false negative voxels are presented in green and false positive voxels are presented in blue.

The best segmentation performance in terms of the Dice coefficient was obtained using focal loss with $p = 0.1$ and $\gamma = 1.0$ as we can see in Fig. 2(f), in which a small increase in false negative rate and a large decrease in false positive rate is observed compared respect to Fig. 2(c).

In Fig. 2(d) a similar number of false negatives and a slight increase in false positives for this slice can be observed compared to Fig. 2(f). This difference is most noticeable in Table 2 which represents the average behavior for all slices and patients. Figure 2(e) shows an increase in false negatives and decrease in false positives compared Fig. 2(c), obtaining a similar Dice coefficient, which indicates that the use of focal loss ($\gamma = 1.0$) does not necessarily imply obtaining a better Dice coefficient if in the training set does not have the adequate representation of the most difficult segmentation voxels, as if it occurs with stratified sampling ($p = 0.1$) in Fig. 2(f).

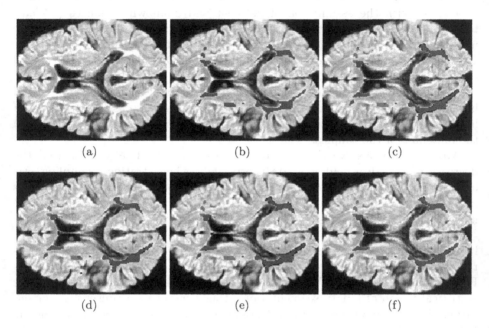

Fig. 2. Automatic consensus segmentations in the axial plane for Patient 2. True positives, false negatives and false positives are presented in red, green and blue, respectively. (a) T2-FLAIR, (b) T2-FLAIR with rater mask, (c–d) binary cross entropy with uniform and stratified sampling with $p = 0.1$, respectively, and (e–f) focal loss with $\gamma = 1.0$ using uniform sampling and stratified sampling with $p = 0.1$, respectively. (Color figure online)

5   Conclusions and Future Work

This work showed empirically that in the task of segmentation of lesions of multiple sclerosis, better results are obtained by focusing the learning of CNN in the most difficult voxels to classify in lesion or non-injury classes. The results indicate that these difficult voxels correspond largely to those located near the edges of the lesions since the focal loss application obtained the best results when a stratified sampling was applied increasing the representation of examples of these areas in the training stage. We can also conclude that the use of this methodology allows to improve the results by maintaining end-to-end training without having to resort to the implementation of architectures and cascade training as in some proposals present in the state of the art [8,11], in which the CNN is retrained with a resampling concentrated on the poorly classified examples.

Although this methodology allows to improve the results without incurring in a higher computational cost, a possible clinical application requires Dice coefficient performance higher than 0.7, due that results higher than this are comparable to segmentations performed by human experts. Considering that the brain, as its structures, tissues and lesions correspond to volumes, we find intuitive that the implementation of this methodology using 3D resampling patches will improve the current results considerably.

As future work, it is proposed to continue working on the development of a loss function that focuses learning on difficult voxels belonging to the edges of the lesions through the use of local voxel information to be classified as well as neighborhood information. We also propose the possibility of extending this methodology based on the approach presented in [10], by performing a prediction of the class of the whole resampling patch instead of predicting the class of the central voxel. So, the architecture of the CNN will adopt the advantages of the 3D approach presented in [12], using the acquired data sets from different acquisition protocols and resonators.

Acknowledgments. This work was supported in part by the Fondecyt Grant 1170123 and ANID PIA/apoyo Project AFB 1800082. Alejandro Veloz is supported by the Fondecyt Grant 1201822. Héctor Allende-Cid is supported by Fondecyt Initiation into Research 11150248.

References

1. Birenbaum, A., Greenspan, H.: Multi-view longitudinal CNN for multiple sclerosis lesion segmentation. Eng. Appl. Artif. Intell. **65**, 111–118 (2017)
2. Carass, A., et al.: Longitudinal multiple sclerosis lesion segmentation: resource and challenge. NeuroImage **148**, 77–102 (2017)
3. Chollet, F., et al.: Keras (2015). https://keras.io
4. Danelakis, A., Theoharis, T., Verganelakis, D.A.: Survey of automated multiple sclerosis lesion segmentation techniques on magnetic resonance imaging. Comput. Med. Imaging Graph. **70**, 83–100 (2018). https://doi.org/10.1016/j.compmedimag.2018. 10.002. http://www.sciencedirect.com/science/article/pii/S0895611118303227
5. Giorgio, A., Stefano, N.D.: Effective utilization of MRI in the diagnosis and management of multiple sclerosis. Neurol. Clin. **36**(1), 27–34 (2018). https:// doi.org/10.1016/j.ncl.2017.08.013. http://www.sciencedirect.com/science/article/ pii/S0733861917301007, multiple Sclerosis
6. Goldenberg, M.M.: Multiple sclerosis review. P & T Peer-Rev. J. Formul. Manag. **37**(3), 175–184 (2012). https://www.ncbi.nlm.nih.gov/pmc/articles/PMC3 351877/, https://www.ncbi.nlm.nih.gov/pubmed/22605909, multiple Sclerosis
7. Jenkinson, M., Bannister, P., Brady, M., Smith, S.: Improved optimization for the robust and accurate linear registration and motion correction of brain images. NeuroImage **17**(2), 825–841 (2002). https://doi.org/10.1006/nimg.2002. 1132. http://www.sciencedirect.com/science/article/pii/S1053811902911328

8. Kazancli, E., Prchkovska, V., Rodrigues, P., Villoslada, P., Igual, L.: Multiple sclerosis lesion segmentation using improved convolutional neural networks. In: Proceedings of the 13th International Joint Conference on Computer Vision, Imaging and Computer Graphics Theory and Applications - Volume 4 VISAPP: VISAPP, pp. 260–269. INSTICC, SciTePress (2018). https://doi.org/10.5220/0006540902600269

9. Lin, T., Goyal, P., Girshick, R., He, K., Dollár, P.: Focal loss for dense object detection. IEEE Trans. Pattern Anal. Mach. Intell. **42**(2), 318–327 (2020). https://doi.org/10.1109/TPAMI.2018.2858826

10. Roy, S., Butman, J.A., Reich, D.S., Calabresi, P.A., Pham, D.L.: Multiple sclerosis lesion segmentation from brain MRI via fully convolutional neural networks (2018)

11. Valverde, S., et al.: Improving automated multiple sclerosis lesion segmentation with a cascaded 3D convolutional neural network approach. NeuroImage **155**, 159–168 (2017)

12. Valverde, S., et al.: One-shot domain adaptation in multiple sclerosis lesion segmentation using convolutional neural networks. NeuroImage Clin. **21**, 101638 (2019). https://doi.org/10.1016/j.nicl.2018.101638. http://www.sciencedirect.com/science/article/pii/S2213158218303863

13. Warfield, S.K., Zou, K.H., Wells, W.M.: Simultaneous truth and performance level estimation (STAPLE): an algorithm for the validation of image segmentation. IEEE Trans. Med. Imaging **23**(7), 903–921 (2004). https://doi.org/10.1109/TMI.2004.828354

14. Zhou, T., Ruan, S., Canu, S.: A review: deep learning for medical image segmentation using multi-modality fusion. Array **3–4**, 100004 (2019). https://doi.org/10.1016/j.array.2019.100004. http://www.sciencedirect.com/science/article/pii/S2590005619300049

B-Mode Ultrasound Breast Anatomy Segmentation

João F. Teixeira[1,3]([✉]) [iD], António M. Carreiro[1], Rute M. Santos[4] [iD],
and Hélder P. Oliveira[2,3] [iD]

[1] Faculty of Engineering of the University of Porto, Porto, Portugal
jpfteixeira.eng@gmail.com
[2] Faculty of Sciences of the University of Porto, Porto, Portugal
[3] INESC TEC, Porto, Portugal
helder.f.oliveira@inesctec.pt
[4] Escola Superior de Tecnologia da Saúde de Coimbra, Coimbra, Portugal

Abstract. Breast Ultrasound has long been used to support diagnostic
and exploratory procedures concerning breast cancer, with an interesting
success rate, specially when complemented with other radiology information. This usability can further enhance visualization tasks during pre-
treatment clinical analysis by coupling the B-Mode images to 3D space, as
found in Magnetic Resonance Imaging (MRI) per instance. In fact, Lesions
in B-mode are visible and present high detail when comparing with other
3D sequences. This coupling, however, would be largely benefited from the
ability to match the various structures present in the B-Mode, apart from
the broadly studied lesion. In this work we focus on structures such as
skin, subcutaneous fat, mammary gland and thoracic region. We provide
a preliminary insight to several structure segmentation approaches in the
hopes of obtaining a functional and dependable pipeline for delineating
these potential reference regions that will assist in multi-modal radiological data alignment. For this, we experiment with pre-processing stages
that include Anisotropic Diffusion guided by Log-Gabor filters (ADLG)
and main segmentation steps using K-Means, Meanshift and Watershed.

Among the pipeline configurations tested, the best results were found
using the ADLG filter that ran for 50 iterations and H-Maxima suppression of 20% and the K-Means method with $K = 6$. The results present
several cases that closely approach the ground truth despite overall having larger average errors. This encourages the experimentation of other
approaches that could withstand the innate data variability that makes
this task very challenging.

Keywords: Segmentation · Rigid registration · Minimum path ·
B-mode Ultrasound · Anatomical structures · Breast cancer

This work was funded by the ERDF - European Regional Development Fund through
the Norte Portugal Regional Operational Programme (NORTE 2020), under the POR-
TUGAL 2020 Partnership Agreement and through the Portuguese National Innovation Agency (ANI) as a part of project BCCT.Plan–NORTE-01-0247-FEDER-01768
and by Fundação para a Ciência e a Tecnologia (FCT) within PhD grants number
SFRH/BD/135834/2018.

© Springer Nature Switzerland AG 2020
A. Campilho et al. (Eds.): ICIAR 2020, LNCS 12132, pp. 193–201, 2020.
https://doi.org/10.1007/978-3-030-50516-5_17

1   Introduction

Breast cancer is the leading cause of cancer in women worldwide. The mortality rate has decreased in recent years [15] and even Conservative Treatment (BCCT) has reached Mastectomy prognosis. Ultrasound imaging (US) is deeply ingrained into the breast cancer treatment procedure. The information gathered from US and other exam modalities enables to accurately locate and characterize the lesions, which in turn can improve the understanding of their surroundings and help in the treatment decision process. This enhanced perception of the physiology of the individual breast can be of particular assistance in clinical, pre-surgical discussions and planning, specially when coupled with 3D references, as provided by some other exams such as e.g. Magnetic Resonance Imaging (MRI).

These kinds of fusion approaches, however, require some kind of common reference and, as such, accurate anatomical landmarks would be of use to produce them. There is considerable work on segmenting lesions on B-Mode US, although, without accurate know positioning of the acquisition probe, this only provides limited information about the accurate placement of those slices into a 3D reference space. Further structures must be outlined on US images, to help this placement and yet, there seems to be a lack in the literature concerning the segmentation of the remaining anatomy of the breast.

With this in mind, we intend to further the developments in automated segmentation of US anatomical structures, with the ultimate ambition of enabling the maturation of multimodal registration applications.

The main goal of this paper is, therefore, to present an experimental application of segmentation models to tackle the issue of US segmentation of the anatomical layered structures, such as skin, subcutaneous fat, mammary gland and thoracic region.

1.1   Literary Review

B-Mode Ultrasound images have essentially two focuses of research: noise removal or attenuation and lesion segmentation and classification. US images are mainly affected by inherent imaging artifacts called speckle noise [22]. The performance of segmentation methods depends not only on their internal workings but also on the cascading effect of pre-processing techniques used for reducing speckle. Several filtering approaches aimed at reducing speckle noise have been proposed in the literature. Popular techniques include Adaptive Filters such as the Median filter [14], Anisotropic Diffusion Filters (SRAD) [23], and interference-based speckle filtering followed by anisotropic diffusion (ISFAD) [6]. Other approaches include Multi-channel Decomposition of US images using Log–Gabor filters (ADLG) [9]. Furthermore, a few more approaches have also attracted attention because of their considerable performance in speckle reduction despite being an algorithm of high complexity. Such method encompass Non-local means filters, like Optimized Bayesian Non-Local Means filter (OBNLM) [7], which uses a Bayesian framework to adapt the Non-Local (NL) Means filter by introducing the Pearson distance as a relevant measure for patch comparison or a modified NL-means filter (MNL) [12] that uses maximum likelihood estimation as the first step. Finally, Multi-scale filters which are produced by applying a single scale method on sub-images obtained with the Laplacian pyramid [2] or Wavelet Decomposition [1], represent relevant pre-processing alternatives due to their multi-scale analysis despite their potential computational intensity.

For segmenting B-mode ultrasound images, various techniques have been developed and used, encompassing multiple levels of complexity. Some are based on classical thresholding techniques such as Otsu's method [18]. Aggregation techniques also have been experimented on US data, such as Region Growing, adding pixels of similar characteristics or fulfilling a aggregation particular condition [19]. Further models have been proposed based on the iterative adaptation to certain regions [4]. Moreover, watershed based methods have also been used for ultrasound segmentation [12]. Clustering based methods are also often used in ultrasound segmentation tasks, by splitting pixels into groups based on their characteristics or multiple layers of features [13]. Lastly, Machine Learning approaches, such as Artificial Neural Networks [16], have also been employed. Despite all this research, the segmentation of the several breast layers in B-mode Ultrasound images has scarcely been studied. To the best of our knowledge, there has been only another related study [5] in which the mammary gland was being segmented, and no other work proposed to approach segmenting skin, subcutaneous fat, mammary gland and thoracic region altogether.

2   B-Mode Ultrasound Anatomy Segmentation Method

This preliminary work explores a 3 step pipeline for anatomical structures' segmentation, illustrated in the diagram of Fig. 1. After the pre-processing and segmentation stages, the resulting image is subjected to an algorithm which finds the shortest paths connecting one margin of the image to the other. This helps to assign the segments found in the segmentation step to the respective layer or anatomical structure.

Fig. 1. Developed pipeline

A few different approaches were explored on both pre-processing and segmentation modules. The performance of each pipeline was also analyzed in a grid search fashion, over the validation set, in order to find the most suitable parameters.

2.1   Pre-processing

The pre-processing module accepts an US image that only contains the sensed data and transforms it using one of two alternate noise removal techniques: an adaptive median filter, for baseline purposes, or an Anisotropic Diffusion filter guided by Log-Gabor texture descriptors (ADLG) [9]. During the systematic experiments, for a few particular pipeline configurations, this module was also bypassed, i.e. the sensed data was feed directly to the segmentation module.

The parameter sets experimented for the optimization step using the validation dataset included a 9×9 kernel for the adaptive median, used as a baseline, and a number of different iterations for the ADLG (50, 100, 200, 500 and 1000).

2.2   Segmentation

The developed pipeline comprises 3 methods: K-means clustering [3], Meanshift clustering [10] and Watershed segmentation [17]. The K-means approach was specifically preceded by a H-Maxima suppression step in which the percentage parameter, along with the number of clusters (K) are optimized. Respectively, the values studied were 0%, 20% and 40%, and the range of $K = 2$ to $K = 6$. In the case of Meanshift clustering, the parameters adjusted were its bandwidth, and the weight ratio between position and intensity features, given as a concatenated input. The bandwidth encompassed the values 0.02, 0.05 and 0.1, while for the weight ratio were 0.0625, 0.125, 0.25 and 1. Finally, the Watershed algorithm was preceded by a H-Minima transform suppression step. Its suppression percentage was also varied during the validation portion of the work (0%, 5%, 10% and 20%).

2.3   Evaluation

The quality of the methods needs to be objectively assessed. In order to do that we use the Sørensen-Dice (DICE) coefficient [8,20], which ranges from 0 to 1, to estimate the area similarity between the GT and segmentation results, and the Average Distance (AD), measured in pixels, to evaluate the proximity of the detected contours of the structures [21].

3   Results and Discussion

3.1   Dataset

This study uses a B-Mode Ultrasound dataset (Fig. 2a) provided by the BCCT.plan project[1] containing data from 10 patients with breast cancer. Each image depicts a part of the breast and contains the typical layers of anatomical structures, as well as, at least one lesion, despite its segmentation not being

---

[1] BCCT.plan - 3D tool for planning breast cancer conservative treatment - NORTE-01-0247-FEDER-017688.

the focus of this work. Each image provides the respective annotation for the available structures, obtained by an expert radiology technician (>5 years). The annotated layers include Skin, Subcutaneous Fat, Mammary Gland and Thorax, as presented in Fig. 2b.

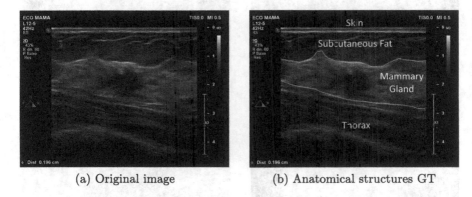

(a) Original image   (b) Anatomical structures GT

Fig. 2. Ultrasound image and respective ground truth (GT)

The images of this dataset are composed by the acquisition window (ROI) and by auxiliary information that improves the physician's understanding of the radiology data. In order to reduce the probability of errors in the segmentation module, it is important to discard this information.

Prior to feeding any data to the pipeline, the images are cropped, as illustrated in Fig. 3. For this we employ the ADLG filter with 500 iterations, to smooth the original image (Fig. 3a) whilst preserving its edges (Fig. 3b). After that, the resulting image undergoes a morphological opening to remove small objects surrounding the ROI (Fig. 3c). Finally, we find the linear ROI boundaries based on the outputs of a Hough Transform, and proceed to extract only the inner contents of that ROI, using the intensity data from the original image (Fig. 3d).

The database is divided into an image set used for parametric tuning (25 images) and a test set (75 images). Each set is randomly selected from the pool of all patients. As mentioned in Sect. 2, an experimental methodology based on grid search is conducted on the tuning set, leading to the best performing parameter configurations being used on the models over the test set.

3.2   Experimental Results

During the parametric tuning stage we optimized each pipeline variant for its DICE score. For the K-Means approach, the best configuration and parameterization obtained was employing the ADLG filter with 50 iterations, with an H-max suppression of 20% and $K = 6$. For the Meanshift approach, the best parameter set involved the use of the ADLG filter with 500 iteration and the

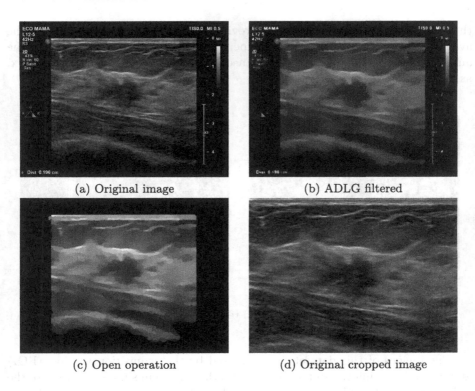

(a) Original image   (b) ADLG filtered

(c) Open operation   (d) Original cropped image

Fig. 3. Pre-pipeline cropping steps

Meanshift bandwidth of 0.1 and Weight ratio of 0.125. Finally, the Watershed approach reached best overall results using the unprocessed image and no minima suppression (0%).

Concerning the application of the test set, the results suggest that K-means clustering exceeds, on average, the other methods tested for the segmentation of these anatomical structures. Table 1 summarizes the quantifiable results. Additionally, the results show that this method exceeds both Meanshift clustering and Watershed in the segmentation of the following structures: Skin, Mammary gland and Thoracic region. However, it is very closely outshined by the Watershed's results regarding Subcutaneous Fat segmentation.

Watershed is sensible to noise and, thus, prone to over-segmentation, which might explain undesired results and unexpected margin-to-margin paths. Concerning that it presented its best results for a largely unaltered image, it suggests that the pre-processing methods experimented were not the most suited for this particular approach and data. Conversely, the Meanshift implementation used leverages K-means' cluster centroids as initial seeds, which may inadvertently induce some limitations to the ability to accurately discern some layers. This might explain the reason why K-means outperforms Meanshift for every anatomical structure detection and segmentation.

Table 1. Test set segmentation average and standard deviation results

	DICE			Average Distance		
	K-means	Mean shift	Watershed	K-means	Mean shift	Watershed
Skin	0.61 (0.22)	0.48 (0.27)	0.52 (0.25)	3.1 (5.5)	13.8 (24.5)	3.4 (5.1)
Subcut. fat	0.49 (0.29)	0.44 (0.29)	0.51 (0.26)	17.4 (17.2)	29.6 (34.2)	17.2 (20.9)
Mam. gland	0.49 (0.32)	0.44 (0.32)	0.32 (0.29)	23.2 (24.8)	38.9 (54.3)	41.4 (37.0)
Thorax	0.71 (0.20)	0.64 (0.22)	0.66 (0.22)	13.1 (17.0)	15.0 (13.3)	21.3 (20.3)

Metrics units: DICE - [0,1] (max is best); Average Distance - in pixels (min is best)

4   Conclusion

Biomedical imaging plays an important role in breast cancer diagnosis and treatment. Breast ultrasound images can provide relevant information and contribute towards a multimodal analysis, leveraging the patients continued well-being on their recovery outcome from cancer treatment. The methods contemplated on this paper involve K-Means, Meanshift and Watershed, as focal algorithms, in which the K-Means with $K = 6$ presented the best overall results concerning the target structures. This configuration also incorporated a pre-processing stage that used an ADLG filter that ran for 50 iterations and H-Maxima suppression of 20%. By proposing a preliminary framework for anatomical structures detection and segmentation, such as skin, subcutaneous fat, mammary gland and thoracic region, in breast ultrasound images, we hope to contribute to that research.

References

1. Ahmad, S., Bolic, M., Dajani, H., Groza, V., Batkin, I., Rajan, S.: Measurement of heart rate variability using an oscillometric blood pressure monitor. IEEE Trans. Instrum. Meas. **59**(10), 2575–2590 (2010). https://doi.org/10.1109/TIM.2010.2057571
2. Aiazzi, B., Alparone, L., Baronti, S.: Multiresolution local-statistics speckle filtering based on a ratio Laplacian pyramid. IEEE Trans. Geosci. Remote Sens. **36**(5), 1466–1476 (1998). https://doi.org/10.1109/36.718850
3. Arthur, D., Vassilvitskii, S.: k-Means++: the advantages of careful seeding. In: Proceedings of the Eighteenth Annual ACM-SIAM Symposium on Discrete Algorithms, pp. 1027–1035. Society for Industrial and Applied Mathematics (2007)
4. Boukerroui, D., Baskurt, A., Noble, J.A., Basset, O.: Segmentation of ultrasound images - multiresolution 2D and 3D algorithm based on global and local statistics. Pattern Recogn. Lett. **24**(4), 779–790 (2003). https://doi.org/10.1016/S0167-8655(02)00181-2
5. Braz, R., Pinheiro, A.M.G., Moutinho, J., Freire, M.M., Pereira, M.: Breast ultrasound images gland segmentation. In: 2012 IEEE International Workshop on Machine Learning for Signal Processing, pp. 1–6, September 2012. https://doi.org/10.1109/MLSP.2012.6349748

6. Cardoso, F.M., Matsumoto, M.M.S., Furuie, S.S.: Edge-preserving speckle texture removal by interference-based speckle filtering followed by anisotropic diffusion. Ultrasound Med. Biol. **38**(8), 1414–1428 (2012). https://doi.org/10.1016/j.ultrasmedbio.2012.03.014
7. Coupe, P., Hellier, P., Kervrann, C., Barillot, C.: Bayesian non local means-based speckle filtering. In: 2008 5th IEEE International Symposium on Biomedical Imaging From Nano to Macro, pp. 1291–1294, May 2008. https://doi.org/10.1109/ISBI.2008.4541240
8. Dice, L.R.: Measures of the amount of ecologic association between species. Ecology **26**(3), 297–302 (1945). https://doi.org/10.2307/1932409
9. Flores, W.G., Pereira, W.C.A., Infantosi, A.F.C.: Breast ultrasound despeckling using anisotropic diffusion guided by texture descriptors. Ultrasound Med. Biol. **40**(11), 2609–2621 (2014). https://doi.org/10.1016/j.ultrasmedbio.2014.06.005
10. Fukunaga, K., Hostetler, L.: The estimation of the gradient of a density function, with applications in pattern recognition. IEEE Trans. Inf. Theory **21**(1), 32–40 (1975). https://doi.org/10.1109/tit.1975.1055330
11. Gómez, W., Leija, L., Alvarenga, A.V., Infantosi, A.F.C., Pereira, W.C.A.: Computerized lesion segmentation of breast ultrasound based on marker-controlled watershed transformation. Med. Phys. **37**(1), 82–95 (2010). https://doi.org/10.1118/1.3265959
12. Guo, Y., Wang, Y., Hou, T.: Speckle filtering of ultrasonic images using a modified non local-based algorithm. Biomed. Signal Process. Control **6**(2), 129–138 (2011). https://doi.org/10.1016/j.bspc.2010.10.004. Special Issue: The Advance of Signal Processing for Bioelectronics
13. Hassan, M., Chaudhry, A., Khan, A., Iftikhar, M.A., Kim, J.Y.: Medical image segmentation employing information gain and fuzzy c-means algorithm. In: 2013 International Conference on Open Source Systems and Technologies, pp. 34–39, December 2013. https://doi.org/10.1109/ICOSST.2013.6720602
14. Huang, T., Yang, G., Tang, G.: A fast two-dimensional median filtering algorithm. IEEE Trans. Acoust. Speech Signal Process. **27**(1), 13–18 (1979). https://doi.org/10.1109/TASSP.1979.1163188
15. Jameson, J.L., et al.: Harrison's Principles of Internal Medicine. McGraw-Hill Education / Medical, New York (2018). OCLC: 990065894
16. K. D. Marcomini, A.A.O.C., Schiabel, H.: Application of artificial neural network models in segmentation and classification of nodules in breast ultrasound digital images. Int. J. Biomed. Imaging 13 (2016). https://doi.org/10.1155/2016/7987212
17. Meyer, F.: Topographic distance and watershed lines. Sig. Process. **38**(1), 113–125 (1994). https://doi.org/10.1016/0165-1684(94)90060-4
18. Otsu, N.: A threshold selection method from gray-level histograms. IEEE Trans. Syst. Man Cybern. **9**(1), 62–66 (1979). https://doi.org/10.1109/TSMC.1979.4310076
19. Shan, J., Cheng, H.D., Wang, Y.: A novel automatic seed point selection algorithm for breast ultrasound images. In: 2008 19th International Conference on Pattern Recognition, pp. 1–4, December 2008. https://doi.org/10.1109/ICPR.2008.4761336
20. Sørensen, T.J.: A method of establishing groups of equal amplitude in plant sociology based on similarity of species content and its application to analyses of the vegetation on danish commons. Biol. Skar. **5**, 1–34 (1948)
21. Taha, A., Hanbury, A.: Metrics for evaluating 3D medical image segmentation: analysis, selection, and tool. BMC Med. Imaging **15**(1) (2015). https://doi.org/10.1186/s12880-015-0068-x

22. Thijssen, J.M.: Ultrasonic speckle formation, analysis and processing applied to tissue characterization. Pattern Recogn. Lett. **24**(4), 659–675 (2003). https://doi.org/10.1016/S0167-8655(02)00173-3
23. Yu, Y., Acton, S.T.: Speckle reducing anisotropic diffusion. IEEE Trans. Image Process. **11**(11), 1260–1270 (2002). https://doi.org/10.1109/TIP.2002.804276

Enhancing the Prediction of Lung Cancer Survival Rates Using 2D Features from 3D Scans

Tahira Ghani and B. John Oommen$^{(\boxtimes)}$

School of Computer Science, Carleton University, Ottawa K1S 5B6, Canada
tahira.ghani@carleton.ca, oommen@scs.carleton.ca

Abstract. The survival rate of cancer patients depends on the type of cancer, the treatments that the patient has undergone, and the severity of the cancer when the treatment was initiated. In this study, we consider adenocarcinoma, a type of lung cancer detected in chest Computed Tomography (CT) scans on the entire lung, and images that are "sliced" versions of the scans as one progresses along the thoracic region. Typically, one extracts 2D features from the "sliced" images to achieve various types of classification. In this paper, we show that the 2D features, in and of themselves, can be used to also yield fairly reasonable predictions of the patients' survival rates if the underlying problem is treated as a *regression* problem. However, the fundamental contribution of this paper is that we have discovered that there is a strong correlation between the *shapes* of the 2D images at successive layers of the scans and these survival rates. One can extract features from these successive images and augment the basic features used in a 2D classification system. These features involve the area at the level, and the mean area along the z-axis. By incorporating additional *shape*-based features, the error involved in the prediction decreases drastically – by almost an order of magnitude. The results we have obtained deal with the cancer treatments done on 60 patients (Understandably, it is extremely difficult to obtain training and testing data for this problem domain! Thus, both authors gratefully acknowledge the help given by Drs. Thornhill and Inacio, from the University of Ottawa, in providing us with domain knowledge and expertise for understanding and analyzing the publicly-available dataset.) at varying levels of severity, and with a spectrum of survival rates. For patients who survived up to 24 months, the average relative error is as low as 9%, which is very significant.

Keywords: Medical image processing · Lung cancer treatment · Prediction of survival rates

---

B. J. Oommen—*Chancellor's Professor*; *Life Fellow: IEEE* and *Fellow: IAPR*. This author is also an *Adjunct Professor* with the University of Agder in Grimstad, Norway.

A. Campilho et al. (Eds.): ICIAR 2020, LNCS 12132, pp. 202–215, 2020.
https://doi.org/10.1007/978-3-030-50516-5_18

1 Introduction

The past few decades have boasted significant progress in diagnostic imaging in the domain of healthcare and medicine. This has, in turn, enhanced the process of exposing internal structures hidden by skin and bones through, for example, radiology. The spectrum of medical imaging techniques include, but are not limited to, X-rays, Magnetic Resonance Images (MRIs), ultrasounds, endoscopies, etc., and a wide range of pathological phenomena can now be detected.

The application of these techniques can be refined depending on the subsystem of the body that is under consideration (i.e., cardiovascular, respiratory, abdominal, etc.), *and* the associated imaging technique. Cancer, as a subcategory of these pathologies, constitutes over 100 different types.

Aspects of Lung Cancer: Apart from folklore, the statistics about cancer are disheartening. The American Cancer Society (ACS) estimated their annual statistics for 2018 based on collected historical data [1]. Lung cancer is now the second leading type of cancer for newly diagnosed patients, behind breast cancer, and it has the highest mortality rate out of all cancer sites. The ACS projected 234,030 new cases of lung cancer. They also forecasted that 154,050 deaths would be caused by lung cancer. However, cancers diagnosed at an early phase, such as Stage 1, can be treated with surgery and radiation therapy with an 80% success rate. Late diagnosis, typically, implies a lower survival rate.

Radiomics is the field of study that extracts quantitative measures of tumour phenotypes from medical images using data-characterization algorithms. These features are explored to uncover disease characteristics that are not visible to the naked eye, but which can then be used for prognosis and treatment plans. Many researchers have worked on engineering the feature sets through a radiomic analysis [2,3]. Our goal, however, is that of predicting the survival rate of lung cancer patients *once they have been diagnosed,* and the result of this study is to demonstrate that a lot of this information resides in the *3D shape* of the tumour. We hope that our study can provide insight into the cancer's severity, and also aid in formulating the treatment plans so as to increase the chances of survival.

1.1 Contributions of This Paper

The contributions of this paper can be summarized as follows:

- Although the diagnosis of lung cancer has been extensively studied, the correlation of the survival times to the tumour's size/shape is relatively unexplored. Our first major contribution is to show that by a regression analysis, we can predict the survival times based on various features of the tumour. Predicting the survival times can essentially aid medical professionals to judge the severity of the cancer, and to design treatment plans accordingly.
- Our features are 2D features obtained from various scan slices. We show that these features, by themselves, yield impressive estimates of survival times.

– The most significant contribution of this paper is the discovery of a *distribution* for the shapes of the scans as they are processed sequentially. From these images, we can obtain relatively simple indices that relate to some geometric features of the sequence that can be used for classification and regression. By augmenting the original 2D feature set with these, we have been able to obtain significantly improved estimations for the survival rates.
– While these results have been proven to be relevant for our lung cancer scenario, we believe that these phenomena are also valid for other tumour-based cancers, and hope that other researchers can investigate the relevance of the same hypothesis for *their* application domains.

After acknowledging the source of the data and briefly surveying the field, we shall discuss the feature extraction process, and analyze the individual features against the "TumourDepth", for any correlation of the measures at successive layers. Section 5 summarizes all the regression results for the various feature sets, and Sect. 6 concludes the paper.

1.2 Data Source

We have used the publicly available data from The Cancer Imaging Archive[1] (TCIA), a service which hosts an archive of data for de-identified medical images of cancer. The dataset used for this work is the "LungCT-Diagnosis" data [15] on TCIA, uploaded in 2014. The set consists of CT scans for 61 patients that have been diagnosed with adenocarcinoma, a type of lung cancer, with the number of images totalling up to about 4,600 over all the scans. However, considering only images that have the presence of a cancer nodule, the count reduces to approximately 450 images. With healthcare data, we are, of course, constrained to work with what we have. As we will see, it suffices for the purpose of regression analyses. Throughout the experiments and results explained in this paper, the condensed dataset of 450 images has been consistently split into training and testing data with a 70% and 30% split respectively, with the guarantee that there was no overlap between the two subsets. The dataset also includes the clinical metadata, where the survival time of the patient associated with each scan, is listed.

1.3 Literature Review

Feature extraction schemes in biomedical applications have been found to be specific to the medical context of the goal at hand, i.e., the features that are used vary based on the type of image being processed, as well as the pathological focus (lesions and nodules, texture variance, organ size and wall thickness, etc.). However, it may be beneficial to derive a wide variety of features, and to then reduce the set to those which prove to be most relevant [16]. This can be done through the application of feature elimination or feature selection techniques.

---

[1] More information can be found at https://www.cancerimagingarchive.net/.

The goal of forming a descriptor vector in the context of nodules and texture, is a task of local feature extraction. Chabat *et al.* [17] and Kim *et al.* [18] aimed to classify obstructive lung diseases based on texture patches as Regions of Interests (ROIs). For each ROI, a statistical descriptor was calculated to describe the CT attenuation characteristics.

However, Kim *et al.* [18] also included other features such as the co-occurrence matrix and the top-hat transform. Additionally, they incorporated measurements to depict shapes in the ROIs, such as circularity and aspect ratios, which were not well represented by statistical measures extracted for the textures. Similarly, the authors of [19] expanded the aforementioned list with morphological features to describe the shapes of the nodules in the ROIs.

The aforementioned features, however, were used in the context goal of classification of a medical pathology. We adapt the research goal of Grove *et al.* [15] by computing and analyzing features that are indicative of the *severity* of the cancerous nodule, whereby severity can be considered as being synonymous with the survival time of a patient after the diagnosis of the cancer. Whereas their study was, for the most part, a hypothesis-based testing methodology, ours is more explorative in nature with the goal of finding descriptive quantitative measurements. It is important to note that our focus is heavily on the construction of the feature set, rather than the customization of regression models that have been used as testing thresholds.

2   Background

2.1   Computed Tomography (CT) Scans

The most common radiological imaging technique incorporates CT scans where X-ray beams are used to take measurements or images (i.e., "slices") from different angles, as shown in Fig. 1, as the patient's body moves through the scanner. Depending on the section thickness and the associated reconstruction parameters, a scan can range anywhere from 100 to over 500 sections or images [5].

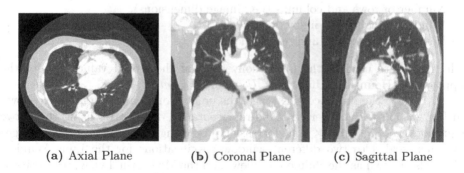

(a) Axial Plane   (b) Coronal Plane   (c) Sagittal Plane

Fig. 1. Planes captured in a Computed Tomography (CT) scan.

The scan records different levels of density and tissues which can then be reconstructed to, non-invasively, create a 3D of the human body.

High-resolution Computed Tomography (HRCT) is specifically used in detecting and diagnosing diffuse lung diseases [4] and cancerous nodules, due to its sensitivity and specificity. It enables the detection and analysis of feature aspects such as morphological lesion characterization, nodule size measurement and growth, as well as attenuation characteristics.

Hounsfield Units: CT numbers are captured and represented as Hounsfield Units (HUs), which serve as a universal standardized dimensionless metric as:

$$HU = 1000 \times \frac{\mu - \mu_{water}}{\mu_{water} - \mu_{air}}, \quad \text{where,} \tag{1}$$

- μ_{water} is the linear attenuation coefficient of water, and
- μ_{air}: is the linear attenuation coefficient of air.

Although measured in HUs, CT scans and other medical imaging reports are saved in the standard Digital Imaging and Communications in Medicine (DICOM) format. There are many DICOM viewer applications for observing and analyzing medical scans specifically.

2.2 The Preprocessing Stage

Working with the images of Chest CT scans means processing data in the DICOM format. This format groups information in the datasets, enabling the attachment of patient and pixel data, as well as technical data, such as the corresponding encoding schemes and window measurements, through attribute tags.

The DICOM images are expressed as 16-bit integer values, where the stored attribute tags specify the default, and include:

- Slice thickness;
- Number of rows and columns (i.e., image dimensions);
- Window centre and window width;
- Rescale intercept, m, and rescale slope[2], b.

The latter are used in the linear conversion of the stored value, SV, to the appropriate Hounsfield Unit (HU), U, where $U = mSV + b$.

As part of the preprocessing stage, the images, originally displayed as shown in Fig. 2a, are first converted to be representative of the HUs using the above equation (i.e., $U = mSV + b$) for easier processing and visibility. The images are then scaled to two different window specifications for the Lung Window view, also known as the Pulmonary view, and the Mediastinal view, respectively.

---

[2] It is important to note that more often than not, the rescale slope was valued at 1. Indeed, we have not encountered a dataset which has a different rescale slope value.

The Lung Window view, shown in Fig. 2b, displays the texture in the lung and is attained by adjusting the window centre and window width, $[C, W]$, parameters to $[-500, 1400]$. The Mediastinal view, shown in Fig. 2c, is attained by adjusting the parameters to $[40, 380]$.

Nodule Segmentation: Rather than segmenting the entire lung region as our Region of Interest (ROI), in this research, we segmented only the cancerous nodule in the "slices" where the presence of the tumour was observed. Similar to the topic of lung segmentation, there is an abundance of published work discussing the automation of so-called nodule segmentation and extraction [9, 10] and [11].

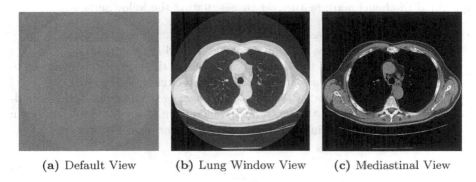

(a) Default View   (b) Lung Window View   (c) Mediastinal View

Fig. 2. Different Chest CT views based on window parameter adjustments.

We made masks of the tumours using the ImageJ software[3]. This was done by manually tracing a contour around the nodule on the images where it was present, filling the shape as "white", and clearing the background to "black". The images that did not contain the nodule were cleared to a "black" background. The CT scans were reviewed, and the segmentation of cancer tumours were validated by a clinical doctor from the Ottawa Heart Institute.

3   Nodule-Based Feature Extraction

When computing and compiling a feature set, we considered the scans in a 2D aspect, where each image or "slice" was treated as a single observation. For instance, if a nodule in a scan ranged over 10 images, each image was treated as a single observation or record, and amounted to 10 records in the dataset. Thus, over 60 scans, we extracted approximately 475 images that contained a nodule.

---

[3] The ImageJ Software is a Java-based image processing program developed at the National Institutes of Health and the Laboratory for Optical and Computational Instrumentation.

3.1 Texture and Shape-Based Features

To adequately analyze the work done for this research, we first created a benchmark of prediction results based on the feature set used in the texture analysis phase of the prior algorithm. This feature set, now referred to as the "Benchmark" feature set, consisted of the Haralick values [13] computed from the Mahotas Python library, which constituted a 12-dimensional vector.

The hypothesis that we worked with in this research was that *irregularities* in a tumour's *shape* are indicative of it being cancerous. Thus, to further investigate the characterizing aspects of the tumour (when it concerns the survival rates), we modified the benchmark feature set by computing additional features relevant to the *shape* of the tumour, with the goal of being able to measure the shape's "regularity". We shall reference these as the "2D Shape" feature set. We appended the benchmark feature set by calculating the following:

- The tumour's area, measured in pixels from its mask (see Fig. 3a);
- Width and height of the tumour's smallest bounding rectangle;
- Mean Squared Error of the boundary pixel from the center (with respect to the radius of the tumour's minimum enclosing circle (see Fig. 3b));
- Moment values of the vector, formed by calculating the distance of the boundary of the tumour from the centre in $10°$ increments for a full $360°$ (as shown in Fig. 3c), where k defines the kth central moment as in Eq. (2):

$$m_k = \frac{1}{n} \sum_{i=1}^{n} (x_i - \bar{x})^k. \tag{2}$$

In the above, these reduce to the following for specific values of k:
- Variance (σ^2), where for $k = 2$, $\sigma^2 = m_2$.
- Skewness (S), where for $k = 3$ and s, the standard deviation, $S = \frac{m_3}{s^3}$.
- Kurtosis (K), where for $k = 4$ and s, the standard deviation, $K = \frac{m_4}{s^4}$.

In our study, we also included σ^2, S and K as additional features in the vector.

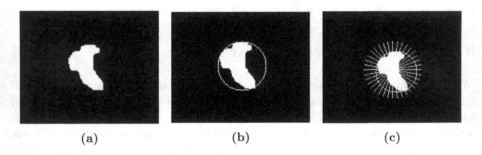

(a) (b) (c)

Fig. 3. Figures explaining the process of calculating the tumour's shape features.

4   Feature Set Analysis

Analyzing the Shape Features: We are now in a position to explain how we further analyzed *these* shape features, and to consider the values of these quantities against the tumour's depth. A tumour's 3D shape will, generally, start small and increase in size as it approaches the centre of the nodule and then decrease in size again. To observe the trend of the tumour's progression in size, we placed the shape features, beginning with "Area", against the "TumourDepth", and determined whether the average area and depth of the nodule displayed a correlation with the target variable. We then replaced the "Area" feature with the computed "MeanArea" given by Eq. (3):

$$MA = \sum_{i=1}^{i=n}(td_i * \frac{a_i}{\sum_{i=1}^{i=n} a_i}), \quad \text{where,} \tag{3}$$

- MA: "MeanArea";
- n: Number of slices which contain the tumour;
- td_i: Tumour's depth at slice i;
- a_i: Tumour's area at slice i.

This process was repeated for all the shape features and appended as new features after removing the original shape feature. This feature set will be referred to as the "Averaged 2D Shape" feature set.

Correlating the Shape Features Against Survival Times: To better understand how the shape features correlate against the survival times, we analyzed the averaged "Area" feature further by plotting curves of the tumour's area in relation to the depth of the image in the tumour, in 6-month bins of "SurvivalTime". Figures 4a, 4b and 4c display plots of the curve for the $[6, 12)$, $[12, 18)$ and $[18, 24)$ months bounds respectively, where in the first case, the average area occurred at 59.1% tumour depth[4]. The overall results are summarized in Table 1, where the area is plotted against the percentage of the tumour depth for each 6-month bin present in the data. A notable observation is the steady trend of the decreasing "Area" mean over the first four bins (i.e., those that have a survival time within 2 years of diagnosis). Given this trend, we hypothesised that we could also test the regression models for this survival time frame.

To verify this hypothesis, we performed a regression analysis on a subset of the data which has a "SurvivalTime" of 24 months or less. Subsets were taken of all of the aforementioned feature sets, and processed (with these augmented features) to recalculate the errors by testing the corresponding regression models.

---

[4] The term "Tumour depth" refers to the level (or z-axis) along the tumour as it progresses down the thoracic region.

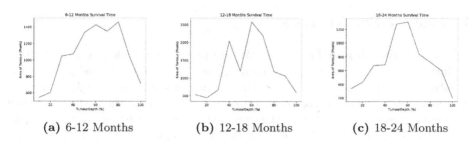

(a) 6-12 Months   (b) 12-18 Months   (c) 18-24 Months

Fig. 4. Area vs. TumourDepth Plots for tumours for patients with a survival rate of 6–12 months, 12–18 months and 18–24 months.

Table 1. Averages of the tumour's "Area" over bins of 6-Month durations.

"SurvivalTime" Bin	Average area
(0, 6]	0.685
(6, 12]	0.591
(12, 18]	0.579
(18, 24]	0.555
(24, 30]	0.589
(30, 36]	0.566
(36, 42]	0.679
(42, 48]	0.671
(48, 54]	0.585
(54, 60]	0.546

5 Results and Discussions

5.1 Model Evaluation

To evaluate the performance of the tested regression models, we utilized two measures, namely the Mean Absolute Error (MAE), measured in months, and the Mean Relative Error (MRE), both of which are defined in Eqs. (4) and (5) respectively:

$$MAE = \frac{1}{n} \sum_{i=1}^{n} |y_i - z_i|, \quad \text{and} \tag{4}$$

$$MRE = \frac{1}{n} \sum_{i=1}^{n} \frac{|y_i - z_i|}{z_i}, \quad \text{where:} \tag{5}$$

- n is the number of test-set data points,
- y_i is the predicted value (i.e., the expected survival time in months), and
- z_i is the true value (i.e., the survival time in months).

The MAE is the average difference between the true values and the predicted values. It provides an overall measure of the distance between the two values, but it does not indicate the direction of the data (i.e., whether the result is an under or over-prediction). Furthermore, this is also seen to be a scale-dependent measure, as the computed values are heavily dependent on the scale of the data, and can be influenced by outliers present in the data [14]. In order to circumvent the scale-dependency, we also computed the MRE which introduces a relativity factor by normalizing the absolute error by the magnitude of the true value. This means that the MRE should, generally, consist of values in the range $[0, 1]$.

As mentioned earlier, all the regression tests were done on the data with a 70% to 30% split of the data for training and testing, respectively.

5.2   Regression Results

We used the `scikit-learn` machine learning library to implement the basic models for regression on the 2D Benchmark feature set. When analyzing the recorded metrics, we emphasize that we were attempting to minimize the error and maximize the accuracy of all the tested models. Also, in the interest of uniformity, we consistently report the results obtained by invoking the three best schemes, namely Linear Regression, kNN Regression and Gradient Boosting[5, 6].

Table 2. Performance of the regression models on the "Benchmark" feature set.

Model	MAE (months)	MRE (%)
Linear regression	12.30	0.76
kNN regression	12.13	0.79
Gradient boosting	12.49	0.77

As can be seen from the MRE indices recorded in Table 2, the error seems to be relatively high when compared to the results of the binary classification attained for this feature set (discussed in [12]). While such a performance may be undesirable, it is certainly not surprising when we consider the following:

1. **Difference of context**: This feature set was initially compiled with the goal of a binary classification task between "Normal" and "Abnormal" lung texture. Processing the entire region belonging to the lung parenchyma enables the visibility of stark differences in texture which can correspond to one of the two aforementioned classes.

---

[5] In the interest of conciseness, we do not discuss the details of the Machine Learning methods invoked. We assume that the reader is aware of them. Additional results, which also detail the results of other methods, are included in the thesis of the first author [12], and not given here in the interest of space.

[6] The thesis publication can be found on the Carleton University website, linked as follows: https://doi.org/10.22215/etd/2019-13731.

2. **Patch size**: The prior algorithm processed a fixed patch size of 37×37 pixels, whereas, in this research, we are processing varying patch sizes depending on the size of the bounding box of the tumour in a 2D image.
3. **Tumour isolation**: In this research, we considered only the tumours and not the texture around them in the 3D matrix compilations. Thus, any pixels that were included in the bounding box but not seen to be a part of the tumour were reduced to zero (i.e., black) and ignored from the feature computations. This could surely affect the variability seen in different tumours.
4. **Correlation of depth and survival time**: Although the depth of the image with regards to the tumour would yield different feature measurements, all slices throughout the tumour would correspond to the same value for the target value since it belonged to a single scan. This would, apparently, reduce the correlation between the features and the target variable.

Table 3. Performance of regression models with shape features

Model	MAE (months)	MRE (%)
Linear Regression	12.13	0.70
kNN Regression	14.48	0.89
Gradient Boosting	8.82	0.54

To further support our initial hypothesis, we ran the same regression models (as we earlier did for the baseline results) with the benchmark feature set and the new shape features, i.e., the 2D Shape feature set. The results that we obtained and the respective error values, are listed in Table 3. As can be seen, the Linear Regression models improved in both the MAE and MRE measures. While the kNN Regression digressed with a larger error, Gradient Boosting performed significantly better – with an improvement of 23% (Table 4).

Table 4. Performance of regression models with averaged 2D Shape feature set

Model	MAE (months)	MRE (%)
Linear regression	11.95	0.70
kNN regression	14.52	0.86
Gradient boosting	7.40	0.45

For the "Averaged 2D Shape" feature set, the Linear Regression displayed a minor improvement in the MAE, while the MRE remained the same. On the other hand, the kNN regression digressed for the MAE measure but improved for the MRE by about 3%. The Gradient Boosting scheme, however, had the best improvement with a decrease of approximately 15% in both the MAE and MRE, confirming the advantage of considering the nodule in its 3-dimensional entirety, compared to the previous 2D feature sets, and to methods that incorporated minimal changes in the Linear and kNN regression schemes. Again, even superior results were obtained on a subset of the data which had a "SurvivalTime" of 24 months or less. Table 5 displays the results of the regression models with respect to each feature set. The results of the MAE improved by at least 60% for all feature sets and regression models. The most notable improvement was with the Gradient Boosting scheme for these modified features, where we obtained a total improvement of almost 70% in the MRE, bringing the absolute error down to 1.29%. This is, by all metrics, remarkable.

Table 5. Regression results with data subset: 'SurvivalTime' less than 24 months

Feature Set	Evaluation	Regression model		
	Metric	Linear	kNN	Gradient Boosting
Benchmark	MAE	4.27	5.22	4.56
	MRE	0.35	0.40	0.37
2D Shape	MAE	3.53	4.60	1.79
	MRE	0.27	0.35	0.12
Average 2D Shape	MAE	3.23	5.33	1.24
	MRE	0.26	0.41	0.09

6   Conclusion

In this paper, we discussed the domain of healthcare imaging for diagnostics and the implementation of radiomics on CT scans, in particular, to predict the survival rates of lung cancer patients. We first tested the feature set that was used in a prior algorithm, where the goal was to examine regression models.

By modifying the existing features by including shape descriptors, which were focused on the cancer nodule itself, we were able to obtain improvements in the best regression models. These new features were then analyzed against the "TumourDepth"', from which a strong correlation between the images at successive layers of the scan was discovered. Further investigating the shape aspects of the tumour, we observed the progression of the "Area"-based feature values *versus* tumour progressions in bins of "SurvivalTime", and were able to observe a notable trend for survival rates up to 24 months. By performing a regression testing for the data within this subset yielded an MRE of as low as 9%, and we even obtained a total improvement of almost 70% in the MRE.

With regard to future work, the feature sets that have been explored, should also be tested on other collections of CT scans to ensure consistency. We also believe that similar methods can be used for other tumour-based cancers.

References

1. Siegel, R.L., Miller, K.D., Jemal, A.: Cancer statistics. CA Cancer J. Clin. **68**, 7–30 (2018)
2. Paul, R., Hawkins, S.H., Schabath, M.B., Gillies, R.J., Hall, L.O., Goldgof, D.B.: Predicting malignant nodules by fusing deep features with classical radiomics features. J. Med. Imaging **5**, 011021 (2018)
3. Fan, L., et al.: Radiomics signature: a biomarker for the preoperative discrimination of lung invasive adenocarcinoma manifesting as a ground glass nodule. Eur. Radiol. **29**, 889–897 (2019)
4. Elicker, B.M., Webb, W.R.: Fundamentals of High-Resolution Lung CT. Wolters Kluwer (2013)
5. Al Mohammad, B., Brennan, P.C., Mello-Thoms, C.: A review of lung cancer screening and the role of computer-aided detection. Clin. Radiol. **72**, 433–442 (2017)
6. Armato III, S.G., Sensakovic, W.F.: Automated lung segmentation for thoracic CT: impact on computer-aided diagnosis. Acad. Radiol. **11**, 1011–1021 (2004)
7. Zhou, S., Cheng, Y., Tamura, S.: Automated lung segmentation and smoothing techniques for inclusion of juxtapleural nodules and pulmonary vessels on chest CT images. Newblock Biomed. Sig. Process. Control **13**, 62–70 (2014)
8. Singadkar, G., Mahajan, A., Thakur, M., Talbar, S.: Automatic lung segmentation for the inclusion of juxtapleural nodules and pulmonary vessels using curvature based border correction. J. King Saud Univ. Comput. Inf. Sci. (2018)
9. Zhao, B., Gamsu, G., Ginsberg, M.S., Jiang, L., Schwartz, L.H.: Automatic detection of small lung nodules on CT utilizing a local density maximum algorithm. J. Appl. Clin. Med. Phys. **4**, 248–260 (2003)
10. Armato III, S.G., Giger, M.L., MacMahon, H.: Automated detection of lung nodules in CT scans. Med. Phys. **28**, 1552–1561 (2001)
11. Messay, T., Hardie, R.C., Tuinstra, T.R.: Segmentation of pulmonary nodules in computed tomography using a regression neural network approach and its application to the lung image database consortium and image database resource initiative dataset. Med.l Image Anal. **22**, 48–62 (2015)
12. Ghani, T.: Feature Engineering with Radiomics for Optimal Prediction of Survival Rates of Adenocarcinoma Patients. MCS thesis, Carleton University, Ottawa (2019)
13. Haralick, R.M., Shanmugam, K., Dinstein, H.: Textural features for image classification. IEEE Trans. Syst. Man Cybern. **6**, 610–621 (1973)
14. Chen, C., Twycross, J., Garibaldi, J.M.: A new accuracy measure based on bounded relative error for time series forecasting. PloS ONE **12**, e0174202 (2017)
15. Grove, O., et al.: Quantitative computed tomographic descriptors associate tumour shape complexity and intratumor heterogeneity with prognosis in lung adenocarcinoma. PloS ONE **10**, e0118261 (2015)
16. Hall, E.L., Kruger, R.P., Dwyer, S.J., Hall, D.L., McLaren, R.W., Lodwick, G.S.: A survey of preprocessing and feature extraction techniques for radiographic images. IEEE Trans. Comput. **20**, 1032–1044 (1971)
17. Chabat, F., Yang, G.Z., Hansell, D.M.: Obstructive Lung diseases: texture classification for differentiation at CT. Radiology **228**, 871–877 (2003)

18. Kim, N., Seo, J.B., Lee, Y., Lee, J.G., Kim, S.S., Kang, S.H.: Development of an automatic classification system for differentiation of obstructive lung disease using HRCT. J. Digit. Imaging **22**, 136–148 (2009)
19. Demir, O., Camurcu, A.Y.: Computer-aided detection of lung nodules using outer surface features. Bio-Med. Mater. Eng. **26**, S1213–S1222 (2015)

Lesion Localization in Paediatric Epilepsy Using Patch-Based Convolutional Neural Network

Azad Aminpour[1], Mehran Ebrahimi[1(✉)], and Elysa Widjaja[2]

[1] Ontario Tech University, Oshawa, ON L1G 0C5, Canada
{azad.aminpour,mehran.ebrahimi}@ontariotechu.net
[2] The Hospital for Sick Children (SickKids), Toronto, ON M5G 1X8, Canada
elysa.widjaja@sickkids.ca

Abstract. Focal Cortical Dysplasia (FCD) is one of the most common causes of paediatric medically intractable focal epilepsy. In cases of medically resistant epilepsy, surgery is the best option to achieve a seizure-free condition. Pre-surgery lesion localization affects the surgery outcome. Lesion localization is done through examining the MRI for FCD features, but the MRI features of FCD can be subtle and may not be detected by visual inspection. Patients with epilepsy who have normal MRI are considered to have MRI-negative epilepsy. Recent advances in machine learning and deep learning hold the potential to improve the detection and localization of FCD without the need to conduct extensive pre-processing and FCD feature extraction. In this research, we apply Convolutional Neural Networks (CNNs) to classify FCD in children with focal epilepsy and localize the lesion. Two networks are presented here, the first network is applied on the whole-slice of the MR images, and the second network is taking smaller patches extracted from the slices of each MRI as input. The patch-wise model successfully classifies all healthy patients (13 out of 13), while 12 out of 13 cases are correctly identified by the whole-slice model. Using the patch-wise model, we identified the lesion in 17 out of 17 MR-positive subjects with coverage of 85% and for MR-negative subjects, we identify 11 out of 13 FCD subjects with lesion coverage of 66%. The findings indicate that convolutional neural network is a promising tool to objectively identify subtle lesions such as FCD in children with focal epilepsy.

Keywords: Focal Cortical Dysplasia · Convolutional Neural Network · Deep learning · Patch-based

1 Introduction

Epilepsy is a common neuroulogical disorder which has devastating consequences on children's quality of life. It has an incidence rate of 4 to 9 in 1000 per year in paediatric patients [8,9]. Despite advances in the treatment of epilepsy, approximately 30% of patients with epilepsy continue to have seizures refractory to

© Springer Nature Switzerland AG 2020
A. Campilho et al. (Eds.): ICIAR 2020, LNCS 12132, pp. 216–227, 2020.
https://doi.org/10.1007/978-3-030-50516-5_19

medications [17,21,23]. Prolonged uncontrolled epilepsy has detrimental effects on the neurodevelopment of a child due to brain injury [20].

For patients with medically intractable epilepsy surgery may be the best option to achieve seizure freedom [6,24,26]. In this case, a comprehensive pre-surgical evaluation to detect the lesion responsible for epilepsy is necessary. Magnetic Resonance Imaging (MRI) plays an essential role in the pre-surgical workup of children with epilepsy in order to identify an underlying lesion responsible for the epilepsy [7,21]. The detection of a lesion on MRI varies in the literature and ranges from 30% to 85% of patients with refractory focal epilepsy [10]. Failure to identify a lesion on MRI will result in a lower likelihood of epilepsy surgery, increased use of invasive electroencephalography monitoring for surgical planning, and lower odds of seizure-free surgery outcome. For instance, the surgical outcome is considerably better in patients with an identifiable lesion than patients in whom the lesion cannot be identified.

Focal cortical dysplasia (FCD) is a brain malformation and one of the most common lesions responsible for medically intractable focal epilepsy in children. The incidence of FCD in those with intractable partial epilepsy ranges from 10% to 12% in combined paediatric and adult series, and up to 26% in paediatric patients alone [13]. The MRI features of FCD are frequently subtle, and may not be detected by visual assessment in up to 50% to 80% of patients with medically intractable epilepsy. The MRI features of FCD include cortical thickening, abnormal signal in the white and/or gray matter, blurring of the gray-white matter junction, and abnormal sulcal pattern [2,16]. Patients with MRI-visible FCD are considered to have MR-positive FCD. These patients may have one or more of these FCD features identifiable on the MRI, while others may demonstrate more. In some patients FCD may not be detected on MRI; these patients are considered to have MR-negative epilepsy [2,16].

A neuroradiologist performs most pre-surgery lesion localization through examining the patient's MRI scan. However, visual assessment of MRI is subjective and highly dependent on the expertise of the observer. Therefore, there is a need for a more advanced and objective tool for analyzing the MRI data. Image processing algorithms offer the potential to detect subtle structural changes, which may not be identifiable on visual inspection of MRI. Existing image processing methods to identify FCD based on the MRI features are limited in their ability as they are mostly based on extracting and detecting specific features of FCD. Recently, artificial intelligence (AI) techniques based on deep neural networks have emerged and been applied to many fields, including various areas of medicine. Deep learning methods' success motivates us to apply deep learning-based techniques to provide the neuroradiologists with means for lesion localization.

2   Literature Review

Due to the importance of FCD detection for pre-surgery evaluation, numerous works [1–5,11,12,14–16,18] have been proposed to help neuroradiologists better detect and localize FCD. The published works are mostly considering adult cases while a few studies [1, 16] have concentrated on identifying FCD in children with focal epilepsy.

Most of the methods mentioned above are based on extracting MRI features which are fed into a classifier. Morphometry features are among the most common features that have been used, for example, Martin et al. in 2015 [18] utilized voxel-based morphometry (VBM) along with surface-based morphometry (SBM) for better identification of FCD. Other works considered morphometric features and textural features together [4, 5]. Besson et al. performed an automatic detection by utilizing a two-step classification, a neural network trained on manual labels followed by a fuzzy k-Nearest Neighbor classifier (fkNN) to remove false positive clusters [5]. In another research, textural and structural features were used to develop a two-stage Bayesian classifier [2]. Most studies utilizing computer aided tools to identify a lesion in patients with medically intractable epilepsy have been conducted in adults with fully developed brain. The immature brain in children may present additional challenges for detecting FCD on images.

There is limited research evaluating paediatric FCD, [1, 16] are two of the most recent ones. Adler et al. utilized morphometric and structural features for training a neural network. Post-processing methods such as the "doughnut" method - which quantifies local variability in cortical morphometry/MRI signal intensity, and per-vertex interhemispheric asymmetry has also been developed in [1]. Kulaseharan et al. in 2019 [16] combined morphometric features and textural analysis using Gray-Level Co-occurrence Matrices (GLCM) on MRI sequences, to detect FCD in paediatric cases. A modified version of the 2-Step Bayesian classifier method proposed by Antel et al. [2] is applied and trained on textural features derived from T1-weighted (T1-w), T2-weighted (T2-w), and FLAIR (Fluid Attenuated Inversion Recovery) sequences. The authors correctly identified 13 out of 13 of healthy subjects and localized lesions in 29 out of 31 of the MR-positive FCD cases with 63% coverage of the complete extent of the lesion. The method also co-localized lesions in 11 out of 23 of the MR-negative FCD cases with coverage of 31%.

A common feature of all these studies for FCD detection is that they are based on extracted MRI features where, through analysis software and image processing techniques, a feature vector is constructed for a patch or a voxel in the MR sequence. Later each feature vector is classified using a classifier. In this research, we propose to implement a CNN, which is applied directly on the MR sequences. More recently, two studies have utilized a deep learning approach to detect FCD [3, 12].

Gill et al. in 2018 [12] trained two networks on patches extracted from patients' MR sequences with histologically-validated FCD. The algorithm was trained on multimodal MRI data from a single site and evaluated on independent data from seven sites (including the first site) worldwide for a total of 107 subjects. Both networks had a similar structure where they included three stacks of convolution and max-pooling layers with 48, 96 and 2 feature channels and

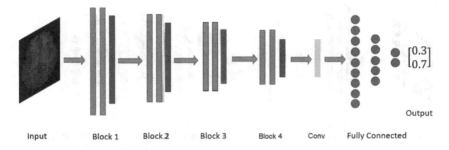

Fig. 1. Whole-slice network architecture, blue rectangles are convolutional layers, and green ones are pooling layers, columns of circles are representing fully connected layers. The light blue rectangle is a 1 × 1 convolutional layer while the other convolutional layers are 3 × 3. (Color figure online)

$3 \times 3 \times 3$ kernels. The first network is trained to maximize recognized lesional voxels, while the second network is trained to reduce the number of misclassified voxels (i.e., removing false positives while maintaining optimal sensitivity) [12]. Later in 2019, David et al. [3] developed a CNN for FCD detection. The proposed network is trained on T1 weighted and FLAIR. In this work to overcome the limited number of training data, authors utilized a conditional Generative Adversarial Network (cGAN) to compose synthetic data where it is used in training the CNN along with the real data [3]. They trained and evaluated their model on 42 cases with FCD and 56 healthy controls. The method classified 54 out of 56 healthy control cases as normal and detected 39 out of 42 FCD cases as FCD.

In this study, our aim was to develop a model where it takes an MRI and decides whether it has FCD or not, and in the case of FCD, it will localize the region of abnormal tissue. Our work is different from previous studies with deep learning on two main parts. First, we are only considering paediatric FCD which means we train and evaluate our model only on paediatric patients. Gill et al. [12] are considering both adult and paediatric data and David et al. [3] are focusing on adult cases. Second, both works have considered subject classification where they report if an input case has FCD or not. In this work, in addition to subject-wise classification, we localize the lesion in the FCD subjects.

3   Method

We apply a specific type of deep learning networks known as Convolutional Neural Networks (CNN) to classify MRI of FCD patients and localize the responsible lesion in the image. As processing the two-dimensional data will be faster than the three-dimensional ones, we will consider presenting slices of the 3D MR data as 2D inputs to our network. To this end, we choose 2D slices extracted from the MRI volume as the first set of inputs and later, each slice was partitioned into smaller patches that was used as the second set of inputs. At the end, each

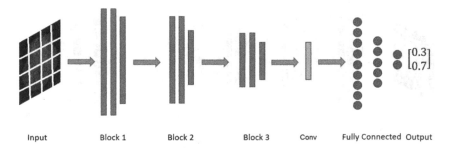

| Input | Block 1 | Block 2 | Block 3 | Conv | Fully Connected | Output |

Fig. 2. Patch-wise Network Architecture, blue rectangles are convolutional layers, and columns of circles are representing fully connected layers. The light blue rectangle is a 1×1 convolutional layer while the other convolutional layers are 3×3. (Color figure online)

volume was given to a neuroradiologist in order to detect and segment the lesion. We proposed two networks based on the two sets of inputs.

The first proposed model is a two-class classifier that takes two-dimensional slices out of 3D MR data and classifies the slice as FCD or healthy. We trained the network on our data where a slice is labelled as FCD if at least one pixel is lesional; otherwise, it is considered to be healthy. The proposed whole-slice CNN has two parts, a feature extraction part and a classification part.

The feature extraction part of the network consists of four blocks of two convolution layers, followed by a Max-Pooling layer. The blocks are followed by one convolutional layer at the end. The classification part consists of three fully connected layers which produces a vector with two elements representing the probability of each class. Therefore in total, there are nine Convolutional layers, four Max-Pooling layers and three Fully connected layers. The architecture is illustrated in Fig. 1. We implemented the Relu activation function after each layer except for the last fully connected layer, where we used Softmax instead. Also, the output of each layer was normalized using batch normalizer and to prevent overfitting, the dropout technique was applied.

In our early experiments, we realized that our model tends to over-fit toward the healthy slices. This was the result of the higher number of healthy slices compared to lesional slices. The unbalanced data distribution is a common problem in FCD detection and segmentation [12]. To overcome this problem, we reduced the number of healthy slices as we were not able to generate more lesional slices. First, we considered slices that contain more than 10% foreground pixels. In other words, we discarded the first few and several end slices containing mostly background/air. Then, we sampled the same number of healthy slices as lesional slices from all healthy slices and performed the training. The sampling process is random; however, it is weighted toward slices, which contain more brain tissue pixels.

One of the problems in the slice level classifier is that the lesion segment is smaller compared to healthy brain tissue. In the process of generating ground truth labels for the training slices, we labelled a slice as lesional if it included at least one lesional pixel. Such labelling process may cause an error in the framework as the network can treat a slice where a large portion of it is healthy as lesional. To overcome this problem, we partitioned each slice into smaller patches and trained the network on the new training set.

The patch-wise network architecture as shown in Fig. 2 is similar to the whole-slice network. We have three blocks of convolutional layers followed by three fully connected layers. In the patch-wise network, we are using convolution layers instead of Max-pooling layers at the end of each block. These layers are equipped with a filter of size 2×2 that slides by a stride of size two, which produces an output half the size of the input. The advantage of the convolution layer over the Max-pooling layer is that the network is able to learn filters' weights for "better" downsampling [19, 25]. Overall, the network has ten convolution layers and three fully connected layers. We utilized ReLu activation on all layers except the last layer, which is using Softmax. Furthermore, we normalized the output of each layer using batch normalizer and to prevent over-fitting, we applied dropout on fully connected layers.

In the patch-wise network, we label each patch with two labels, FCD or healthy. Again, we labelled a patch as FCD if at least one pixel is lesional. As we partitioned each slice into smaller patches, we ended up with patches only containing background/air or a small portion of healthy tissue, which we omitted from our training. We controlled the healthy patch selection with one parameter, percent of valued pixels relative to air/background. Here we selected patches with more than 20% valued pixels.

In our evaluation, we faced a similar challenge as the whole-slice classifier where our model was predicting every patch as healthy. Therefore, we tune the healthy patch selection parameter and change it to 50% (patches with more than 50% valued pixels). Also, we performed sampling when selecting the healthy patches for training. The sampling is random but with higher probability of patches with greater number of valued pixels. The next parameter to tune is the patch size which we explored 100×100, 50×50, and 25×25. We picked 50×50 patch sizes as it outperformed others in our preliminary experiments. Patch size was selected independent of the lesion size as the lesion segment has various sizes across all subjects. In the partitioning step, we examine both stride and non-stride version of the patch extraction. The non-stride partitioning simply means to partition the slice into patches of the same size where they don't overlap. For the case of stride, we also consider an overlap between patches to take into account every possible local information in the slice. We applied stride of 25 on the 50×50 patch selection and trained both networks on their corresponding dataset. The stride of 25 ensures that we have covered every point twice without increasing redundant training data. Larger strides will result in missing portions of the image and smaller strides will generate a lot of patches covering the same area which makes the training time longer.

4 Result

Our dataset contains T1 MRI volumes of paediatric subjects. It includes 30 FCD diagnosed subjects (17 MR-positive and 13 MR-Negative) and 13 healthy controls. The data is in the form of DICOM images and required several preprocessing steps including brain extraction, re-orientation, and resizing to a standard size. The brain extraction part is done utilizing the BET (Brain Extraction Tool) algorithm proposed by Smith et al. in 2002 [22]. Since CNN is taking images of the same size, every slice in our data needed to be resized to a standard size. We zero-padded the smaller slices while cutting background/air part of the larger slices to a size of 400 × 300. Due to the sensitive nature of our data, we avoided shrinking or stretching the data to resize them. At the end, 'clean' data was presented to a neuroradiologist to segment the lesion in FCD subjects, which is used as ground truth in our training and evaluation.

In both whole-slice and patch-wise models, each network was trained and evaluated using the leave-one-out technique. This means the network was trained on all of the available data except for the one used for evaluation. We repeated the experiment 30 times, which is the same as the number of FCD subjects. The output of each network is a label that identifies if the input slice or patch is lesional or healthy. Then based on the ratio of predicted lesional patches (or slices) over the number of predicted healthy patches (or slices), we will decide if the subject is lesional or healthy. We have quantified our results using lesional sensitivity, which is the true positive (TP) rate and lesional specificity, which is the true negative (TN) rate. Here true positive means number of healthy patches (or slices) classified as healthy, and true negative refers to the number of lesional patches (or slices) classified as lesional. We also reported subject-wise sensitivity and specificity which indicates the number of healthy or FCD subjects classified as healthy or FCD respectively.

The result of the whole-slice network is presented in Table 1. The model identified 12 out of 13 healthy control subjects. In the case of FCD subjects, the whole slice network failed to classify one of the MR-Positive cases while localized the lesion in the other 16 subjects with an average coverage of 54%. For MR-Negative cases, 11 out of 13 subjects were detected with an average coverage of 56%. The result for the patch-wise network is presented in Tables 2 and 3. Each row represents one experiment where we train the network using patches extracted from slices within all subjects except for the subject relating to its row.

Table 1. Whole-slice two class classifier results

	MR-positive	MR-negative
Subject-wise sensitivity	0.92	0.92
Subject-wise specificity	0.94	0.85
Lesional sensitivity	0.59	0.72
Lesional specificity	0.54	0.56

Table 2. 2D patch-wise two class classifier results for lesional subjects

Test sample	Train accuracy	Test accuracy	Sensitivity	Specificity	Subject label
MR-positive subjects					
4	99%	99%	1.00	0.87	1
5	99%	99%	1.00	0.70	1
8	99%	99%	1.00	0.81	1
10	99%	99%	1.00	0.89	1
13	99%	99%	1.00	0.85	1
18	99%	99%	1.00	0.86	1
20	99%	99%	1.00	0.92	1
21	99%	99%	1.00	0.85	1
22	99%	99%	1.00	0.87	1
23	99%	99%	1.00	0.86	1
24	99%	99%	1.00	0.84	1
26	99%	99%	1.00	0.88	1
27	99%	99%	1.00	0.78	1
29	99%	99%	1.00	0.89	1
35	99%	99%	1.00	0.86	1
36	99%	99%	1.00	0.86	1
39	99%	99%	1.00	0.86	1
Average	99%	99%	1.00	0.85	
MR-negative subjects					
1	99%	99%	1.00	0.78	1
2	99%	99%	1.00	0.67	1
3	99%	97%	1.00	0.55	0
7	99%	98%	1.00	0.47	0
9	99%	95%	1.00	0.66	1
11	99%	99%	1.00	0.75	1
12	99%	99%	1.00	0.74	1
16	99%	99%	1.00	0.69	1
17	99%	94%	1.00	0.63	1
19	99%	95%	1.00	0.64	1
31	99%	94%	1.00	0.54	1
32	99%	95%	1.00	0.64	1
38	99%	99%	1.00	0.83	1
Average	99%	97%	1.00	0.66	

The first column is the test subject, the second and third columns are training and test accuracy, which are the percentage of correctly predicted patches in training and evaluation set. We cannot solely rely on training/test accuracy as

Table 3. 2D patch-wise two class classifier results for healthy controls. We cannot report the lesional specificity as we do not have any lesional patches in our data. Zeros in the last column mean that the test subject is being classified as healthy.

Test sample	Train accuracy	Test accuracy	Sensitivity	Specificity	Subject label
30	99%	99%	1.00	N/A	0
51	99%	99%	0.99	N/A	0
52	99%	99%	1.00	N/A	0
53	99%	99%	1.00	N/A	0
54	99%	99%	1.00	N/A	0
55	99%	99%	1.00	N/A	0
56	99%	99%	1.00	N/A	0
57	99%	99%	1.00	N/A	0
58	99%	99%	1.00	N/A	0
60	99%	99%	1.00	N/A	0
61	99%	99%	1.00	N/A	0
62	99%	99%	1.00	N/A	0
63	99%	99%	1.00	N/A	0
Average	99%	99%	1.00		

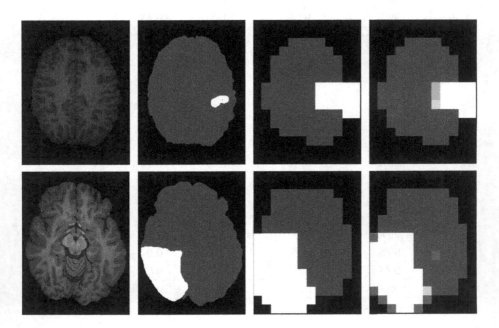

Fig. 3. Patch-wise model output; from left to right, columns represent the original slice, pixel-wise ground-truth, patch-wise ground-truth, patch-wise network prediction.

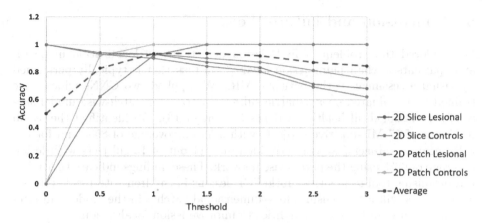

Fig. 4. Subject class label threshold, horizontal axis shows the threshold and the vertical axis indicates the accuracy for lesional and control subjects.

a measure of how well the network is performing since high numbers may be a result of high number of correctly classified healthy patches. Two complete the table, lesional sensitivity and specificity are reported in the next two columns.

The results for FCD approved patients are presented in Table 2. The first block is the MR-positive subjects, where we can see the network detected all 17 out of 17 cases with 85% coverage. The second portion is the MR-negative subjects' results. The model detected 11 out of 13 cases with a coverage of 66%. The large variation of the specificity (47%–83%) in Table 2 for MR-negative subjects can be attributed to lack of visual features which makes the training and evaluation for MR-Negative subjects more difficult. In Table 3, the results for 13 healthy control cases are reported where we successfully classified all 13 as healthy. The patch-wise network output along with the pixel-level and patch-level ground truth is illustrated in Fig. 3. Each row represents a patient, the top row is a patient with MR-positive FCD and the bottom is an MR-negative FCD approved subject. First image in each row is the actual slice, next one is the pixel-wise ground truth provided by a neuroradiologist, followed by the patch ground truth extracted from the pixel-wise ground truth, and the network prediction. As it is obvious the network's output localized the lesion in both subjects with acceptable accuracy.

In both models, we consider a subject prediction class as healthy or lesional by finding the number of predicted lesional patches (or slices) over the number of predicted healthy patches (or slices). If this ratio is larger than one percent, we consider the subject as lesional, otherwise it is considered as healthy. Selecting the right threshold is important as it decides whether a patient has FCD or not. We examined several options for each network and, to stay consistent in all experiments we plotted the results in Fig. 4. Lesional accuracy here is the average of MR-positive and MR-negative accuracy. The red curve in the graph is the average for all accuracies given the specific threshold. The curve is maximum at one. Hence, we chose one as a universal threshold for all of our experiments.

5 Conclusion and Future Work

We explored the problem of FCD lesion detection and localization in paediatric patients, a challenging lesion localization task that is typically performed by visual assessment of the patients' MRI. We applied two CNNs which were trained and evaluated on paediatric subjects' data. Our patch-wise model successfully classified all healthy patients (13 out of 13). We identified the lesion in 17 out of 17 MR-positive subjects with a lesion coverage of 85% and for the MR-negative subjects, we correctly identified 11 out of 13 subjects with lesion coverage of 66% using the patch-wise network. These findings indicate that CNN is a promising tool to identify subtle FCD lesions in children with focal epilepsy. Future work will include using three-dimensional patches in the model and integrating additional imaging modalities to improve lesion localization.

Acknowledgments. This research was conducted with the support of EpLink – The Epilepsy Research Program of the Ontario Brain Institute (OBI). The opinions, results and conclusions are those of the authors and no endorsement by the OBI is intended or should be inferred. This research was also supported in part by an NSERC Discovery Grant for M.E. AA would like to acknowledge Ontario Tech university for a doctoral graduate international tuition scholarship (GITS). The authors gratefully acknowledge the support of NVIDIA Corporation for the donation of GPUs used in this research through its Academic Grant Program.

References

1. Adler, S., et al.: Novel surface features for automated detection of focal cortical dysplasias in paediatric epilepsy. NeuroImage: Clin. **14**, 18–27 (2017)
2. Antel, S.B., et al.: Automated detection of focal cortical dysplasia lesions using computational models of their MRI characteristics and texture analysis. Neuroimage **19**(4), 1748–1759 (2003)
3. Bastian, D., et al.: Conditional generative adversarial networks support the detection of focal cortical dysplasia. Organization for Human Brain Mapping (2019, Abstract Submission)
4. Bernasconi, A., et al.: Texture analysis and morphological processing of magnetic resonance imaging assist detection of focal cortical dysplasia in extra-temporal partial epilepsy. Ann. Neurol.: Off. J. Am. Neurol. Assoc. Child Neurol. Soc. **49**(6), 770–775 (2001)
5. Besson, P., Bernasconi, N., Colliot, O., Evans, A., Bernasconi, A.: Surface-based texture and morphological analysis detects subtle cortical dysplasia. In: Metaxas, D., Axel, L., Fichtinger, G., Székely, G. (eds.) MICCAI 2008. LNCS, vol. 5241, pp. 645–652. Springer, Heidelberg (2008). https://doi.org/10.1007/978-3-540-85988-8_77
6. Bourgeois, M., Di Rocco, F., Sainte-Rose, C.: Lesionectomy in the pediatric age. Child's Nerv. Syst. **22**(8), 931–935 (2006)
7. Centeno, R.S., Yacubian, E.M., Sakamoto, A.C., Ferraz, A.F.P., Junior, H.C., Cavalheiro, S.: Pre-surgical evaluation and surgical treatment in children with extratemporal epilepsy. Child's Nerv. Syst. **22**(8), 945–959 (2006)

8. Chin, R.F., Neville, B.G., Peckham, C., Bedford, H., Wade, A., Scott, R.C., et al.: Incidence, cause, and short-term outcome of convulsive status epilepticus in childhood: prospective population-based study. Lancet **368**(9531), 222–229 (2006)

9. Coeytaux, A., Jallon, P., Galobardes, B., Morabia, A.: Incidence of status epilepticus in french-speaking switzerland (EPISTAR). Neurology **55**(5), 693–697 (2000)

10. Colombo, N., et al.: Focal cortical dysplasias: MR imaging, histopathologic, and clinical correlations in surgically treated patients with epilepsy. Am. J. Neuroradiol. **24**(4), 724–733 (2003)

11. Gill, R.S., et al.: Automated detection of epileptogenic cortical malformations using multimodal MRI. In: Cardoso, M.J., et al. (eds.) DLMIA/ML-CDS -2017. LNCS, vol. 10553, pp. 349–356. Springer, Cham (2017). https://doi.org/10.1007/978-3-319-67558-9_40

12. Gill, R.S., et al.: Deep convolutional networks for automated detection of epileptogenic brain malformations. In: Frangi, A.F., Schnabel, J.A., Davatzikos, C., Alberola-López, C., Fichtinger, G. (eds.) MICCAI 2018. LNCS, vol. 11072, pp. 490–497. Springer, Cham (2018). https://doi.org/10.1007/978-3-030-00931-1_56

13. Hader, W.J., et al.: Cortical dysplastic lesions in children with intractable epilepsy: role of complete resection. J. Neurosurg.: Pediatrics **100**(2), 110–117 (2004)

14. Huppertz, H.J., et al.: Enhanced visualization of blurred gray-white matter junctions in focal cortical dysplasia by voxel-based 3D MRI analysis. Epilepsy Res. **67**(1–2), 35–50 (2005)

15. Hutton, C., De Vita, E., Ashburner, J., Deichmann, R., Turner, R.: Voxel-based cortical thickness measurements in MRI. Neuroimage **40**(4), 1701–1710 (2008)

16. Kulaseharan, S., Aminpour, A., Ebrahimi, M., Widjaja, E.: Identifying lesions in paediatric epilepsy using morphometric and textural analysis of magnetic resonance images. NeuroImage: Clin. **21**, 101663 (2019)

17. Kwan, P., Brodie, M.J.: Early identification of refractory epilepsy. N. Engl. J. Med. **342**(5), 314–319 (2000)

18. Martin, P., Bender, B., Focke, N.K.: Post-processing of structural MRI for individualized diagnostics. Quant. Imaging Med. Surg. **5**(2), 188 (2015)

19. Nazeri, K., Aminpour, A., Ebrahimi, M.: Two-stage convolutional neural network for breast cancer histology image classification. In: Campilho, A., Karray, F., ter Haar Romeny, B. (eds.) ICIAR 2018. LNCS, vol. 10882, pp. 717–726. Springer, Cham (2018). https://doi.org/10.1007/978-3-319-93000-8_81

20. Pitkänen, A., Sutula, T.P.: Is epilepsy a progressive disorder? Prospects for new therapeutic approaches in temporal-lobe epilepsy. Lancet Neurol. **1**(3), 173–181 (2002)

21. Rastogi, S., Lee, C., Salamon, N.: Neuroimaging in pediatric epilepsy: a multi-modality approach. Radiographics **28**(4), 1079–1095 (2008)

22. Smith, S.M.: Fast robust automated brain extraction. Hum. Brain Mapp. **17**(3), 143–155 (2002)

23. Snead III, O.C.: Surgical treatment of medically refractory epilepsy in childhood. Brain Dev. **23**(4), 199–207 (2001)

24. Spencer, S., Huh, L.: Outcomes of epilepsy surgery in adults and children. Lancet Neurol. **7**(6), 525–537 (2008)

25. Springenberg, J.T., Dosovitskiy, A., Brox, T., Riedmiller, M.: Striving for simplicity: the all convolutional net. arXiv preprint arXiv:1412.6806 (2014)

26. Wyllie, E.: Surgical treatment of epilepsy in children. Pediatr. Neurol. **19**(3), 179–188 (1998)

Deep Learning Models for Segmentation of Mobile-Acquired Dermatological Images

Catarina Andrade[1]($\boxtimes$), Luís F. Teixeira[2], Maria João M. Vasconcelos[1], and Luís Rosado[1]

[1] Fraunhofer Portugal AICOS, 4200-135 Porto, Portugal
{catarina.andrade,maria.vasconcelos,luis.rosado}@fraunhofer.pt
[2] FEUP/INESC, 4200-465 Porto, Portugal
luisft@fe.up.pt

Abstract. With the ever-increasing occurrence of skin cancer, timely and accurate skin cancer detection has become clinically more imperative. A clinical mobile-based deep learning approach is a possible solution for this challenge. Nevertheless, there is a major impediment in the development of such a model: the scarce availability of labelled data acquired with mobile devices, namely macroscopic images. In this work, we present two experiments to assemble a robust deep learning model for macroscopic skin lesion segmentation and to capitalize on the sizable dermoscopic databases. In the first experiment two groups of deep learning models, U-Net based and DeepLab based, were created and tested exclusively in the available macroscopic images. In the second experiment, the possibility of transferring knowledge between the domains was tested. To accomplish this, the selected model was retrained in the dermoscopic images and, subsequently, fine-tuned with the macroscopic images. The best model implemented in the first experiment was a DeepLab based model with a MobileNetV2 as feature extractor with a width multiplier of 0.35 and optimized with the soft Dice loss. This model comprehended 0.4 million parameters and obtained a thresholded Jaccard coefficient of 72.97% and 78.51% in the Dermofit and SMARTSKINS databases, respectively. In the second experiment, with the usage of transfer learning, the performance of this model was significantly improved in the first database to 75.46% and slightly decreased to 78.04% in the second.

Keywords: Skin lesion segmentation · Macroscopic images · Dermoscopic images · Convolution neural networks · Knowledge transfer

1 Introduction

Skin cancer is the most prevalent malignancy worldwide [5]. Among the carcinogenic skin lesions, malignant melanoma is less common yet the most lethal, with a worldwide morbidity of 60.7 thousand people in 2018 [5]. Timely and accurate

© The Author(s) 2020
A. Campilho et al. (Eds.): ICIAR 2020, LNCS 12132, pp. 228–237, 2020.
https://doi.org/10.1007/978-3-030-50516-5_20

skin cancer detection is clinically highly relevant since the estimated 5-year survival rate for malignant melanoma drops from over 98% to 23% if detected when the metastases are distant from the origin point [20]. Nonetheless, this effective diagnosis raises another paradigm: the clinical presentation of most common cutaneous cancers is, every so often, identical to benign skin lesions.

The mobile technological advancement and the ubiquitous adoption of smartphones associated with the high performance of deep learning algorithms have the potential to improve skin cancer triage with the creation of an algorithm which can match or outperform the visual assessment of skin cancer. Convolutional neural networks have been the staple method used in the skin lesion segmentation challenge. Most methods are based on modifications of encoder-decoder architecture of the U-Net [15]. From small changes, as modifying the number of input channels and loss optimization [11] to the addition of recurrent layers and residual units [1], Sarker et al. [18] developed a U-Net with an encoder path consisting of four pre-trained dilated residual networks and a pyramid pooling block.

Nevertheless, the scarce availability of labelled data acquired with mobile devices, namely macroscopic images may prove to be a major impediment for the creation of such a method. Habitually, the cutaneous skin lesions are diagnosed by skin lesion surface microscopy (dermoscopy) which allows for the visualization of the subsurface skin structures which are usually not visible to the naked eye. This compelled the creation of several dermoscopic databases of substantial size. Withall, the direct inference between the macroscopic and dermoscopic domain is not advisable due to their paradoxical characteristics and challenges namely, the acquisition of images with the dermoscope generates several structures, colours and artefacts which are not detectable in the macroscopic image. The polarized light that permits the visualization of these characteristics eliminates the surface glare of the skin, which is abundantly common in the macroscopic setting. Additionally, structures clearly visible in dermoscopic images like pigmented network, streaks, dots, globules, blue-whitish veil or vascular patterns are usually less noticeable or even imperceptible in macroscopic images. Furthermore, the flat outward aspect in the dermoscopic images, caused by the direct contact of the dermoscope with the skin is paradoxical with the visual depth normally present in the macroscopic images. In fact, even for the diagnosis, there are rules and methods specific for each domain [9].

This work aims to evaluate the possibility of designing a deep learning algorithm for segmenting the lesion in macroscopic images which would operate fully in the mobile environment. This involves creating a fast and lightweight algorithm with expert-level accuracy to be integrated into the mobile environment. To assemble such a model, we explored the capitalization of the sizable dermoscopic databases and designed two separate experiments.

2   Methodology

2.1   Databases and Problem Definition

As there were several databases available which provided matching binary segmentation masks, it was possible to assemble two distinct datasets: the dermoscopic (set D) and the macroscopic (set M). The set D was constituted by the combined images of all ISIC Challenges (2016 [7], 2017 [4] and 2018 [3,21]) and the PH2 database [13] and set M was comprised of the Dermofit image Library [12] and the SMARTSKINS database [22].

For the PH2 and Dermofit image Library, an 80/20 slip was used for the creation of training/validation and test sets. In the case of the SMARTSKINS database, a 50/50 split was used due to the database the small size. Considering the three ISIC challenges had duplicate images in the different years, the training datasets of ISIC 2016, ISIC 2017 and ISIC 2018 and the validation of ISIC 2017 were combined and the duplicates removed. Subsequently, the image instances of the test dataset of ISIC 2017 were also eliminated from the combined dataset and reserved as a test subset. When the databases were classified, the division was structurally made to maintain an equal percentage of each class in the training/validation and test sets. The characterization and the slitting of the databases in each dataset are summarised in Table 1.

Table 1. Overview of available segmentation databases and separation into train/validation and test subsets.

Set	Database	No. images (type)	No. train/val.	No. test
Set M	Dermofit	1300 (MD)	1036	264
	SMARTSKINS	80 (MP)	39	41
Set D	ISIC	2594 (De)	1994	600
	PH2	200 (De)	160	40

No. - Number; De - Dermoscopic images; MD - Macroscopic images acquired with a digital camera; MP - Macroscopic images acquired with a mobile phone;

As it can be observed in Table 1, the size of set D is almost double of the size of set M, which lead to the creation of two experiments. In the first experiment, a comparative study, using exclusively the set M, was performed with two major groups of deep Learning models, U-Net and DeepLab based. From this study, a model was to be chosen to be used in the following experiment. The second experiment tested the possibility of transferring the knowledge from the dermoscopic to the macroscopic domain was tested. This was accomplished by re-training the chosen model in the first experiment with the set D and subsequent fine-tuning of the model with set M.

2.2   Deep Learning Models for Semantic Segmentation

For implementation, we adopt the Tensorflow API r1.15 in Python 3.7.3 on three NVIDIA Tesla V100 PCIe GPU module, two with 16GB and the other with 32 GB. Initially, the standard image resizing (512×512 pixels) and standardization were performed as a preprocessing stage. As for the training protocol, we employ a batch size of 4, a 90/10 partition for the training/validation subsets and the Adam optimizer to perform stochastic optimization with a cyclic learning rate (CLR) [19]. The cycle used was the cosine annealing variation with periodic restarts [8] associated with early stopping and model checkpointing.

U-Net Based Models. As a baseline model, we implemented a classical U-Net [15] with the addition of dropout layers (0.2) and zero-padding. The second model was an Attention U-Net [14] (AttU-Net) which main modification lies in the addition of an Attention Gate (AG) at the skip connection of each level. The third U-Net based model trained was the R2U-Net [1] which is a recurrent residual convolutional neural network based on U-Net. The last model implemented was a combination of the two aforementioned models (AttR2U-Net) [10]. The optimal number of initial feature channels was also analysed. This model parameter can be used to decrease the model complexity, however, it can also downgrade the performance of the model. This hypothesis was tested, using the values 16, 32 and 64, to ascertain the fidelity of this technique.

DeepLab Based Models. The second proposed approach was the state-of-the-art DeepLabv3+ model [2]. Initially, the original modified Aligned Xception encoder was used [2]. However, due to its considerable size which is not suited to the mobile environment, the MobileNetV2 as in [17] was also implemented. Primarily, all models were tested with randomly initialized weights and then two different sets of pre-trained weights were used: one pre-trained in Cityscapes and the other on the Pascal VOC 2012. Furthermore, some encoder specific experiments were also performed. To the modified Aligned Xception encoder, two output strides (OS), which refer to the spatial resolution ratio between the input and the output images, were tested: 8 and 16. In the case of the MobileNetV2, the variation of width multiplier (α) was analyzed. This hyperparameter allows the manipulation of the input width of a layer, which can lead to the reduction or augmentation of the models by a ratio roughly of α^2.

Model Optimization. The selection of a suitable loss function for the challenge at hand is pivotal to reach the appropriate capacity of the model. In total, five losses were tested. Initially, cross-entropy (CE) was chosen as the standard loss. Then four losses were tested: the soft Dice coefficient (DI) loss $(1 - DI)$ and soft Jaccard coefficient (JA) loss $(1 - JA)$, and the logarithmic combinations with the cross-entropy ($CE - \log JA$, $CE - \log DI$). These losses use a soft variant of the DI or JA, which uses the predicted probabilities instead of a binary mask, to decrease the effect of the class imbalance amidst each sample.

Performance Assessment. In the first experiment all the image segmentation models were trained on set M. Here, all the model configurations were tested and selected based on the results on the validation dataset. The measures used to evaluate the performance of the models were the ones used in the ISIC challenge of 2018 [3]: threshold JA (TJA), JA, DI, accuracy (AC), sensitivity (SE) and specificity (SP). The decision of the best model was a balance between model complexity and TJA, as it was the scoring metric of the ISIC 2018 challenge. The threshold on which these metrics were taken was inferred with the result of a JA analysis performed on the validation subset. Considering the sigmoid nonlinearity used in the last layer, the binarization threshold was estimated with the intent of maximizing the resultant JA. For this purpose, the JA was evaluated in 50 thresholds within the [0.5, 1] range. In the second experiment, the model was evaluated in the test subset and compared with the results of the selected model of the first experiment.

3   Experimental Results

3.1   First Experiment

The best performing configurations for each model architecture of the U-Net based models and each encoder of the Deeplab based models are summarized in Table 2.

Regarding the best performing U-Net based models, the addition of the AG proved to be quite disadvantageous. The models with this extension reach underperforming results leading to a decrease in 2% in JA and TJA when compared with the classical U-Net structure. Besides, the requirement of a higher number of ifC, 64 instead of 32 needed in the baseline U-Net elucidates to the inefficiency of the networks with the AG in learning representative features. On the other hand, the addition of the recurrent residual unit leads to 3% improvement in JA and 4% in TJA when compared to the Classical U-Net. Contrarily to the AttU-Net, the R2U-Net reached higher performance with the lowest ifC tested, 16. However, the number of parameters of this network increases significantly when compared with the other networks. Consequentially, the R2U-Net with 16 ifC and the classical U-Net with 32 ifC have a similar number parameter, around 7M. When adding the recurrent residual units with the AG the results are entirely consistent with the aforementioned conclusions. Essentially, the AttR2U-Net result improves in comparison to the classical U-Net however it decreases 1% the JA when comparing with the R2U-Net.

Table 2. Performance metrics for each model architecture used in the U-Net approach best configuration and for each encoder used in the DeepLab approach, evaluated in the validation subset of set M.

U-Net based models										
Model architecture	L	ifC	P	T	TJA	JA	DI	AC	SE	SP
U-Net	$1-JA$	32	8 M	0.52	78.44	80.80	89.24	93.16	**90.42**	94.65
AttU-Net	$1-DI$	64	32 M	0.50	76.55	78.78	87.91	93.27	88.27	95.69
R2U-Net	$1-JA$	16	6 M	0.50	**83.02**	**83.02**	**90.60**	**94.39**	90.05	**96.48**
AttR2U-Net	$CE-\log JA$	64	96 M	0.50	82.08	82.08	90.00	93.98	89.38	96.18
Deeplab based models										
Encoder	L	PreT	P	T	TJA	JA	DI	AC	SE	**SP**
Xception, OS = 16	$1-JA$	Pascal	41 M	0.50	**85.37**	**85.37**	**92.00**	**95.54**	91.46	97.05
MobileNetV2, $\alpha=1.0$	$1-JA$	Pascal	2 M	0.55	84.80	84.80	91.70	95.18	**91.60**	96.55
MobileNetV2, $\alpha=0.35$	$1-DI$	None	0.4 M	0.50	82.85	82.85	90.52	94.65	88.31	**97.27**

L - Loss Function; ifC - intial feature channels; P - Model Parameters; T - binarization threshold used in the output probability map; PreT - Pretraining of weights;

Concerning the DeepLab approach, an overall conclusion can be drawn from the robustness of the combination of the soft JA loss function $(1 - JA)$ and the pre-trained model in Pascal VOC 2012 which lead to the top-performing results for each of the encoders. Concerning the output stride of $OS = 16$ surpasses in terms of performance the $OS = 8$, meaning a denser feature extraction in the last layers of the model decoder is not suitable for skin lesion segmentation. Not surprisingly, the addition of the inverted residual depthwise separable convolution of the MobileNetV2 encoder leads to a dramatical reduction of model complexity. In fact, there is almost a reduction of approximately nineteen times fewer parameters and with the loss of less than 1% in all of the performance metrics. This result prompted a second study, which focuses only on the effects on the reduction of the MobileNetV2. Thus, several models with various α and no pre-training of weights were implemented and optimized with the five designed loss functions. The best result from this study is summarised in Table 2 (Deeplab based models, row 3).

Pertaining to the loss function the results are quite consistent. For each model architecture, the stochastic optimization performed by a loss function, which takes into account the soft variation of JA and DI, leads to improved results. The soft DI and soft JA losses yielded the best results in all the models except the AttR2U-Net. Therefore, the use of a loss function which takes into consideration the measure of overlap between two samples is an effective way of reducing the class imbalance between the surrounding skin and the lesion pixels.

Based on the aforementioned approaches and experiments, one model was chosen to be evaluated in the test subset of set M. The main rationale behind this selection was choosing the model which offers the best a balance between two desirable but usually incompatible aspects features: performance and model complexity. The selected model was the reduced Mobile DeepLab with $\alpha = 0.35$ and optimized with soft DI loss function mainly due to its reduced size and its JA and TJA values above the 80% threshold, which is above the interobserver agreement and the visual correctness threshold [4].

3.2   Second Experiment

After the selection of the reduced Mobile DeepLab, the model was retrained with set D. The same training procedure and network parameters were used. Subsequently, the model trained with set D was fine-tunned with set M. The obtained results of both experiments, evaluated in the test subset of the SMARTSKINS and Dermofit, are presented in Table 3. From this table, it is possible to infer that the fine-tuning, with the macroscopic data, of the pre-trained model, on set D, leads to a 2.49% TJA improvement in digital-acquired images (Dermofit) and a slight decrease of less than 0.48% in the mobile-acquired images (SMARTSKINS).

For the SMARTSKINS database, there is no standard used for the slitting of the database into train-validation-test subset. Therefore, the comparison with the models in the literature might not be as equitable as desired. Nevertheless, the reduced Mobile DeepLab attains in the first experiment the performance of

Table 3. Comparison of the performance metrics of the reduced Mobile DeepLab from the two proposed experiments, evaluated in the test subset of the SMARTSKINS and Dermofit.

Test subset	Experiment	T	TJA	JA	DI	AC	SE	SP
SMARTSKINS	Exp 1	0.50	**78.51**	**82.64**	**90.14**	**98.96**	95.40	**99.15**
	Exp 2	0.56	78.04	82.21	89.89	98.90	**96.05**	99.09
Dermofit	Exp 1	0.50	72.97	80.26	88.26	93.51	87.56	**96.86**
	Exp 2	0.56	**75.46**	**81.03**	**88.79**	**93.78**	**89.68**	96.13

T - binarization threshold used in the output probability map;

82.64% in JA, 90.14% in DI and 99.15% in AC. These values set a new state-of-the-art performance in the SMARTSKINS database which previously was of 81.58% in JA [16], 83.36% in DI [6] and 97.38% in AC [16].

Figure 1 presents several examples of the predicted segmentation mask of the model trained in each experiment compared with the ground truth label (GT). The model in both experiments shows highly satisfactory results when the lesion is pigmentated with high contracts with the skin (Fig. 1, row 1). The presence of lesions with dysplastic form and uneven pigmentation can lead to the underperformance of the model of the second experiment (Fig. 1, row 2). The model of the second experiment outperforms the other in the presence of dark hair and lesions with other moles near (Fig. 1, row 3). Both experiments seem to underperform when the lesion presents red regions amidst the normal skin and vascularizations near the lesion border (Fig. 1, row 4).

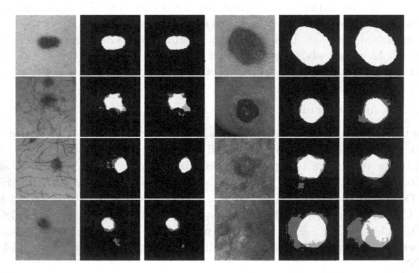

Fig. 1. Examples of successful and failed segmentation results on the SMARTSKINS (left) and Dermofit (right) test subset. In the comparison images: yellow - true positives; red - false positives; green - false negatives; black - true negatives; (Color figure online)

4   Conclusion and Future Work

The yielded results show considerable potential in the use of models with decreased complexity and size. Altogether, the selected network had less than half a million parameters and a decrease in performance of TJA of 3% when compared with a model with approximately 41 M parameters.

When comparing the two experiments it can be inferred that the knowledge transfer between the dermoscopic and macroscopic domains still resulted in an overall improvement of the model. Despite the slight decrease in performance on the SMARTSKINS dataset, the improvement in the Dermofit dataset is significantly larger. It should be noted that the Dermofit dataset has more variety of skin lesion classes, including non-pigmented lesions that are not present in the SMARTSKINS dataset, thus we can assume that the fine-tuning procedure brought an overall model improvement. Nevertheless, there's still room for improvements, namely further experiments should be done in order to effectively take advantage of the sizable dermoscopic datasets.

Acknowledgements. This work was done under the scope of project "DERM.AI: Usage of Artificial Intelligence to Power Teledermatological Screening" and supported by national funds through 'FCT–Foundation for Science and Technology, I.P.', with reference DSAIPA/AI/0031/2018.

References

1. Alom, M.Z., Hasan, M., Yakopcic, C., Taha, T.M., Asari, V.K.: Recurrent residual convolutional neural network based on U-Net (R2U-Net) for medical image segmentation (2018)
2. Chen, L.-C., Zhu, Y., Papandreou, G., Schroff, F., Adam, H.: Encoder-decoder with atrous separable convolution for semantic image segmentation. In: Ferrari, V., Hebert, M., Sminchisescu, C., Weiss, Y. (eds.) ECCV 2018. LNCS, vol. 11211, pp. 833–851. Springer, Cham (2018). https://doi.org/10.1007/978-3-030-01234-2_49
3. Codella, N., et al.: Skin lesion analysis toward melanoma detection 2018: a challenge hosted by the international skin imaging collaboration (ISIC) (2019)
4. Codella, N.C.F., et al.: Skin lesion analysis toward melanoma detection: a challenge at the 2017 international symposium on biomedical imaging (ISBI), hosted by the international skin imaging collaboration (ISIC). In: 2018 IEEE 15th International Symposium on Biomedical Imaging (ISBI 2018), April 2018
5. Ferlay, J., et al.: Estimating the global cancer incidence and mortality in 2018: globocan sources and methods. Int. J. Cancer 144(8), 1941–1953 (2019)
6. Fernandes, K., Cruz, R., Cardoso, J.S.: Deep image segmentation by quality inference. In: 2018 International Joint Conference on Neural Networks (IJCNN), pp. 1–8. IEEE (2018)
7. Gutman, D., et al.: Skin lesion analysis toward melanoma detection: a challenge at the international symposium on biomedical imaging (ISBI) 2016, hosted by the international skin imaging collaboration (ISIC) (2016)
8. Huang, G., Li, Y., Pleiss, G., Liu, Z., Hopcroft, J.E., Weinberger, K.Q.: Snapshot ensembles: train 1, get M for free. arXiv preprint arXiv:1704.00109 (2017)

9. Korotkov, K., Garcia, R.: Computerized analysis of pigmented skin lesions: a review. Artif. Intell. Med. **56**(2), 69–90 (2012)
10. LeeJunHyun: Pytorch implementation of U-Net, R2U-Net, attention U-Net, attention R2U-Net (2019). https://github.com/LeeJunHyun/Image_Segmentation. Accessed 1 Jan 2020
11. Lin, B.S., Michael, K., Kalra, S., Tizhoosh, H.R.: Skin lesion segmentation: U-nets versus clustering. In: 2017 IEEE Symposium Series on Computational Intelligence (SSCI), pp. 1–7. IEEE (2017)
12. Ltd., E.I.: Dermofit image library - edinburgh innovations (2019). https://licensing. eri.ed.ac.uk/i/software/dermofit-image-library.html. Accessed 11 June 2019
13. Mendonça, T., Ferreira, P.M., Marques, J.S., Marcal, A.R., Rozeira, J.: PH^2-a dermoscopic image database for research and benchmarking. In: 2013 35th Annual International Conference of the IEEE Engineering in Medicine and Biology Society (EMBC), pp. 5437–5440. IEEE (2013)
14. Oktay, O., B., et al.: Attention U-Net: learning where to look for the pancreas. arXiv preprint arXiv:1804.03999 (2018)
15. Ronneberger, O., Fischer, P., Brox, T.: U-Net: convolutional networks for biomedical image segmentation. In: Navab, N., Hornegger, J., Wells, W.M., Frangi, A.F. (eds.) MICCAI 2015. LNCS, vol. 9351, pp. 234–241. Springer, Cham (2015). https://doi.org/10.1007/978-3-319-24574-4_28
16. Rosado, L., Vasconcelos, M.: Automatic segmentation methodology for dermatological images acquired via mobile devices. In: Proceedings of the International Joint Conference on Biomedical Engineering Systems and Technologies, vol. 5, pp. 246–251 (2015)
17. Sandler, M., Howard, A., Zhu, M., Zhmoginov, A., Chen, L.C.: MobileNetV 2: inverted residuals and linear bottlenecks. In: 2018 IEEE/CVF Conference on Computer Vision and Pattern Recognition, June 2018
18. Sarker, M.M.K., et al.: SLSDeep: skin lesion segmentation based on dilated residual and pyramid pooling networks. In: Frangi, A.F., Schnabel, J.A., Davatzikos, C., Alberola-López, C., Fichtinger, G. (eds.) MICCAI 2018. LNCS, vol. 11071, pp. 21–29. Springer, Cham (2018). https://doi.org/10.1007/978-3-030-00934-2_3
19. Smith, L.N.: Cyclical learning rates for training neural networks. In: 2017 IEEE Winter Conference on Applications of Computer Vision (WACV), pp. 464–472. IEEE (2017)
20. American Cancer Society: Cancer facts and figures 2019 (2019)
21. Tschandl, P., Rosendahl, C., Kittler, H.: The HAM10000 dataset, a large collection of multi-source dermatoscopic images of common pigmented skin lesions. Sci. Data **5**, 180161 (2018)
22. Vasconcelos, M.J.M., Rosado, L., Ferreira, M.: Principal axes-based asymmetry assessment methodology for skin lesion image analysis. In: Bebis, G., et al. (eds.) ISVC 2014. LNCS, vol. 8888, pp. 21–31. Springer, Cham (2014). https://doi.org/ 10.1007/978-3-319-14364-4_3

Semi-automatic Tool to Identify Heterogeneity Zones in LGE-CMR and Incorporate the Result into a 3D Model of the Left Ventricle

Maria Narciso[1]([✉])(iD), António Ferreira[2]([✉])(iD), and Pedro Vieira[1]([✉])(iD)

[1] FCT Nova, Caparica, Portugal
m.narciso@campus.fct.unl.pt
[2] Cardiology Department, Hospital de Santa Cruz,
Centro Hospitalar Lisboa Ocidental, Lisbon, Portugal

Abstract. Fatal scar-related arrhythmias are caused by an abnormal electrical wave propagation around non conductive scarred tissue and through viable channels of reduced conductivity. Late gadolinium enhancement (LGE) cardiovascular magnetic resonance (CMR) is the gold-standard procedure used to differentiate the scarred tissue from the healthy, highlighting the dead cells. The border regions responsible for creating the feeble channels are visible as gray zones. Identifying and monitoring (as they may evolve) these areas may predict the risk of arrhythmias that may lead to cardiac arrest. The main goal of this project is the development of a system able to aid the user in the extraction of geometrical and physiological information from LGE images and the replication of myocardial heterogeneities onto a three-dimensional (3D) structure, built by the methods described by our team in another publication, able to undergo electro-physiologic simulations. The system components were developed in MATLAB R2019b the first is a semi-automatic tool, to identify and segment the myocardial scars and gray zones in every two-dimensional (2D) slice of a LGE CMR dataset. The second component takes these results and assembles different sections while setting different conductivity values for each. At this point, the resulting parts are incorporated into the functional 3D model of the left ventricle, and therefore the chosen values and regions can be validated and redefined until a satisfactory result is obtained. As preliminary results we present the first steps of building one functional Left ventricle (LV) model with scarred zones.

Keywords: Gray zone · Ischaemia · Arrhythmia · Heart computational model

1 Introduction

Coronary artery disease, characterized by blockages in the coronary arteries, frequently results in infarcted (dead) myocardial tissue, therefore incapable of conducting electrical signal.

© Springer Nature Switzerland AG 2020
A. Campilho et al. (Eds.): ICIAR 2020, LNCS 12132, pp. 238–246, 2020.
https://doi.org/10.1007/978-3-030-50516-5_21

Ischemic injuries create not only non excitable portions of myocardial tissue, constituting the scar core, but also damaged portions where the electrical conductivity is low but not zero, these areas are termed gray zones. Although they can also stand alone, often these gray zones constitute the peri-infarct tissue - the scar borders. If they happen to cross the dead scarred zones and reach conductive or semi conductive tissue on the other side, conductive channel (CC) are formed, critical in the ventricular fibrillation as explained in [4].

In clinical decision-making, LGE is the gold-standard to identify and locate scars, allowing to evaluate the scar and gray zone extensions as well as identify the presence of CC. Monitoring the gray zone evolution is critical to help identify the optimal timing and candidates for placement of implantable cardioverter-defibrillator [3]. The conditions may deteriorate or improve due to the fact that damaged myocardium can recover function following coronary revascularization.

In the following chapters, we report the development of a system, using MAT-LAB R2019b, capable of assembling 3D models of left ventricular myocardium heterogeneities based on patient CMR data. The output of this system allows subsequent virtual eletrophysiological studies to be conducted. Semi-automatic histogram and segmentation methods based on signal intensity (SI) thresholds are used to identify the heterogeneity zones and corresponding conductivity assignment are based on values described in the literature. All the development is made on top of a structure, built by the methods detailed in a previous publication from our team [7], able to be used in electro-physiologic simulations. An overview of the whole system can be seen in Fig. 1.

2   Methods

2.1   Setup

In order to build and test our application, anonymized 3D CMR datasets in DICOM format from a LGE exam were used as input. The images passed trough the transformations described in detail in a previous publication to obtain the important data in regard of this paper. Therefore the input data of our system is a set of 2D slices representing the cardiac short-axis view (SAX), alongside with a mask identifying the left ventricle myocardium, healthy and damaged tissue alike, the pixels within are used in steps described in the following sections.

2.2   Segmentation

Our approach to the segmentation of the tissue heterogeneities is two-fold: in the first category the user slides a pointer through the masked myocardium histogram, selecting a pixel intensity level above which, all the pixels are labelled as dead tissue, the same process is repeated to select a value bellow which all the pixels are labelled as healthy. All the pixels with intensities between the two selected thresholds are labelled as gray zone (GZ). Henceforward the myocardium is divided in three parts: GZ, healthy tissue (H) and scarred dead

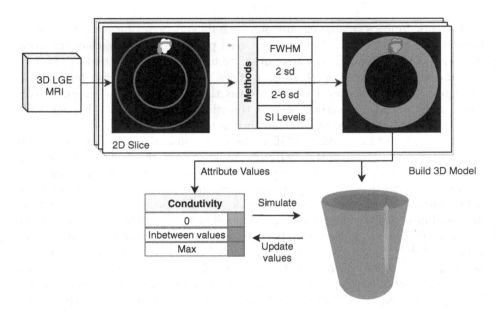

Fig. 1. Overview of the system being developed, starting from the 3D CMR segmentation, passing through the heterogeneities analysis and conductivity values assignment until the 3D structure generation.

tissue (S). The user may decide that the thresholds have the same value and the GZ will be disregarded.

The second category of our approach is to let the user chose between previously validated SI threshold algorithms: full width half max (FWHM), 2–6 standard deviation (sd) and 6 sd, to identify the S and GZ. [1,2,5]

Considering the histogram for the pixels inside the masked myocardium, using the FWHM, the infarct area is made by all the pixels which intensities are above 50% of the maximum intensity. Using this method, the S zones are distinguished from the H, but if GZ are present they will be considered part of S as well. They may be distinguished on a later step.

To apply the 2–6 sd and sd methods, the mean intensity of the healthy tissue is calculated from the largest contiguous area of myocardium with no visually apparent enhanced areas or artefacts, selected during a previous step that leads to the masked myocardium.

The 2–6 sd method recognize all the pixels, of the masked myocardium, with intensities bellow the mean plus 2 sd to be healthy, pixels with intensity values between the mean plus 2 sd to the mean plus 6 sd as gray zones and pixels with intensities above the mean value plus 6 sd to be scars.

The 6 sd method works similarly but the pixels that are recognised as part of gray zone have intensities between the mean and the mean value plus 6 sd .

For both types of segmentation, the next (optional) step is to take the GZ, or in FWHM case the S, and split it into a number of levels the user decides, limited by the quantity of unique values in the region. Defining more sub-regions within the GZ.

This strategy is applied to every 2D slice, both parts of the approach and all the methods can be applied to evaluate the results, however to conclude the segmentation the same procedure must be applied to the whole set. The results are stored in a 3D matrix where every zone is identified by a distinct integer number.

2.3 Conductivity Assignment

Considering an electrical propagation bidomain model, the myocardium cells have extracellular medium between them, thus the intra and extra cellular conductivities must be defined for the H zones. The assigned values will be based on the ones listed in [8]. After any chosen division, the myocardium has at least two different portions, H and S. S zones conductivity is set to zero in every direction. H zones intra conductivities are set to 3.75 S/m in the longitudinal direction (along the tissue fibres) and 2.14 S/m transversal and normal and the extra cellular conductivities 4.69 S/m and 0.47 S/m respectively.

For the first approximation, the conductivity values will be chosen in the following way: the number of intervals is selected (it may be only one) and the conductivities for each will be set by a simple linear regression function, trained by one vector with all the intensities of the other zones (H and S) and one with the corresponding conductivities. The function will then attribute a conductivity value to each pixel intensity of the GZ. The chosen value for the whole sub-interval (when working with more than one are) is the median value of all the conductivities set to this sub-interval.

The outset values may be changed later on, analysing the results from electrophysiological simulations run on the generated model will give an understanding of the accuracy level of the modelled tissue.

The in-between values assignment function should also be improved based on the results until a satisfactory approximation is obtained.

2.4 Output Data Assembling

From now on, the term *region* will be use to refer to the groups of adjacent voxels that are labelled as the same zone. Distinct parts of the muscle can have the same level of heterogeneity, therefore the zone is the same but the region is different. The regions are generated by reading the 3D matrix, where the zones are stored and assorting the adjacent voxels that have the same value to a region.

The resulting regions representing the heterogeneities volumes have to be translated into a format suited for simulation software. Our team chose to work with the open source software CHASTE (Cancer, Heart and Soft Tissue Environment) [6] to solve and visualize the propagation of excitation waves. In this program the simulation settings are read from a extensible markup language (xml) file where several physiologic parameters, including conductivities, can be defined. The heterogeneities regions can be defined in the parameters file, as parallelepipedal volumes, specifying a pair of a upper and lower 3D coordinates.

Since the regions are not rectangular shapes, each one has to be divided into smaller rectangular shapes. As the parallelepipedal structures are grouped and can form larger ones, these last are the ones validated to be part of the final structure, this is important to minimize the number of nodes read by the simulation software, if each voxel was simply assigned to one region, the parametrization file would become unreadable. In Fig. 2, a region of a single slice divided into cubes - as the slice has thickness - is exemplified.

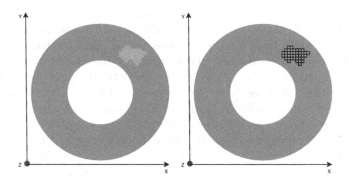

Fig. 2. Splitting a region of one slice into small cubes with the original voxels dimensions.

Figure 3 exemplifies how a region that spans through two consecutive slices [left] is split into three cubes [right], which corners coordinates will be registered into the parametrization file.

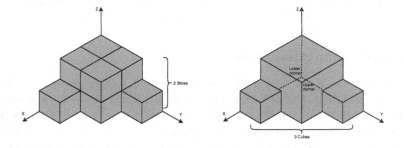

Fig. 3. Example of how eight cubes, four at each consecutive slice, are reduced to one in the final structure.

For every region block, the pairs of upper and lower coordinates and the designated conductivity are written to a xml file formatted to be compatible with CHASTE. Once again following the methods from our previous publication, a volumetric model of the LV is generated. Our system gives the possibility to define a series of physiologic parameters that are also included in the xml.

3   Preliminary Results

Using as input a LGE CMR we first tested the identification of enhanced zones, meaning scarred tissue that may include gray zones. In Fig. 4 three segmentation methods on a the SAX slice. The manual segmentation was accomplished by dragging two lines, one for each threshold, on top of the histogram, visualizing the segmentation result on real-time and stopping when the highlighted are was clearly an enhanced area.

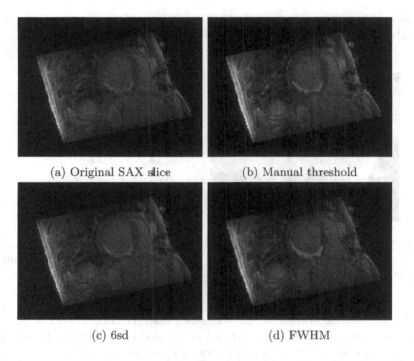

(a) Original SAX slice   (b) Manual threshold

(c) 6sd   (d) FWHM

Fig. 4. Scarred myocardium zone (in red) segmented by different methods b) User input thresholds, c) 6 sd and d) FWHM (Color figure online)

Selecting only one gray zone we obtain the area highlighted in blue in Fig. 5 [left] and three levels of longitudinal conductivity as seen in 5 [right].

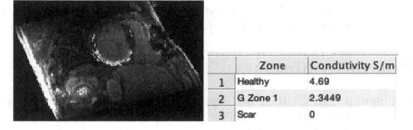

	Zone	Condutivity S/m
1	Healthy	4.69
2	G Zone 1	2.3449
3	Scar	0

Fig. 5. Gray zone for the given slice highlighted in light blue. The corresponding conductivity values to this area and the other two are presented on the table on the right. (Color figure online)

Taking the same gray zone, and telling our system we want to split it into two sub-zones, the image on the left of Fig. 6 is returned.

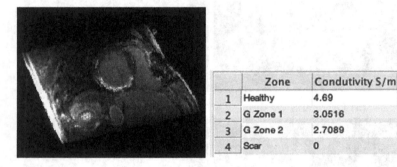

	Zone	Condutivity S/m
1	Healthy	4.69
2	G Zone 1	3.0516
3	G Zone 2	2.7089
4	Scar	0

Fig. 6. Two gray zones for the given slice highlighted in light blue and blue. The corresponding conductivity values to these areas and the other two are presented on the table on the right. (Color figure online)

The conductivity values for the 'G Zone 1' and 'G Zone 2' are calculated by the linear regression function described in the previous chapter, the other two values are pre-setted but as stated previously can be updated if the simulation results indicate that they should.

To simplify the rest of our initial testing, the area that resulted of the manual segmentation on Fig. 4b was exported to the CHASTE parametrization file, with a conductivity of zero mS/cm in every direction.

Figure 7 gives a 3D perception of the scarred area from inside the ventricle. The slices upfront represent the base of the LV. This image was generated only as an visual aid to understand the extension of the area on the final volume.

Fig. 7. 3D view of the LV myocardial surface represented in blue and the scarred zone in orange (Color figure online)

Both the parametrization file with the scarred zone near the base and the generated LV model were used in CHASTE to visualize the electrical wave propagation. On Fig. 8 the LV wall around the scar is being depolarized - by a stimuli we defined - while the scarred area maintains its potential at zero.

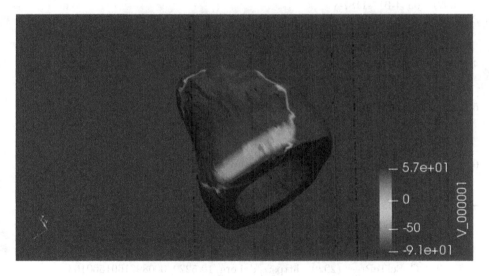

Fig. 8. Signal propagation on the LV wall (in red) with a dead zone (Color figure online)

4   Conclusion

The system described in this paper, may be a valuable asset in the assembling of personalized heart computational models. Offering the possibility to include heterogeneities adds a very important component in the study of arrhythmia mechanisms and risk assessment. In the future we aim to extend the reported methods to include the right ventricle, therefore obtaining a more faithful model. Another crucial goal is to compare the simulation results with real electrical maps from the same patient that provided the CMR, the evaluation of the results using this data will then be used to fed the method described in the Sect. 2.3 therefore new and more accurate conductivity values could be defined and the assignment function well improved. Overall, although this project is still at an early stage and needs several improvements, our approach proven to be worthwhile as an add-on to the parallel cardiac simulations on our ongoing research.

References

1. Flett, A.S., et al.: Evaluation of techniques for the quantification of myocardial scar of differing etiology using cardiac magnetic resonance. JACC Cardiovascular Imag. **4**(2), 150–156 (2011). https://doi.org/10.1016/j.jcmg.2010.11.015
2. Gho, J.M.I.H., et al.: A systematic comparison of cardiovascular magnetic resonance and high resolution histological fibrosis quantification in a chronic porcine infarct model. Int. J. Cardiovascular Imag. **33**(11), 1797–1807 (2017). https://doi.org/10.1007/s10554-017-1187-y
3. Ghugre, N.R., et al.: Evolution of gray zone after acute myocardial infarction: influence of microvascular obstruction. J. Cardiovascular Magnetic Reson. **13**(S1), 151 (2011). https://doi.org/10.1186/1532-429x-13-s1-p151
4. Lin, L.Y., et al.: Conductive channels identified with contrast-enhanced MR imaging predict ventricular tachycardia in systolic heart failure. JACC Cardiovascular Imag. **6**(11), 1152–1159 (2013). https://doi.org/10.1016/j.jcmg.2013.05.017
5. Mikami, Y., et al.: Accuracy and reproducibility of semi-automated late gadolinium enhancement quantification techniques in patients with hypertrophic cardiomyopathy. J. Cardiovasc. Magn. Resonance **16**(1), 85 (2014). https://doi.org/10.1186/s12968-014-0085-x
6. Mirams, G.R., et al.: Chaste: an open source C++ library for computational physiology and biology. PLoS Comput. Biol. **9**(3), e1002970 (2013). https://doi.org/10.1371/journal.pcbi.1002970
7. Narciso., M., Sousa., A.I., Crivellaro., F., de Almeida., R.V., Ferreira., A., Vieira., P.: Left ventricle computational model based on patients three-dimensional MRI. In: Proceedings of the 13th International Joint Conference on Biomedical Engineering Systems and Technologies, vol. 2 BIOIMAGING: BIOIMAGING, pp. 156–163. INSTICC, SciTePress (2020). https://doi.org/10.5220/0008961601560163
8. Sachse, F.B., Moreno, A.P., Seemann, G., Abildskov, J.A.: A model of electrical conduction in cardiac tissue including fibroblasts. Ann. Biomed. Eng. **37**(5), 874–889 (2009). https://doi.org/10.1007/s10439-009-9667-4

Analysis of Histopathology Images

A Deep Learning Based Pipeline
for Efficient Oral Cancer Screening
on Whole Slide Images

Jiahao Lu[1](ID), Nataša Sladoje[1](ID), Christina Runow Stark[2],
Eva Darai Ramqvist[3], Jan-Michaél Hirsch[4](ID), and Joakim Lindblad[1](✉)(ID)

[1] Centre for Image Analysis, Department of IT, Uppsala University, Uppsala, Sweden
joakim@cb.uu.se
[2] Department of Orofacial Medicine at Södersjukhuset, Folktandvården Stockholms
Län AB, Stockholm, Sweden
[3] Department of Clinical Pathology and Cytology, Karolinska University Hospital,
Stockholm, Sweden
[4] Department of Surgical sciences, Uppsala University, Uppsala, Sweden

Abstract. Oral cancer incidence is rapidly increasing worldwide. The
most important determinant factor in cancer survival is early diagnosis. To facilitate large scale screening, we propose a fully automated
pipeline for oral cancer detection on whole slide cytology images. The
pipeline consists of fully convolutional regression-based nucleus detection, followed by per-cell focus selection, and CNN based classification.
Our novel focus selection step provides fast per-cell focus decisions at
human-level accuracy. We demonstrate that the pipeline provides efficient cancer classification of whole slide cytology images, improving over
previous results both in terms of accuracy and feasibility. The complete source code is made available as open source (https://github.com/
MIDA-group/OralScreen).

Keywords: CNN · Whole slide imaging · Big data · Cytology ·
Detection · Focus selection · Classification

1 Introduction

Cancers in the oral cavity or the oropharynx are among the most common malig-
nancies in the world [27,30]. Similar as for cervical cancer, visual inspection of
brush collected samples has shown to be a practical and effective approach for
early diagnosis and reduced mortality [25]. We, therefore, work towards introducing screening of high risk patients in General Dental Practice by dentists
and dental hygienists. Computer assisted cytological examination is essential for
feasibility of this project, due to large data and high involved costs [26].

This work is supported by: Swedish Research Council proj. 2015-05878 and 2017-04385,
VINNOVA grant 2017-02447, and FTV Stockholms Län AB.

A. Campilho et al. (Eds.): ICIAR 2020, LNCS 12132, pp. 249–261, 2020.
https://doi.org/10.1007/978-3-030-50516-5_22

Whole slide imaging (WSI) refers to scanning of conventional microscopy glass slides to produce digital slides. WSI is gaining popularity among pathologists worldwide, due to its potential to improve diagnostic accuracy, increase workflow efficiency, and improve integration of images into information systems [5]. Due to the very large amount of data produced by WSI, typically generating images of around 100,000 × 100,000 pixels with up to 100,000 cells, manipulation and analysis are challenging and require special techniques. In spite of these challenges, the advantage to reproduce the traditional light microscopy experience in digital format makes WSI a very appealing choice.

Deep learning (DL) has shown to perform very well in cancer classification. An important advantage, compared to (classic) model-based approaches, is absence of need for nucleus segmentation, a difficult task typically required for otherwise subsequent feature extraction. At the same time, the large amount of data provided by WSI makes DL a natural and favorable choice. In this paper we present a complete fully automated DL based segmentation-free pipeline for oral cancer screening on WSI.

2 Background and Related Work

A number of studies suggest to use DL for classification of histology WSI samples, [1,2,18,28]. A common approach is to split tissue WSIs into smaller patches and perform analysis on the patch level. Cytological samples are, however, rather different from tissue. For tissue analysis the larger scale arrangement of cells is important and region segmentation and processing is natural. For cytology, though, the extra-cellular morphology is lost and cells are essentially individual (especially for liquid based samples); the natural unit of examination is the cell.

Cytology generally has slightly higher resolution requirements than histology; texture is very important and accurate focus is therefore essential. On the other hand, auto-focus of slide scanners works much better for tissue samples being more or less flat surfaces. In cytology, cells are partly overlapping and at different z-levels. Tools for tissue analysis rarely allow z-stacks (focus level stacks) or provide tools for handling such. In this work we present a carefully designed complete chain of processing steps for handling cytology WSIs acquired at multiple focus levels, including cell detection, per-cell focus selection, and CNN based classification.

Malignancy-associated changes (MACs) are subtle morphological changes that occur in histologically normal cells due to their proximity to a tumor. MACs have been shown to be reproducibly measured via image cytometry for numerous cancer types [29], making them potentially useful as diagnostic biomarkers. Using a random forest classifier [15] reliably detected MACs in histologically normal (normal-appearing) oropharyngeal epithelial cells located in tissue samples adjacent to a tumor and suggests to use the approach as a noninvasive means of detecting early-stage oropharyngeal tumors. Reliance on MAC enables using patient-level diagnosis for training of a cell-level classifier, where *all* cells of a patient are assigned the same label (either cancer or healthy)[32]. This hugely reduces the burden of otherwise very difficult and laborious manual annotation on a cell level.

Cell Detection: State-of-the-art object detection methods, such as the R-CNN family [7,8,23] and YOLO [20–22], have shown satisfactory performance for natural images. However, being designed for computer vision, where perspective changes the size of objects, we find them not ideal for cell detection in microscopy images. Although appealing to learn end-to-end the classification directly from the input images, s.t. the network jointly learns region of interest (RoI) selection *and* classification, for cytology WSIs this is rather impractical. The classification task is very difficult and requires tens of thousands of cells to reach top performance, while a per-cell RoI detection is much easier to train (much fewer annotated cell locations are needed), requires less detail and can be performed at lower resolution (thus faster). To jointly train localization and classification would require the (manual) localization of the full tens of thousands of cells. Our proposal, relying on patient-level annotations for the difficult classification task, reaches good performance using only around 1000 manually marked cell locations. Methods for detecting objects with various size and the bounding boxes also cost unnecessary computation, since all cell nuclei are of similar size and bounding box is not of interest in diagnosis. Further, these methods tend to not handle very large numbers of small and clustered objects very well [36].

Many DL-based methods specifically designed for the task of nucleus detection are similar to the framework summarized in [12]: first generate a probability map by sliding a binary patch classifier over the whole image, then find nuclei positions as local maxima. However, considering that WSIs are as large as 10 giga-pixels, this approach is prohibitively slow. U-Net models avoid the sliding window and reduce computation time. Detection is performed as segmentation where each nucleus is marked as a binary disk [4]. However, when images are noisy and with densely packed nuclei, the binary output mask is not ideal for further processing. We find the regression approach [16,33,34], where the network is trained to reproduce fuzzy nuclei markers, to be more appropriate.

Focus Selection: In cytological analysis, the focus level has to be selected for each nucleus individually, since different cells are at different depth. Standard tools (e.g., the microscope auto-focus) fail since they only provide a large field-of-view optimum, and often focus on clumps or other artifacts. Building on the approaches of Just Noticeable Blur (JNB) [6] and Cumulative Probability of Blur Detection (CPBD) [19], the Edge Model based Blur Metric (EMBM) [9] provides a no-reference blur metric by using a parametric edge model to detect and describe edges with both contrast and width information. It claims to achieve comparable results to the former while being faster.

Classification: Deep learning has successfully been used for different types cell classification [10] and for cervical cancer screening in particular [35]. Convolutional Neural Networks (CNNs) have shown ability to differentiate healthy and malignant cell samples [32]. Whereas the approach in [32] relies on manually selected free lying cells, our study proposes to use automatic cell detection. This allows improved performance by scaling up the available data to *all* free lying cells in each sample.

3 Materials and Methods

3.1 Data

Three sets of images of oral brush samples are used in this study. **Dataset 1** is a relatively small Pap smear dataset imaged with a standard microscope. **Dataset 2** consist of WSIs of the same glass slides as Dataset 1. **Dataset 3** consist of WSIs of liquid-based (LBC) prepared slides. All samples are collected at Dept. of Orofacial Medicine, Folktandvården Stockholms län AB. From each patient, samples were collected with a brush scraped at areas of interest in the oral cavity. Each scrape was either smeared onto a glass (Datasets 1 and 2) or placed in a liquid vial (Dataset 3). All samples were stained with standard Papanicolau stain. Dataset 3 was prepared with Hologic T5 ThinPrep Equipment and standard non-gynecologic protocol. **Dataset 1** was imaged with an Olympus BX51 bright-field microscope with a 20×, 0.75 NA lens giving a pixel size of $0.32\,\mu m$. From 10 Pap smears (10 patients), free lying cells (same as in "Oral Dataset 1" in [32]) are manually selected and $80 \times 80 \times 1$ grayscale patches are extracted, each with one centered in-focus cell nucleus. **Dataset 2**: The same 10 slides as in Dataset 1 were imaged using a NanoZoomer S60 Digital slide scanner, 40×, 0.75 NA objective, at 11 z-offsets ($\pm 2\,\mu m$, step-size $0.4\,\mu m$) providing RGB WSIs of size $103936 \times 107520 \times 3$, $0.23\,\mu m$/pixel. **Dataset 3** was obtained in the same way as Dataset 2, but from 12 LBC slides from 12 other patients.

Slide level annotation and reliance on MAC appears as a useful way to avoid need for large scale very difficult manual cell level annotations. Both [15] and [32] demonstrate promising results for MAC detection in histology and cytology. In our work we therefore aim to classify cells based on the patient diagnosis, i.e., all cells from a patient with diagnosed oral cancer are labeled as cancer.

3.2 Nucleus Detection

The nucleus detection step aims to efficiently detect each individual cell nucleus in WSIs. The detection is inspired by the Fully Convolutional Regression Networks (FCRNs) approach proposed in [33] for cell counting. The main steps of the method are described below, and illustrated on an example image from Dataset 3, Fig. 1.

Training: Input is a set of RGB images $I_i, i = 1 \ldots K$, and corresponding binary annotation masks B_i, where each individual nucleus is indicated by one centrally located pixel.

Each ground truth mask is dilated by a disk of radius r [4], followed by convolution with a 2D Gaussian filter of width σ. By this, a fuzzy ground truth is generated. A fully convolutional network is trained to learn a mapping between the original image I (Fig. 1a) and the corresponding "fuzzy ground truth", D (Fig. 1b). The network follows the architecture of U-Net [24] but with the final softmax replaced by a linear activation function.

Inference: A corresponding density map D' (Fig. 1c) is generated (predicted) for any given test image I. The density map D' is thresholded at a level T and centroids of the resulting blobs indicate detected nuclei locations (Fig. 1d).

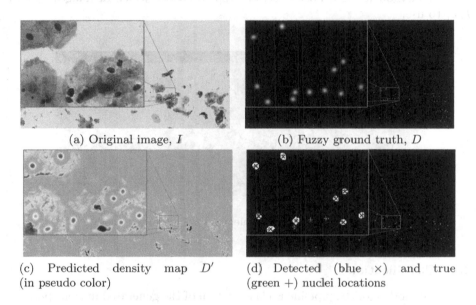

(a) Original image, I

(b) Fuzzy ground truth, D

(c) Predicted density map D' (in pseudo color)

(d) Detected (blue ×) and true (green +) nuclei locations

Fig. 1. A sample image at different stages of nucleus detection (Color figure online)

3.3 Focus Selection

Slide scanners do not provide sufficiently good focus for cytological samples and a focus selection step is needed. Our proposed method utilizes N equidistant z-levels acquired of the same specimen. Traversing the z-levels, the change between consecutive images shows the largest variance at the point where the specimen moves in/out of focus. This novel focus selection approach provides a clear improvement over the Edge Model based Blur Metric (EMBM) proposed in [9].

Following the Nucleus detection step (which is performed at the central focus level, $z = 0$) we cut out a square region for each detected nucleus at all acquired focus levels. Each such cutout image is filtered with a small median filter of size $m \times m$ on each color channel to reduce noise. This gives us a set of images P_i, $i = 1, \ldots, N$, of an individual nucleus at the N consecutive z-levels. We compute the difference of neighboring focus levels, $P'_i = P_{i+1} - P_i$, $i = 1, \ldots, N-1$. The variance, σ_i^2, is computed for each difference image P'_i:

$$\sigma_i^2 = \frac{1}{M} \sum_{j=1}^{M} \left(p'_{ij} - \mu_i \right)^2 , \text{ where } \mu_i = \frac{1}{M} \sum_{j=1}^{M} p'_{ij} ,$$

M is the number of pixels in P_i', and p_{ij}' is the value of pixel j in P_i'. Finally the level l corresponding to the largest σ_i^2 is selected,

$$l = \underset{i=1,\dots,N-1}{\arg\max} \ \sigma_i^2 .$$

To determine which of the two images in the pair P_i' is in best focus, we use the EMBM method [9] as a post selection step to choose which of images P_l and P_{l+1} to use.

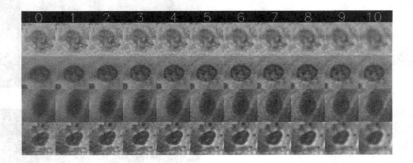

Fig. 2. Example of focus sequences for experts to annotate

3.4   Classification

The final module of the pipeline is classification of the generated nucleus patches into two classes – cancer and healthy. Following recommendation from [32], we evaluate ResNet50 [11] as a classifier. We also include the more recent DenseNet201 [13] architecture. In addition to random (Glorot-uniform) weight initialization, we also evaluate the two architectures using weights pre-trained on ImageNet.

Considering that texture information is a key feature for classification [15,31], the data is augmented without interpolation. During training, each sample is reflected with 50% probability and rotated by a random integer multiple of 90°.

4   Experimental Setup

4.1   Nucleus Detection

The WSIs at the middle z-level ($z = 0$) are used for nucleus detection. Each WSI is split into an array of $6496 \times 3360 \times 3$ sub-images using the Open Source tool ndpisplit [3]. The model is trained on 12 and tested on 2 sub-images (1014 resp. 119 nuclei) from Dataset 3. The manually marked ground truth is dilated by a disk of radius $r = 15$. All images, including ground truth masks, are resized to 1024×512 pixels, using area weighted interpolation. A Gaussian filter, $\sigma = 1$, is applied to each ground truth mask providing the fuzzy ground truth D.

Each image is normalized by subtracting the mean and dividing by the standard deviation of the training set. Images are augmented by random rotation in the range $\pm 30°$, random horizontal and vertical shift within 30% of the total scale, random zoom within the range of 30% of the total size, and random horizontal and vertical flips. Nucleus detection does not need the texture details, so interpolation does not harm. To improve stability of training, batch normalization [14] is added before each activation layer. Training is performed using RMSprop with mean squared error as loss function, learning rate $\alpha = 0.001$ and decay rate $\rho = 0.9$. The model is trained with mini-batch size 1 for 100 epochs, the checkpoint with minimum training loss is used for testing.

Performance of nucleus detection is evaluated on Dataset 3. A detection is considered correct if its closest ground truth nucleus is within the cropped patch *and* that ground truth nucleus has no closer detections (s.t. one true nucleus is paired with at most one detection).

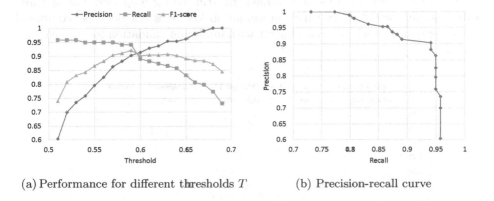

(a) Performance for different thresholds T (b) Precision-recall curve

Fig. 3. Results of nucleus detection

4.2 Focus Selection

100 detected nuclei are randomly chosen from the two test sub-images (Dataset 3). Every nucleus is cut to an $80 \times 80 \times 3$ patch for each of the 11 z-levels. For EMBM method the contrast threshold of a salient edge is set to $c_T = 8$, following [9].

To evaluate the focus selection, 8 experts are asked to choose the best of the 11 focus-levels for each of the 100 nuclei (Fig. 2). The median of the 8 assigned labels is used as true best focus, l_{GT}. A predicted focus level l is considered accurate enough if $l \in [l_{GT} - 2, l_{GT} + 2]$.

4.3 Classification

The classification model is evaluated on Dataset 1 as a benchmark, and then on Dataset 2, to evaluate effectiveness of the nucleus detection and focus selection modules in comparison with the performance on Dataset 1. The model is also run on Dataset 3 to validate generality of the pipeline. Datasets are split on a patient level; no cell from the same patient exists in both training and test sets. On Dataset 1 and 2, three-fold validation is used, following [32]. On Dataset 3, two-fold validation is used. Our trained nucleus detector with threshold $T = 0.59$ (best-performing in Sect. 5.1) is used for Dataset 2 and 3 to generate nucleus patches. Some cells in Dataset 2 and 3 lie outside the $\pm 2\,\mu$m imaged z-levels, and the best focus is still rather blurred. We use the EMBM to exclude the most blurred ones. Cell patches with an EMBM score < 0.03 are removed, leaving 68509 cells for Dataset 2 and 130521 for Dataset 3.

We use Adam optimizer, cross-entropy loss and parameters as suggested in [17], i.e., initial learning rate $\alpha = 0.001$, $\beta_1 = 0.9$, $\beta_2 = 0.999$. 10% of the training set is randomly chosen as validation set.

When using models pre-trained on ImageNet, since the weights require three input channels, the grayscale images from Dataset 1 are duplicated into each channel. Pre-trained models are trained (fine-tuned) for 5 epochs. The learning rate is scaled by 0.4 every time the validation loss does not decrease compared to the previous epoch. The checkpoint with minimum validation loss is saved for testing.

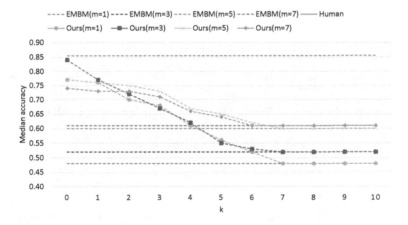

Fig. 4. Accuracy of focus selection

A slightly different training strategy is used when training from scratch. ResNet50 models are trained with mini-batch size 512 for 50 epochs on Dataset 2 and 3, and with mini-batch size 128 for 30 epochs on Dataset 1, since it contains fewer samples. Because DenseNet201 takes larger GPU memory, mini-batch sizes are set to 256 on Dataset 2 and 3. To mitigate overfitting, DenseNet201 models are trained for only 30 epochs on Dataset 2 and 3, and 20 epochs on Dataset 1. When the validation loss has not decreased for 5 epochs, the learning rate is scaled by 0.1. Training is stopped after 15 epochs of no improvement. The checkpoint with minimum validation loss is saved for testing.

5   Results and Discussion

5.1   Nucleus Detection

Results of nucleus detection are presented in Fig. 3. Figure 3a shows Precision, Recall, and F1-score as the detection threshold T varies in $[0.51, 0.69]$. At $T = 0.59$, F1-score reaches 0.92, with Precision and Recall being 0.90 and 0.94 respectively. Using $T = 0.59$, 94,685 free lying nuclei are detected in Dataset 2 and 138,196 in Dataset 3.

The inference takes 0.17 s to generate a density map D' of size 1024×512 on an NVIDIA GeForce GTX 1060 Max-Q. To generate a density map of the same size based on the sliding window approach (Table 4 of [12]), takes 504 s.

5.2   Focus Selection

Performance of the focus selection is presented in Fig. 4. The "human" performance is the average of the experts, using a leave-one-out approach. We plot performance when using EMBM to select among the $2(k + 1)$ levels closest to our selected pair l; for increasing k the method approaches EMBM.

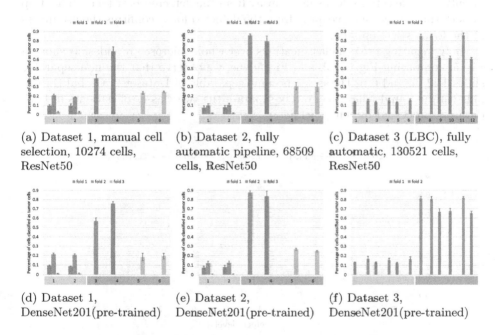

(a) Dataset 1, manual cell selection, 10274 cells, ResNet50

(b) Dataset 2, fully automatic pipeline, 68509 cells, ResNet50

(c) Dataset 3 (LBC), fully automatic, 130521 cells, ResNet50

(d) Dataset 1, DenseNet201(pre-trained)

(e) Dataset 2, DenseNet201(pre-trained)

(f) Dataset 3, DenseNet201(pre-trained)

Fig. 5. Cell classification results per microscope slide; green samples (bars to the left) are healthy, red samples (bars to the right) are from cancer patients. ResNet50 is used for (a)–(c) and DenseNet201 pre-trained on ImageNet is used for (d)–(f). (Color figure online)

It can be seen that EMBM alone does not achieve satisfying performance on this task. Applying a median filter improves the performance somewhat. Our proposed method performs very well on the data and is essentially at the level of a human expert (accuracy 84% vs. 85.5%, respectively) using $k = 0$ and a 3×3 median filter.

5.3 Classification

Classification performance is presented in Table 1 and Fig. 5. The two architectures (ResNet50 and DenseNet201) perform more or less equally well. Pretraining seems to help a bit for the smaller Dataset 1, whereas for the larger Datasets 2 and 3 no essential difference is observed. Results on Dataset 2 are consistently better than on Dataset 1. This confirms effectiveness of the nucleus detection and focus selection modules; by using more nuclei (from the same samples) than those manually selected, improved performance is achieved. The results on Dataset 3 indicate that the pipeline generalizes well to liquid-based images. We also observe that our proposed pipeline is robust w.r.t. network architectures and training strategies of the classification.

In Fig. 6 we plot how classification performance decreases when nuclei are intentionally selected n focus levels away from the detected best focus. The drop in performance as we move away from the detected focus confirms the usefulness of the focus selection step.

If aggregating the cell classifications over whole microscopy slides, as show in Fig. 5, comparing Fig. 5a–5b and Fig. 5d–5e, we observe that the non-separable slides 1, 2, 5, and 6 in Dataset 1 become separable in Dataset 2. Global thresholds can be found which accurately separate the two classes of patients in both datasets processed by our pipeline.

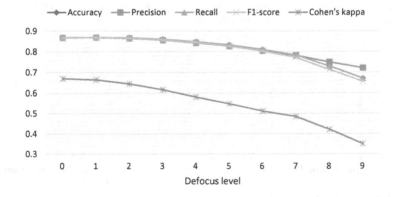

Fig. 6. The impact of defocused testset on Dataset 2, fold 1 (ResNet50)

Table 1. Classification performance. The best F1-score for each dataset is presented in bold.

Dataset	Network	Accuracy	Precision	Recall	F1-score
1	ResNet50	70.5 ± 0.5	63.1 ± 1.2	34.8 ± 1.4	44.8 ± 1.3
	ResNet50 (pre-trained)	72.0 ± 0.9	66.4 ± 2.0	37.5 ± 2.0	**48.0 ± 2.1**
	DenseNet201	70.4 ± 0.5	63.1 ± 1.8	33.8 ± 0.9	44.0 ± 0.7
	DenseNet201 (pre-trained)	70.6 ± 0.7	63.4 ± 1.6	34.3 ± 1.7	44.5 ± 1.8
2	ResNet50	74.4 ± 1.9	83.3 ± 2.9	46.3 ± 3.8	59.5 ± 3.8
	ResNet50 (pre-trained)	74.0 ± 0.1	83.9 ± 0.5	44.6 ± 0.7	58.2 ± 0.5
	DenseNet201	75.4 ± 0.8	84.3 ± 1.5	48.3 ± 1.1	**61.4 ± 1.3**
	DenseNet201 (pre-trained)	73.3 ± 0.7	81.7 ± 2.8	44.4 ± 0.3	57.5 ± 0.6
3	ResNet50	81.6 ± 0.7	71.7 ± 1.2	73.8 ± 0.9	**72.8 ± 1.0**
	ResNet50 (pre-trained)	81.3 ± 1.5	72.1 ± 3.0	71.6 ± 0.6	71.8 ± 1.8
	DenseNet201	81.3 ± 0.5	71.4 ± 0.7	73.0 ± 0.8	72.2 ± 0.7
	DenseNet201 (pre-trained)	81.5 ± 1.3	71.2 ± 2.4	74.5 ± 2.4	72.8 ± 1.9

6 Conclusion

This work presents a complete fully automated pipeline for oral cancer screening on whole slide images; source code (utilizing TensorFlow 1.14) is shared as open source. The proposed focus selection method performs at the level of a human expert and significantly outperforms EMBM. The pipeline can provide fully automatic inference for WSIs within reasonable computation time. It performs well for smears as well as liquid-based slides.

Comparing the performance on Dataset 1, using human selected nuclei and Dataset 2, using computer selected nuclei from the same microscopy slides, we conclude that the presented pipeline can reduce human workload while at the same time make the classification easier and more reliable.

References

1. Campanella, G., et al.: Clinical-grade computational pathology using weakly supervised deep learning on whole slide images. Nat. Med. **25**(8), 1301–1309 (2019)
2. Cruz-Roa, A., et al.: Accurate and reproducible invasive breast cancer detection in whole-slide images: a deep learning approach for quantifying tumor extent. Sci. Rep. **7**, 46450 (2017)
3. Deroulers, C., Ameisen, D., Badoual, M., Gerin, C., Granier, A., Lartaud, M.: Analyzing huge pathology images with open source software. Diagnostic Pathol. **8**(1), 92 (2013)
4. Falk, T., et al.: U-Net: deep learning for cell counting, detection, and morphometry. Nat. Methods **16**(1), 67–70 (2019)
5. Farahani, N., Parwani, A., Pantanowitz, L.: Whole slide imaging in pathology: advantages, limitations, and emerging perspectives. Pathol. Lab Med. Int. **7**, 23–33 (2015)

6. Ferzli, R., Karam, L.: A no-reference objective image sharpness metric based on the notion of just noticeable blur (JNB). IEEE Trans. Image Process. **18**(4), 717–728 (2009)
7. Girshick, R.: Fast R-CNN. arXiv:1504.08083 [cs], September 2015
8. Girshick, R., Donahue, J., Darrell, T., Malik, J.: Rich feature hierarchies for accurate object detection and semantic segmentation. arXiv:1311.2524 [cs], October 2014
9. Guan, J., Zhang, W., Gu, J., Ren, H.: No-reference blur assessment based on edge modeling. J. Vis. Commun. Image Represent. **29**, 1–7 (2015)
10. Gupta, A., et al.: Deep learning in image cytometry: a review. Cytometry Part A **95**(4), 366–380 (2019)
11. He, K., Zhang, X., Ren, S., Sun, J.: Deep residual learning for image recognition. arXiv:1512.03385 [cs], December 2015
12. Höfener, H., Homeyer, A., Weiss, N., Molin, J., Lundström, C., Hahn, H.: Deep learning nuclei detection: a simple approach can deliver state-of-the-art results. Comput. Med. Imag. Graph. **70**, 43–52 (2018)
13. Huang, G., Liu, Z., van der Maaten, L., Weinberger, K.Q.: Densely connected convolutional networks. arXiv:1608.06993 [cs], January 2018
14. Ioffe, S., Szegedy, C.: Batch normalization: accelerating deep network training by reducing internal covariate shift. arXiv:1502.03167 [cs] (2015)
15. Jabalee, J., et al.: Identification of malignancy-associated changes in histologically normal tumor-adjacent epithelium of patients with HPV-positive oropharyngeal cancer. Anal. Cellular Pathol. **2018**, 1–9 (2018)
16. Kainz, P., Urschler, M., Schulter, S., Wohlhart, P., Lepetit, V.: You should use regression to detect cells. In: Navab, N., Hornegger, J., Wells, W.M., Frangi, A.F. (eds.) MICCAI 2015. LNCS, vol. 9351, pp. 276–283. Springer, Cham (2015). https://doi.org/10.1007/978-3-319-24574-4_33
17. Kingma, D., Ba, J.: Adam: a method for stochastic optimization. arXiv:1412.6980 [cs], December 2014
18. Korbar, B., et al.: Deep learning for classification of colorectal polyps on whole-slide images. J. Pathol. Inform. **8**, 30 (2017)
19. Narvekar, N.D., Karam, L.J.: A no-reference image blur metric based on the cumulative probability of blur detection (CPBD). IEEE Trans. Image Process. **20**(9), 2678–2683 (2011)
20. Redmon, J., Divvala, S., Girshick, R., Farhadi, A.: You only look once: unified, real-time object detection. In: Proceedings of IEEE Conference on CVPR, pp. 779–788 (2016)
21. Redmon, J., Farhadi, A.: YOLO9000: better, faster, stronger. In: Proceedings of IEEE Conference on CVPR, pp. 7263–7271 (2017)
22. Redmon, J., Farhadi, A.: YOLOv3: An Incremental Improvement. arXiv:1804.02767 [cs], April 2018
23. Ren, S., He, K., Girshick, R., Sun, J.: Faster R-CNN: towards real-time object detection with region proposal networks. arXiv:1506.01497 [cs], January 2016
24. Ronneberger, O., Fischer, P., Brox, T.: U-Net: convolutional networks for biomedical image segmentation. In: Navab, N., Hornegger, J., Wells, W., Frangi, A. (eds.) Medical Image Computing and Computer Assisted Intervention MICCAI 2015, pp. 234–241. Springer, Cham (2015). https://doi.org/10.1007/978-3-319-24574-4_28
25. Sankaranarayanan, R., et al.: Long term effect of visual screening on oral cancer incidence and mortality in a randomized trial in Kerala. India. Oral Oncol. **49**(4), 314–321 (2013)

26. Speight, P., et al.: Screening for oral cancer—a perspective from the global oral cancer forum. Oral Surg., Oral Med. Oral Pathol. Oral Radiol. **123**(6), 680–687 (2017)
27. Stewart, B., Wild, C.P., et al.: World cancer report 2014. Public Health (2014)
28. Teramoto, A., et al.: Automated classification of benign and malignant cells from lung cytological images using deep convolutional neural network. Inform. Med. Unlocked **16**, 100205 (2019)
29. Us-Krasovec, M., et al.: Malignancy associated changes in epithelial cells of buccal mucosa: a potential cancer detection test. Anal. Quantit. Cytol. Histol. **27**(5), 254–262 (2005)
30. Warnakulasuriya, S.: Global epidemiology of oral and oropharyngeal cancer. Oral Oncol. **45**(4–5), 309–316 (2009)
31. Wetzer, E., Gay, J., Harlin, H., Lindblad, J., Sladoje, N.: When texture matters: texture-focused CNNs outperform general data augmentation and pretraining in oral cancer detection. In: Proceedings of IEEE International Symposium on Biomedical Imaging (ISBI) (2020) forthcoming
32. Wieslander, H., et al.: Deep convolutional neural networks for detecting cellular changes due to malignancy. In: 2017 IEEE International Conference on Computer Vision Workshops (ICCVW), pp. 82–89. IEEE, October 2017
33. Xie, W., Noble, J., Zisserman, A.: Microscopy cell counting and detection with fully convolutional regression networks. Comput. Methods Biomech. Biomed. Eng. Imag. Visual. **6**(3), 283–292 (2018)
34. Xie, Y., Xing, F., Kong, X., Su, H., Yang, L.: Beyond classification: structured regression for robust cell detection using convolutional neural network. In: Navab, N., Hornegger, J., Wells, W.M., Frangi, A.F. (eds.) MICCAI 2015. LNCS, vol. 9351, pp. 358–365. Springer, Cham (2015). https://doi.org/10.1007/978-3-319-24574-4_43
35. Zhang, L., Lu, L., Nogues, I., Summers, R.M., Liu, S., Yao, J.: DeepPap: deep convolutional networks for cervical cell classification. IEEE J. Biomed. Health Inform. **21**(6), 1633–1643 (2017)
36. Zou, Z., Shi, Z., Guo, Y., Ye, J.: Object detection in 20 years: a survey. arXiv:1905.05055 [cs], May 2019

Studying the Effect of Digital Stain Separation of Histopathology Images on Image Search Performance

Alison K. Cheeseman[1], Hamid R. Tizhoosh[2], and Edward R. Vrscay[1]([✉])

[1] Department of Applied Mathematics, Faculty of Mathematics,
University of Waterloo, Waterloo, ON N2L 3G1, Canada
{alison.cheeseman,ervrscay}@uwaterloo.ca
[2] Kimia Lab, University of Waterloo, Waterloo, ON N2L 3G1, Canada
hamid.tizhoosh@uwaterloo.ca

Abstract. Due to recent advances in technology, digitized histopathology images are now widely available for both clinical and research purposes. Accordingly, research into computerized image analysis algorithms for digital histopathology images has been progressing rapidly. In this work, we focus on image retrieval for digital histopathology images. Image retrieval algorithms can be used to find similar images and can assist pathologists in making quick and accurate diagnoses. Histopathology images are typically stained with dyes to highlight features of the tissue, and as such, an image analysis algorithm for histopathology should be able to process colour images and determine relevant information from the stain colours present. In this study, we are interested in the effect that stain separation into their individual stain components has on image search performance. To this end, we implement a basic k-nearest neighbours (kNN) search algorithm on histopathology images from two publicly available data sets (IDC and BreakHis) which are: a) converted to greyscale, b) digitally stain-separated and c) the original RGB colour images. The results of this study show that using H&E separated images yields search accuracies within one or two percent of those obtained with original RGB images, and that superior performance is observed using the H&E images in most scenarios we tested.

Keywords: Digital histopathology · Encoded Local Projections (ELP) · Digital stain separation · Digital image retrieval and classification

1 Introduction

Histopathology, the examination of tissue under a microscope to study biological structures related to disease manifestation, has traditionally been carried out

This research has been supported in part by the Natural Sciences and Engineering Research of Canada (NSERC) in the form of a Doctoral Scholarship (AKC) and a Discovery Grant (ERV).

A. Campilho et al. (Eds.): ICIAR 2020, LNCS 12132, pp. 262–273, 2020.
https://doi.org/10.1007/978-3-030-50516-5_23

manually by pathologists working in a lab. It is only in recent years that the technology has advanced to a point which allows for the rapid digitization and storage of whole slide images (WSIs). Consequently, digitized histopathology images are now widely available for both clinical and research purposes, and computerized image analysis for digital histopathology has quickly become an active area of research [8,11]. In this paper, we focus specifically on content-based image retrieval (CBIR) for histopathology images, which involves finding images which share the same visual characteristics as the query image. The identification and analysis of similar images can assist pathologists in quickly and accurately obtaining a diagnosis by providing a baseline for comparison.

While most radiology images (X-ray, CT, etc.) are greyscale, histopathology images are typically stained with dyes to highlight certain features of the tissue. In order to properly use the relevant colour information, a WSI analysis system should be able to process colour images and determine relevant biological information from the presence of different stain colours. In a previous work [4], we introduced a new frequency-based ELP (F-ELP) image descriptor for histopathology image search, which captures local frequency information and implemented this new descriptor on images which were separated into two colour channels based on chemical stain components using a digital stain separation algorithm. In [4], we found that both the ELP and F-ELP descriptors saw improved search accuracy when applied to the stain-separated images, as opposed to single-channel greyscale images. In this paper, we focus on studying the effectiveness of digital stain separation of histopathology images for image retrieval applications using a number of common handcrafted image descriptors. We compare the results of image retrieval for the stain-separated images to the results of the same experiment conducted on both greyscale and colour (three-channel RGB) images. Experiments are conducted using two publicly available breast cancer histopathology data sets, IDC and BreakHis.

2   Digital Stain Separation

The most common staining protocol for histology slides involves two dyes, namely hematoxylin and eosin (H&E). The hematoxylin stains cell nuclei blue, and eosin stains other structures varying shades of red and pink [16]. The colours which appear in a slide, and the size, shape and frequency at which they appear are all relevant factors a pathologist may assess when making a diagnosis. For this reason, we consider separating the input images into two stain components prior to the computation of an image descriptor.

In this paper, as in [4], we adopt the stain separation method proposed in [12], an extension of the wedge finding method from [10]. Unlike some previous methods for stain separation [16], this method does not require any calibration or knowledge of the exact stain colours, instead it works by using the available image data to estimate an H&E basis. Given that an image search algorithm should ultimately be applied to data from multiple sources, this is a desirable feature for the stain separation algorithm.

Figure 1 shows two sample images from the BreakHis data set which have been separated in their hematoxylin and eosin components. We can see that the algorithm is able to effectively separate the two components of both images, even though the stains appear as noticeably different colours in each image.

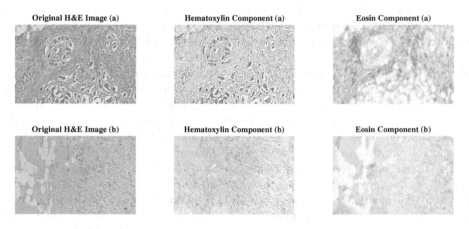

Fig. 1. Two sample images from the BreakHis dataset showing the resulting hematoxylin and eosin components after applying the stain separation algorithm from [12].

3   Proposed Study

The main purpose of this study is to investigate the effectiveness of digital stain separation in isolation from the other parameters of the image search process. To this end, we implement a basic image search algorithm, k-Nearest Neighbours (kNN), using input images that are: a) converted to greyscale (one colour channel), b) separated into their H&E components (two colour channels) by the method described above, and c) the original RGB colour images (three colour channels). For each colour channel of an input image having N total channels, an image descriptor, $\mathbf{h}_i$ is computed, and concatenated to form the final descriptor $\mathbf{h} = [\mathbf{h}_1 \ ... \ \mathbf{h}_N]$, which represents the entire image. We then implement the kNN algorithm for image search using a number of well-known distance functions to determine "nearest" neighbours and four different image descriptors of varying lengths and properties as inputs to the algorithm. The image descriptors and distance functions used are described in the following sections.

3.1   Image Descriptors

Feature extraction, or the computation of compact image descriptors, is an important part of many image analysis tasks, including image retrieval. As such, there has been plenty of research over the years into the design of image descriptors for various imaging applications. Some of the most well known image descriptors include: local binary patterns (LBP) [1], the scale-invariant feature transform (SIFT) [9], speeded-up robust features (SURF) [3], and histograms of oriented gradients (HOG) [6]. Most of these methods, including SIFT, SURF, and HOG, perform well in more traditional applications such as object detection or tracking and face recognition, but perform poorly compared to LBP for the retrieval and classification of histopathology images [2,18]. LBP, which is based on computing binary patterns in local regions of the image, is generally thought to be a better image descriptor for textures, which may explain its superior performance on high resolution histopathology images, which resemble textures more than natural images.

In our study, we implement four image descriptors, including the LBP descriptor, along with the Gabor filter-based GIST descriptor [15] which computes the spatial envelope of a scene, the ELP descriptor (encoded local projections) from [18] which was designed with medical images in mind, and our proposed F-ELP descriptor from [4], designed specifically to be a compact descriptor for histopathology images.

3.2 Distance Functions for Image Search

In any image search algorithm, it is important to properly define what makes two images similar. Typically, that means one must choose an appropriate distance function between image descriptors. Six different distance functions were used in this study to determine the nearest neighbours for the kNN search, including the well-known L_1, L_2, cosine, correlation, and chi-squared metrics. We also consider the Hutchinson (also known as Monge-Kantorovich) distance [13], as it is thought to be a good measure of distance between histograms. In the finite one-dimensional case, the Hutchinson distance can be computed in linear-time using the method from [14].

4   Data Sets and Image Preprocessing

In this study, we used two publicly available data sets containing breast cancer histopathology images: IDC and BreakHis.

Invasive Ductal Carcinoma (IDC) Kaggle Data: The IDC dataset consists of digitized breast cancer slides from 162 patients diagnosed with IDC at the University of Pennsylvania Hospital and the Cancer Institute of New Jersey [5] (Fig. 2). Each slide was digitized at 40x magnification and downsampled to a resolution of $4\,\mu$m/pixel. The dataset provides each WSI split into patches of size 50 px × 50 px in RGB colour space. The supplied data was randomly split into three different subsets of 84 patients for training, 29 for validation and 49 test cases for final evaluation. Ground truth annotation regarding the presence of IDC in each patch was obtained by manual delineation of cancer regions performed by expert pathologists.

Due to their small size, each individual image patch in the IDC data set may not contain both hematoxylin and eosin stains. Since the stain separation algorithm learns the stain colours from the data, both stains must be present in the image for accurate results. To ensure good performance on all image patches, we use the entire WSI to perform stain separation and then split the image back into the original patches to compute image descriptors. One further issue is that the stain separation algorithm used assumes that two (and only two) stain components (H&E in our case) exist in the image. However, some images are observed to have significant discolouration, such as large dark patches, and the introduction of other colours not caused by H&E staining. The prevalence of such artefacts negatively impacts the ability of the stain separation algorithm to provide good results for some patients, so we remove them by searching for images which have minimal variation in the RGB channels across the entire image. A total of 686 patches were flagged and removed from the total data set, all of which contain significant artefacts or discoloration.

Breast Cancer Histopathology Database (Breakhis): The Breast Cancer Histopathology Database (BreakHis) [17] was built as a collaboration between researchers at the Federal University of Parana (UFPR) and the P&D Laboratory - Pathological Anatomy and Cytopathology, in Parana, Brazil (Fig. 3). To date, it contains 9,109 images of breast tumour tissue from 82 patients using four different magnification factors: 40×, 100×, 200×, and 400×. The images are provided in PNG format (3-channel RGB, 8-bit depth/channel) and are 700×460 px. The data is divided into two classes, benign tumours and malignant tumours, with class labels provided by pathologists from the P&D Laboratory. Within each class, further labelling is provided to indicate tumour types. The data set consists of four histologically distinct benign tumours and four malignant tumour types. These additional intra-class labels are not used in the current study.

For this particular study, we use only the images taken at 40× magnification. This subset of the data contains 1,995 images, of which 652 are benign and 1,370 are malignant. Using code provided by the authors of [17], the data is split into a training (70%) and testing (30%) set with the condition that patients in the training set are not used for the testing set. The results presented in this paper are the average of five trials, using the five data folds from [17].

Fig. 2. Sample patches from the IDC data set. The top row shows negative examples (healthy tissue or non-invasive tumour tissue) and the bottom row shows positive examples (IDC tissue).

Fig. 3. Example of image patches from the BreakHis data set. The top two rows show examples of benign tumours and the bottom two rows show malignant tumours.

4.1 Accuracy Calculations

IDC: For consistency with previous works involving the IDC data, we use both the balanced accuracy (BAC) and F-measure (F1), which are defined as follows [5]:

$$\text{BAC} = \frac{\text{Sen} + \text{Spc}}{2}, \quad F1 = \frac{2 \cdot \text{Pr} \cdot \text{Rc}}{\text{Pr} + \text{Rc}}, \tag{1}$$

where Sen is sensitivity, Spc is specificity, Pr is precision and Rc, recall.

BreakHis: For the BreakHis data we compute patient scores and the global recognition rate, which were introduced in [17]. If we let N_P be the number of images of patient P and N_{rec} be the number of images of patient P that are correctly classified, then the patient score for patient P is defined as

$$\text{Patient Score} = \frac{N_{\text{rec}}}{N_P} \tag{2}$$

and the global recognition rate (GRR) as

$$\text{Global Recognition Rate} = \frac{\sum \text{Patient scores}}{\text{Total number of patients}}. \tag{3}$$

In addition to the global recognition rate we also compute the balanced accuracy as defined above in (1).

5   Experiment

In order to evaluate image retrieval performance we implement the kNN algorithm in MATLAB with each set of image descriptors as inputs. The kNN algorithm searches through the training data partition and classifies each image in the test set based on the class of its k nearest neighbours. Since there is no exact metric to quantitatively evaluate image retrieval performance, we measure the accuracy of classification using kNN. In this work, we test the kNN algorithm using three different values for k ($k = 1, 5$ and 15).

Table 1. A list of image descriptors used in this study and the corresponding number of features computed (i.e. the length of the feature vector).

Descriptor	Number of features		
	Greyscale	H&E stains	RGB image
ELP	1024	2048	3072
GIST	512	1024	1536
F-ELP	32	64	96
LBP	18	36	54

Table 1 lists the image descriptors used and their respective lengths on each set of input colour channels. We can see that as we increase the number of input colour channels from one to three, the length of the feature vectors increases. Given that the computation time for the kNN search algorithm has linear dependency on feature vector length [7], it is clear that for a fixed image descriptor, it is desirable to use fewer colour channels, so long as the overall search performance does not suffer significantly.

The length of each image descriptor is dependent on certain parameters of the algorithm. In this work the following parameters are used: the ELP and F-ELP descriptors are implemented with a window size of $w = 9$, the GIST descriptor, by default, divides the image into a 4×4 grid and uses a filter bank of 32 Gabor filters, and the LBP descriptor is computed with a radius of $R = 2$ and $P = 16$ neighbouring pixels. As a result, we have a wide variety of descriptor lengths, from the very short LBP descriptor to the long ELP histogram.

5.1   Comparing Input Image Colour Channels

In this section, we present results which compare the image search performance using greyscale, H&E stain-separated, and RGB images as inputs. For each descriptor, and each set of input colour channels, the best accuracy, taken over all distance functions, is presented. It should be noted here that there are some slight discrepancies between the results presented here for the IDC data set and those in our previous work [4]. This is due to a small error which was found in the code which slightly changes the numerical results, but does not change the overall conclusions of the previous study.

Table 2. The best KNN search ($k = 1$) accuracy for the IDC dataset taken over all distance functions. The top result in each column is highlighted in bold.

Colour channels	ELP		GIST		F-ELP		LBP	
	F1	BAC	F1	BAC	F1	BAC	F1	BAC
Greyscale	0.3987	0.5905	0.5086	0.6500	0.4183	0.6023	0.4625	0.6280
H&E stains	**0.4528**	**0.6235**	**0.5549**	**0.6923**	**0.5565**	**0.6932**	0.5860	0.7130
RGB image	0.4504	0.6219	0.5513	0.6890	0.5419	0.6836	**0.5926**	**0.7187**

Table 3. The best KNN search ($k = 5$) accuracy for the IDC dataset taken over all distance functions. The top result in each column is highlighted in bold.

Colour channels	ELP		GIST		F-ELP		LBP	
	F1	BAC	F1	BAC	F1	BAC	F1	BAC
Greyscale	0.4001	0.6080	0.5598	0.6968	0.4338	0.6208	0.5124	0.6645
H&E stains	**0.4880**	**0.6531**	**0.6052**	**0.7356**	**0.6303**	**0.7406**	0.6618	0.7614
RGB image	0.4832	0.6504	0.5918	0.7270	0.6028	0.7222	**0.6785**	**0.7750**

Table 4. The best KNN search ($k = 15$) accuracy for the IDC dataset taken over all distance functions. The top result in each column is highlighted in bold.

Colour channels	ELP		GIST		F-ELP		LBP	
	F1	BAC	F1	BAC	F1	BAC	F1	BAC
Greyscale	0.3839	0.6069	0.5910	0.7207	0.4218	0.6199	0.5396	0.6825
H&E stains	**0.4943**	**0.6589**	**0.6283**	**0.7570**	**0.6697**	**0.7665**	0.6887	0.7780
RGB image	0.4912	0.6569	0.6147	0.7448	0.6404	0.7462	**0.7125**	**0.7972**

IDC: Tables 2, 3 and 4 show the results for the IDC data set for kNN search with $k = 1, 5$ and 15, respectively. As expected, since coloured images contain relevant information which may be lost when converted to greyscale, we observe that using either the H&E stain separated image or the total RGB image is always an improvement over using the greyscale image. A more interesting comparison comes from looking at the bottom two rows of the tables, comparing the H&E images to the RGB images. We see that generally the F1 scores and balanced accuracies are similar (within one or two percent) for both H&E and RGB images. For all descriptors besides LBP, we actually observe an improved performance using the H&E image over RGB, despite the fact that the input image has less colour channels, and thus the feature vector is shorter.

We also, not surprisingly, observe that as k is increased, the search performance tends to improve, although the jump from $k = 1$ to 5 is quite a bit larger than the jump from 5 to 15. This may indicate that $k = 15$ is near an optimal value for k. We also see that for this particular data set, the LBP descriptor gives the highest accuracy, and the ELP descriptor performs the worst.

BreakHis: Similarly, for the BreakHis data set, Tables 5, 6 and 7 show the best global recognition rates and balanced accuracies for each image descriptor and set of input colour channels. Once again, we observe a general increase in search accuracy when using more than one input colour channel (H&E or RGB) as compared to the greyscale images. In many cases, in addition to the decreased computational cost of using fewer colour channels, for the BreakHis data set we see that there is another benefit to using stain separated images, which is a significant improvement in search performance.

Table 5. The best KNN search ($k = 1$) accuracy for the BreakHis dataset taken over all distance functions. The top result in each column is highlighted in bold.

Colour channels	ELP		GIST		F-ELP		LBP	
	GRR	BAC	GRR	BAC	GRR	BAC	GRR	BAC
Greyscale	0.6584	0.5812	0.6589	0.5602	0.6534	0.5577	0.6874	0.6212
H&E stains	**0.7532**	**0.7047**	**0.7083**	**0.6456**	0.7433	**0.6823**	**0.7397**	**0.6903**
RGB image	0.6604	0.5971	0.6578	0.5787	0.7358	0.6812	0.6689	0.6115

As before, we see that search performance tends to increase with increasing k, but does seem to level off around $k = 15$. Unlike our previous results on the IDC data set, we see here that the best search performance occurs using the ELP descriptor and F-ELP descriptors, while the GIST descriptor fares the worst on the BreakHis data. Due to the higher intra-class variation of the BreakHis data (multiple tumour types for benign and malignant classes), it makes sense that the balanced accuracies are generally lower on this data set.

Table 6. The best KNN search ($k = 5$) accuracy for the BreakHis dataset taken over all distance functions. The top result in each column is highlighted in bold.

Colour channels	ELP		GIST		F-ELP		LBP	
	GRR	BAC	GRR	BAC	GRR	BAC	GRR	BAC
Greyscale	0.6838	0.5844	0.6940	0.5808	0.6802	0.5637	0.6952	0.6049
H&E stains	**0.7602**	**0.7078**	**0.7340**	**0.6324**	0.7557	**0.6915**	**0.7323**	**0.6662**
RGB image	0.6744	0.5931	0.6884	0.6113	0.7480	0.6865	0.6977	0.6335

Table 7. The best KNN search ($k = 15$) accuracy for the BreakHis dataset taken over all distance functions. The top result in each column is highlighted in bold.

Colour channels	ELP		GIST		F-ELP		LBP	
	GRR	BAC	GRR	BAC	GRR	BAC	GRR	BAC
Greyscale	0.7085	0.5898	0.7128	0.5744	0.6957	0.5616	0.7051	0.5979
H&E stains	**0.7660**	**0.7060**	**0.7406**	**0.6286**	0.7737	**0.7033**	**0.7564**	**0.6885**
RGB image	0.6892	0.6008	0.7068	0.6090	**0.7749**	**0.7033**	0.7023	0.6219

5.2   Comparing Distance Functions

In this section, we consider the effect that the choice of distance function has on image search performance for each image descriptor. To do so, we introduce a ranking of each distance function, based on the balanced accuracy result. For a given image descriptor and choice of input colour channels (greyscale, H&E, or RGB) we rank each distance function based on the resulting balanced accuracy as a percentage of the maximum balanced accuracy for that particular search trial. Results presented here show the averaged distance ranking for each distance and each descriptor, averaged over the choice of input colour channels, and the choice of k for the kNN search algorithm. We present results only for the balanced accuracy, as the results for the F1 measure (IDC data) and global recognition rate (BreakHis data) follow similar trends.

IDC: Figure 4 shows the average ranking of all six distance functions tested for each image descriptor on the IDC data. We observe that, in general, the variation in search accuracy caused by the choice of distance function is relatively low, with the exception of the ELP descriptor, where a noticeable variation can be seen. For all but the ELP descriptor, it would be difficult to pinpoint which distance function is the "best" choice. In particular, for the F-ELP and LBP descriptors, the variation in performance across distance functions is almost non-existent.

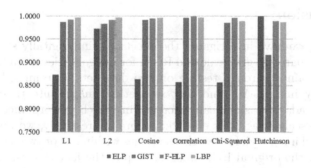

Fig. 4. A comparison of the average ranking of distance functions for each image descriptor on the IDC data set.

BreakHis: In Fig. 5 we show the average rankings of each distance function on the BreakHis data. Once again, the overall change in the accuracy as a result of the choice of distance function is surprisingly low for all descriptors. As with the IDC data, we see the most noticeable variation in performance with the ELP and GIST descriptors. The results for both the IDC and BreakHis data sets do not give any indication that one distance function is necessarily superior for image search even for a fixed image descriptor. Over many tests, we see only one scenario (the ELP descriptor applied to the IDC data) where the choice of distance function significantly impacts the results. Further testing on more data is of course possible, however for good performance and generalisation to unknown data, it would seem that the best choice is simply to use the distance function which can be computed most efficiently.

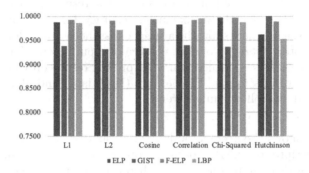

Fig. 5. A comparison of the average ranking of distance functions for each image descriptor on the BreakHis data set.

6   Conclusion

In this paper, we have investigated the effect of using digitally stain separated images, as compared to greyscale and RGB, for image retrieval applications. Our results are obtained through testing on two data sets containing breast cancer histopathology images. We find that separating images into their H&E stain components leads to a significant increase in search performance over simply using the greyscale images, as expected. More interestingly, we find that using H&E separated images yields search accuracies within one or two percent of those obtained with the original RGB images, despite the fact that the H&E images have only two colour channels. In fact, superior performance is observed using the H&E images in most tested scenarios. Given the improved computation speed afforded by using fewer image channels, it is reasonable to conclude that using H&E stain separated images is preferable to using the overall RGB images for image search. As well, ELP appears to benefit from investigations on choosing the distance metric, a factor that should be considered when using this descriptor.

References

1. Ahonen, T., Hadid, A., Pietikainen, M.: Face description with local binary patterns. IEEE Trans. Pattern Anal. Mach. Intell. **28**(12), 2037–2041 (2006)
2. Alhindi, T.J., Kalra, S., Ng, K.H., Afrin, A., Tizhoosh, H.R.: Comparing LBP, HOG and deep features for classification of histopathology images (2018)

3. Bay, H., Ess, A., Tuytelaars, T., Gool, L.V.: Speeded-up robust features (SURF). Comput. Vis. Image Und. **110**(3), 346–359 (2008)
4. Cheeseman, A.K., Tizhoosh, H., Vrscay, E.R.: A compact representation of histopathology images using digital stain separation and frequency-based encoded local projections. In: Karray, F., Campilho, A., Yu, A. (eds.) ICIAR 2019. LNCS, vol. 11663, pp. 147–158. Springer, Cham (2019). https://doi.org/10.1007/978-3-030-27272-2_13
5. Cruz-Roa, A., et al.: Automatic detection of invasive ductal carcinoma in whole slide images with convolutional neural networks. In: Progress in Biomedical Optics and Imaging - Proceedings of SPIE 9041 (2014)
6. Dalal, N., Triggs, B.: Histograms of oriented gradients for human detection. In: 2005 IEEE Computer Society Conference on Computer Vision and Pattern Recognition, San Diego, pp. 886–893 (2005)
7. Friedman, J.H., Bentley, J.L., Finkel, R.A.: An algorithm for finding best matches in logarithmic expected time. ACM Trans. Math. Softw. **3**(3), 209–226 (1977)
8. Gurcan, M.N., Boucheron, L.E., Can, A., Madabhushi, A., Rajpoot, N.M., Yener, B.: Histopathological image analysis: a review. IEEE Rev. Biomed. Eng. **2**(2), 147–171 (2009)
9. Lowe, D.G.: Distinctive image features from scale-invariant keypoints. Int. J. Comput. Vis. **60**(2), 91–110 (2004)
10. Macenko, M., et al.: A method for normalizing histology slides for quantitative analysis. In: Proceedings of International Symposium on Biomedical Imaging, Chicago, pp. 1107–1110 (2009)
11. Madabhushi, A.: Digital pathology image analysis: opportunities and challenges. Imaging Med. **1**(1), 7–10 (2009)
12. McCann, M.T., Majumdar, J., Peng, C., Castro, C.A., Kovacevic, J.: Algorithm and benchmark dataset for stain separation in histology images. In: Proceedings of 2014 IEEE International Conference on Image Processing (ICIP), Paris, pp. 3953–3957 (2014)
13. Mendivil, F.: Computing the Monge–Kantorovich distance. Comput. Appl. Math. **36**(3), 1389–1402 (2016). https://doi.org/10.1007/s40314-015-0303-7
14. Molter, U., Brandt, J., Cabrelli, C.: An algorithm for the computation of the Hutchinson distance. Inf. Process. Lett. **40**(2), 113–117 (1991)
15. Oliva, A., Torralba, A.: Modeling the shape of the scene: a holistic representation of the spatial envelope. Int. J. Comput. Vis. **42**(3), 145–175 (2001). https://doi.org/10.1023/A:1011139631724
16. Ruifrok, A.C., Johnston, D.A.: Quantification of histochemical staining by color deconvolution. Anal. Quant. Cytol. Histol. **23**(4), 291–299 (2001)
17. Spanhol, F.A., Oliveira, L.S., Petitjean, C., Heutte, L.: A dataset for breast cancer histopathological image classification. IEEE Trans. Biomed. Eng. **63**(7), 1455–1462 (2016)
18. Tizhoosh, H.R., Babaie, M.: Representing medical images with encoded local projections. IEEE Trans. Biomed. Eng. **65**(10), 2267–2277 (2018)

Generalized Multiple Instance Learning for Cancer Detection in Digital Histopathology

Jan Hering and Jan Kybic[✉][iD]

Faculty of Electrical Engineering, Czech Technical University in Prague,
Prague, Czech Republic
{jan.hering,kybic}@fel.cvut.cz

Abstract. We address the task of detecting cancer in histological slide images based on training with weak, slide- and patch-level annotations, which are considerably easier to obtain than pixel-level annotations. we use CNN based patch-level descriptors and formulate the image classification task as a generalized multiple instance learning (MIL) problem. The generalization consists of requiring a certain number of positive instances in positive bags, instead of just one as in standard MIL. The descriptors are learned on a small number of patch-level annotations, while the MIL layer uses only image-level patches for training.

We evaluate multiple generalized MIL methods on the H&E stained images of lymphatic nodes from the CAMELYON dataset and show that generalized MIL methods improve the classification results and outperform no-MIL methods in terms of slide-level AUC. Best classification results were achieved by the MI-SVM(k) classifier in combination with simple spatial Gaussian aggregation, achieving AUC 0.962.

However, MIL did not outperform methods trained on pixel-level segmentations.

Keywords: Multiple-Instance Learning · Histopathology image classification · Computer-aided diagnosis

1 Introduction

Training state-of-the-art deep-learning methods in computer-aided diagnosis (CAD) often requires a large image database with pixel-level annotations [9]. When provided with such data, deep-learning computer-aided diagnosis (CAD) methods are the state-of-the-art and can reach or surpass human expert performance. However, obtaining such precise manual annotations is tedious and expensivein terms of time and resources. Therefore, there is a lot of interest in

The project was supported by the Czech Science Foundation project 17-15361S and the OP VVV funded project "CZ.02.1.01/0.0/0.0/16_019/0000765 Research Center for Informatics.".

© Springer Nature Switzerland AG 2020
A. Campilho et al. (Eds.): ICIAR 2020, LNCS 12132, pp. 274–282, 2020.
https://doi.org/10.1007/978-3-030-50516-5_24

methods capable of learning from weak annotations, such as image or patient level labels. This data, e.g. whether a patient is healthy or not, can be often extracted from the hospital information system automatically, with no or very little additional cost.

One popular class of weakly-supervised learning methods is Multiple-Instance Learning (MIL), which considers image as a collection (bag) of instances (pixels or pixel regions) and requires only image-level labels for training [7]. In standard MIL, a bag is positive iff at least one of its instances is positive.

The task of the CAMELYON challenge [9] is to detect metastases in stained breast lymph node images (see examples in Figs. 1 and 2). Each image is to be assigned a score between 0.0 and 1.0 measuring the likelihood of containing a tumor. The best-ranked submissions use a two-stage approach [2]. First a convolutional neural network (CNN) is learned in a fully-supervised manner to classify rectangular patches, yielding a *tumor probability map*. The second, *aggregation stage*, classifies the whole image based on geometrical properties of detected regions in the prediction map [2].

In this work, we use a CNN only to extract patch descriptors, which are then considered as instances for the MIL approach to classify images. The fact that an image (bag) is positive (contains cancer) iff any of its patches (instances) is positive corresponds exactly to the MIL formulation. However, due to the high number of patches and imperfections of the patch (instance) classifier, applying the standard MIL methods is very sensitive to false positive detections. We alleviate this problem by applying the generalized MIL [6] method, increasing the number of required positive instances for positive bags. This correspond to common histopathological guidelines, where the size of the lesion is one of the important factors of the classification.

Existing methods combining MIL and CNN are mainly based on region proposals, like R-CNN [5,11]. However, because of high memory consumption, they have only been applied to much smaller images than ours [4,12].

2   Method

Let us describe the basic sequential blocks of the proposed method—patch descriptor calculation (Sect. 2.2), generalized MIL learning from image-level annotations (Sect. 2.3), and spatial aggregation (Sect. 2.4). We also describe alternative techniques, used as a baseline.

2.1   Patch Extraction

We operate on 256×256 pixel square image patches extracted from the whole slide image (WSI) at the $20\times$ magnification level. We use a random forest classifier on color channels of the down-sampled images to distinguish tissue and background. A patch is used only if it contains at least 80% of tissue. The patch label y_i is set to *tumor* ($y_i = 1$) if at least 60% of its area is tumor tissue and to *normal* ($y_i = -1$) if at most 10% of the tissue are from tumor class. The remaining patches are omitted during training.

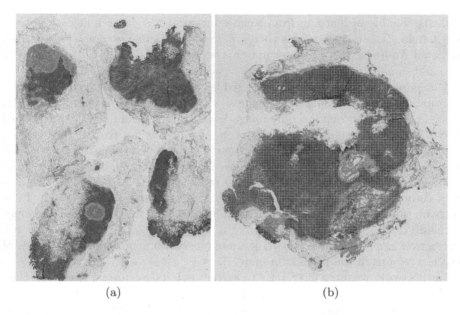

(a)   (b)

Fig. 1. *(a)* Example whole-slide image (WSI) from the CAMELYON'16 dataset. Tumor annotation boundaries are shown in magenta. *(b)* Another WSI with superimposed tissue patch boundaries—green for healthy, red for tumor—based on human expert annotations. Indeterminate (mostly boundary) yellow patches will not be used. Non-tissue patches (not-shown) were determined automatically. (Color figure online)

2.2   Patch Descriptors

The goal of this step is to provide a low dimensional descriptor for each patch. The descriptor is learnt from a limited amount of patch-level labels, which we obtain by aggregating the pixel-level segmentations provided by the CAME-LYON'16 dataset. It would also be possible to ask the expert to annotate the patches directly, which would be much easier than to create full pixel-level annotations. The hope is that even when trained on limited data, the descriptor gives us a useful embedding for the MIL block. Note that the pixel level segmentations are not used directly at all and this is the only place where patch-level segmentations are needed.

We use a VGG'16 deep network, variant D, with 16 weight layers and the binary cross-entropy loss function [13]. We apply implicit color-normalization by adding a color-normalization layer [10]. We used the following augmentation techniques—random crop to the input size of 224×224, $90°$, $180°$ and $270°$ rotations as well as random up-down and left-right flips. The accuracy of the predicted patch labels is shown as 'CNN' in Table 1.

We then insert another fully-connected (FC) layer before the output layer, which reduces the dimensionality at the input of the last layer from 4096 to some much smaller D ($D = 32$ was used in the experiments). This augmented CNN is retrained and the output of the added intermediary layer is used as a patch descriptor f_{CNN}.

As an alternative to the CNN last layer, we have also trained a gradient boosting XGBoost classifier [3] (shown as 'CNN + XGBoost' in Table 1) to predict patch labels from patch descriptors f_{CNN}.

2.3   Generalized MIL

The next block takes the f_{CNN} patch descriptors and produces both patch and image level labels. We have taken two most promising generalized MIL methods based on earlier experiments [6]. Both methods evaluate a scalar patch scoring function $\phi(f_{CNN})$, which is thresholded to obtain the patch labels $\hat{y}_i$. We introduce a parameter k, the minimum number of patches assumed to be positive in a positive image, with $k = 1$ corresponding to the standard MIL formulation [1].

The first method, MI-SVM(k) [6], is an extension of the MI-SVM classifier [1]. It acts iteratively and repeatedly trains an SVM classifier that calculates ϕ using all instances from negative bags and selected instances (the *witnesses*) from positive bags. After each iteration, the set of witnesses is recalculated by taking the top k positive instances from each positive bag. The iteration ends when the set of witnesses does not change.

The second method, MIL-ARF(k) [6], is an extension of MIL-ARF [8]. It implements ϕ using a random forest classifier and applies deterministic annealing. In each iteration, the instances are first classified using the current instance-level classifier. Then the instance labels are modified to enforce at least k positive instances in each positive image and no positive instances in any negative image. The patch labels are randomly perturbed, with probability decreasing as a function of the iteration number. The random forest is incrementally relearned from the updated labels and the process is repeated until convergence.

Finally, image labels are obtained by thresholding the number of positive patches with k. The results of this method are shown in the 'MIL' column in Table 1.

2.4   Spatial Aggregation

The patch-level output from all previously described methods is fed into an aggregator to obtain an image-level prediction. In the simplest but surprisingly efficient case (denoted 'Gaussian' in Table 1), we project patch prediction to pixels to obtain a pixel-level tumor probability map T, apply Gaussian smoothing to obtain $T_\sigma(\mathbf{x}) = G_\sigma * T$ with $\sigma = 2\sqrt{2}$, take a maximum

$$m_\sigma = \max_{\mathbf{x}} T_\sigma(\mathbf{x}) \qquad (1)$$

and threshold, $m_\sigma > \tau$, where τ is the threshold parameter, which can be user-specified or learned from data by cross-validation.

A more sophisticated procedure [2] consists of combining the maximal Gaussian response m_σ for $\sigma \in \sqrt{2}[1, 2, 4]$ with properties of the largest 2-connected component for each binary image $T_\sigma > t$ for thresholds $t \in \{0.5, 0.8\}$. The properties are area, extent, solidity and eccentricity, as well as the mean of T_σ within

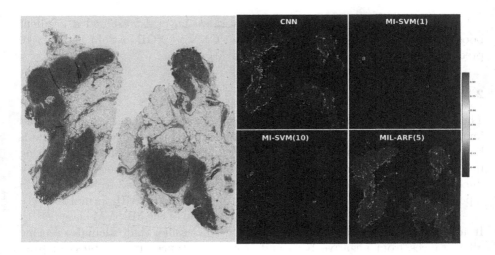

Fig. 2. Example image from the CAMELYON dataset and tumor prediction maps computed by the CNN, the MI-SVM($k = 1$), MI-SVM($k = 10$) and MIL-ARF($k = 5$). The output is scaled between 0.0 and 1.0 (tumor tissue) with the indicated color map. Ground truth annotations are shown as green overlay over the original image. (Color figure online)

the area. A random-forest classifier is trained on the resulting 30 dimensional descriptors. This is denoted as 'RF' in Table 1.

Figure 2 shows the tumor prediction maps (patch-scores) computed by the VGG net (CNN) and the various MIL methods. We see that MI-SVM with $k = 10$ is closest to the ground truth annotation.

3   Experiments and Results

Experiments were performed on all available training (159 healthy, 110 tumor), respective testing (49 healthy, 78 tumor) whole-slide images as provided within the CAMELYON'16 challenge. In all cases, parameters were found using cross-validation, taking out 20% of the training dataset for validation.

We trained the VGG16 network with descriptor dimensionality $D = 32$. We use $k = 1, 5, 10, 15$ for MI-SVM(k) and $k = 1, 5, 10$ (required number of positive instances) for the MIL-ARF(k).

The first experiment evaluates the effect of initialization on the two MIL methods. We have taken a fraction ($l \in \{0, 0.25, 0.5, 1.0\}$) of the patch labels in the training set and used them to initialize the generalized MIL classifiers in their first iterations, initializing the remaining patch labels by the bag labels. We can see in Fig. 3 that unlike MI-SVM, MIL-ARF is very sensitive to this type of initialization and that generalized MI-SVM(k) with $k > 1$ provide robust results even when all instance labels are initialized with bag labels.

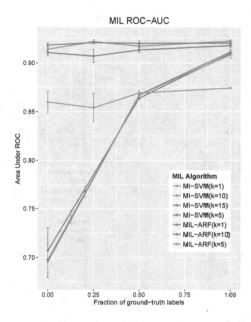

Fig. 3. Bag classification score as a function of the fraction of revealed training labels during initialization. Each line represents the mean ROC-AUC score of the MIL classifier with whiskers denoting ±SD.

The experiment also evaluates the effect of the parameter k (Fig. 3). For MI-SVM(k), the parameter k affects the overall performance significantly. The performance improves with higher values with an optimum around $k = 10$ on our data.

The main results are summarized in Table 1, while the ROC curves are shown in Fig. 4. The pure CNN approach yielded the best patch-level AUC score of 0.973, which resulted in an image-level AUC of 0.941 after the aggregation phase. Plugging-in the supervised XGBoost classifier yielded almost the same results. Also the MIL-ARF(k) for both $k = 1$ and 5 reached similar patch-level AUC.

In terms of image level accuracy, MI-SVM(k) with $k = 10$ performed the best. When considering achieved ROC-AUC for all initialized fractions l together, MI-SVM(k) with $k = 15$ performed in the most consistent way.

Interestingly, combining patch-level predictions from MI-SVM(k) and Gaussian aggregation to obtain image-level results outperformed all other variants.

To evaluate whether using MIL can help reduce the number of required images with detailed (pixel or patch level) annotations, we trained the CNN on patches from 25% of available training images. It turns out that while using MIL helps, it cannot yet compensate for the lack of detailed annotation. After Gaussian spatial aggregation, we get an image-level AUC of 0.831 for the CNN only and 0.838 for MI-SVM($k = 5$), the AUC for the direct output of the MIL ('MIL' score) is 0.815.

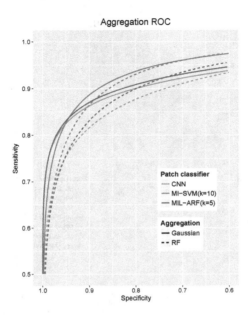

Fig. 4. ROC curves of the spatial aggregation phase. Both *Gaussian* (solid) and *RF* (dashed) [2] aggregation outcomes are shown for the MI-SVM (k = 10), MIL-ARF (k = 5) and the CNN classifiers.

Interestingly, the more sophisticated aggregation procedure [2] never outperformed the simpler but more robust aggregation based on Gaussian smoothing.

Table 1. Patch- and image-level classification scores. The image-level AUC is either a direct output (*MIL*) or the result of spatial aggregation with global features (*Gaussian*) or with a random forest classifier (*RF*) [2]. Best results in each column are shown in bold.

Algorithm	Image-level AUC			Patch-level metrics		
	MIL	Gaussian	RF	AUC	Specificity	Sensitivity
MI-SVM, k=15	0.920	0.945	0.941	0.871	0.991	0.707
MI-SVM, k=10	**0.923**	**0.962**	**0.953**	0.862	**0.992**	0.685
MI-SVM, k=5	0.915	0.955	0.944	0.847	0.990	0.647
MI-SVM, k=1	0.874	0.934	0.904	0.835	0.979	0.529
MIL-ARF, k=1	0.911	0.944	0.931	0.972	0.952	0.916
MIL-ARF, k=5	0.908	0.943	0.931	0.972	0.951	0.917
CNN	*n/a*	0.941	0.935	**0.973**	0.986	0.815
CNN + XGBoost	*n/a*	0.943	0.928	0.972	0.951	0.916

4   Conclusion

We have demonstrated that generalized MIL approaches can boost the performance of fully-supervised methods in the task of classifying histopathology images. We have also shown that it is possible to reach a good level accuracy by training only on a limited amount of patch data. The MI-SVM(k) method was shown to be robust to label initialization.

The direct output of the generalized MIL methods in terms of image-level AUC was lower than for the CNN methods with no MIL, but the high specificity, and thus minimal amount of false positives (see Fig. 2), enabled an important improvement through spatial aggregation, especially in the high specificity regime. The best image-level result (AUC 0.962) is comparable with the pathologist interpreting the slides in the absence of time constraints [2].

References

1. Andrews, S., Tsochantaridis, I., Hofmann, T.: Support vector machines for multiple-instance learning. In: Advances in Neural Information Processing Systems, pp. 561–568 (2002)
2. Bejnordi, B.E., Veta, M., van Diest, P.J., van Ginneken, B., et al.: Diagnostic assessment of deep learning algorithms for detection of lymph node metastases in women with breast cancer. JAMA **318**(22), 2199–2210 (2017). https://doi.org/10.1001/jama.2017.14585
3. Chen, T., Guestrin, C.: XGBoost: a scalable tree boosting system. In: Proceedings of the 22nd ACM SIGKDD International Conference on Knowledge Discovery and Data Mining, pp. 785–794. ACM Press, San Francisco, California, USA (2016). https://doi.org/10.1145/2939672.2939785
4. Das, K., Conjeti, S., Roy, A.G., Chatterjee, J., Sheet, D.: Multiple instance learning of deep convolutional neural networks for breast histopathology whole slide classification. In: 2018 IEEE 15th International Symposium on Biomedical Imaging (ISBI 2018). pp. 578–581. IEEE, Washington, DC, April 2018. https://doi.org/10.1109/ISBI.2018.8363642
5. Durand, T., Thome, N., Cord, M.: WELDON: weakly supervised learning of deep convolutional neural networks. In: 2016 IEEE CVPR, pp. 4743–4752, June 2016. https://doi.org/10.1109/CVPR.2016.513
6. Hering, J., Kybic, J., Lambert, L.: Detecting multiple myeloma via generalized multiple-instance learning. In: Proceedings of SPIE, p. 22. SPIE, March 2018. https://doi.org/10.1117/12.2293112
7. Kandemir, M., Hamprecht, F.A.: Computer-aided diagnosis from weak supervision: a benchmarking study. Comput. Med. Imaging Graph. **42**, 44–50 (2015). https://doi.org/10.1016/j.compmedimag.2014.11.010
8. Leistner, C., Saffari, A., Bischof, H.: MIForests: multiple-instance learning with randomized trees. In: Daniilidis, K., Maragos, P., Paragios, N. (eds.) ECCV 2010. LNCS, vol. 6316, pp. 29–42. Springer, Heidelberg (2010). https://doi.org/10.1007/978-3-642-15567-3_3
9. Litjens, G., et al.: 1399 H&E-stained sentinel lymph node sections of breast cancer patients: the CAMELYON dataset. GigaScience **7**(6), June 2018. https://doi.org/10.1093/gigascience/giy065

10. Mishkin, D., Sergievskiy, N., Matas, J.: Systematic evaluation of convolution neural network advances on the Imagenet. Comput. Vis. Image Underst. **161**, 11–19 (2017). https://doi.org/10.1016/j.cviu.2017.05.007
11. Ren, S., He, K., Girshick, R., Sun, J.: Faster R-CNN: towards real-time object detection with region proposal networks. In: Advances in Neural Information Processing Systems, vol. 28, pp. 91–99. Curran Associates, Inc. (2015)
12. Ribli, D., Horváth, A., Unger, Z., Pollner, P., Csabai, I.: Detecting and classifying lesions in mammograms with Deep Learning. Sci. Rep. **8**(1), 4165 (2018). https://doi.org/10.1038/s41598-018-22437-z
13. Simonyan, K., Zisserman, A.: Very deep convolutional networks for large-scale image recognition. In: International Conference on Learning Representations (2015)

Diagnosis and Screening of Ophthalmic Diseases

A Multi-dataset Approach for DME Risk Detection in Eye Fundus Images

Catarina Carvalho[1]([✉]), João Pedrosa[1], Carolina Maia[2], Susana Penas[2,3],
Ângela Carneiro[2,3], Luís Mendonça[4], Ana Maria Mendonça[1,5],
and Aurélio Campilho[1,5]

[1] Institute for Systems and Computer Engineering,
Technology and Science (INESC TEC), Porto, Portugal
catarina.b.carvalho@inesctec.pt
[2] Centro Hospitalar Universitário São João (CHUSJ), Porto, Portugal
[3] Faculdade de Medicina da Universidade do Porto (FMUP), Porto, Portugal
[4] Hospital de Braga, Braga, Portugal
[5] Faculdade de Engenharia da Universidade do Porto (FEUP), Porto, Portugal

Abstract. Diabetic macular edema is a leading cause of visual loss for
patients with diabetes. While diagnosis can only be performed by optical coherence tomography, diabetic macular edema risk assessment is
often performed in eye fundus images in screening scenarios through the
detection of hard exudates. Such screening scenarios are often associated
with large amounts of data, high costs and high burden on specialists,
motivating then the development of methodologies for automatic diabetic macular edema risk prediction. Nevertheless, significant dataset
domain bias, due to different acquisition equipment, protocols and/or
different populations can have significantly detrimental impact on the
performance of automatic methods when transitioning to a new dataset,
center or scenario. As such, in this study, a method based on residual neural networks is proposed for the classification of diabetic macular edema
risk. This method is then validated across multiple public datasets, simulating the deployment in a multi-center setting and thereby studying the
method's generalization capability and existing dataset domain bias. Furthermore, the method is tested on a private dataset which more closely
represents a realistic screening scenario. An average area under the curve
across all public datasets of 0.891 ± 0.013 was obtained with a ResNet50
architecture trained on a limited amount of images from a single public
dataset (IDRiD). It is also shown that screening scenarios are significantly more challenging and that training across multiple datasets leads
to an improvement of performance (area under the curve of 0.911 ±
0.009).

Keywords: Diabetic macular edema · Eye fundus · Screening ·
Classification

C. Carvalho and J. Pedrosa—Equal contribution.

© Springer Nature Switzerland AG 2020
A. Campilho et al. (Eds.): ICIAR 2020, LNCS 12132, pp. 285–298, 2020.
https://doi.org/10.1007/978-3-030-50516-5_25

1   Introduction

Diabetic macular edema (DME) is a condition characterized by an accumulation of fluid in the macula and is a leading cause of visual loss for patients with diabetes. Although not unique to the diabetic population, the prevalence of this pathology in the diabetic population is notoriously high - up to 42% in type 1 and 53% in type 2 diabetes mellitus patients [38].

While diagnosis of DME can only be performed by optical coherence tomography, the cost and limited availability of this imaging modality makes it prohibitive for screening. Alternatively, the presence of hard-exudates (i.e. lipid residues of serous leakage from damaged capillaries), which are visible in typical eye fundus images, has been found to be good a indicator of DME risk [6]. In other words, presence of exsudates indicates that there is a risk of DME, while the opposite indicates no risk of DME.

Notwithstanding its relevance for early diagnosis of DME, retinopathy screenings produce large amounts of data that need to be assessed by ophthalmologists leading to an unsustainable growth of the costs associated to the acquisition, storing and diagnosis of the data. As an example, in a 2009–2014 retinopathy screening in Portugal [23], only 16% of the screened population (52.739 patients) presented signs of diabetic retinopathy and of those, only 1.4% presented signs of DME. Such high costs and high number of pathology-free patients, have motivated the development of accurate and automatic methods for the classification of DME risk, reducing the amount of data that specialists need to assess.

Most of the methods proposed for DME risk classification follow traditional machine-learning approaches: 1) data pre-processing methods, as image enhancement and anatomical structures removal; 2) detection of exudate candidates using morphological operations for example; 3) extraction of features based on intensity, texture, size, shape or others from the detected candidates; and 4) classification of exudate candidates, using methods such as support-vector machines and naive Bayes classifiers, as true/false exudate. In these works [2,3,9,11,14,20–22,31,36,39], segmentation of the exudates is used for the DME classification. More recently, data-driven approaches based on the use of single or combined convolutional neural networks (CNNs), such as Residual neural networks [1], Deep Convolutional neural networks [30], AlexNet, VGG16, GoogLeNet, and Inception-v3 [19], have been proposed for the DME risk classification.

Despite the promising results in literature for DME risk classification (inferred from exudate segmentation or by direct image classification), ranging from [85%–98%], most studies developed and tested their methods on a limited set of the available public datasets. This fact not only hinders a direct performance comparison between different methods proposed in literature but may also be detrimental to the generalization capabilities of the developed methods. While a given method may have good performance within a certain dataset, it is often the case that test on a different dataset (in which significant domain bias could have been introduced due to the use of different retinal scanners, acquisition conditions, patient population or demographic factors [40]) produces significantly worse performance. This limited generability can be even more significant when transitioning from a research scenario to clinical screening, where image quality is typically degraded and a more diverse population in terms of lesions and anatomical features is found.

As such, and considering the nature of the screening of eye pathologies (spanning multiple centers, populations and acquisition protocols), such domain bias is expected to occur across different acquisition centers and the developed methods should consider the use of domain-invariant predictors and thorough validation across multiple datasets.

This work proposes a method, based on a residual neural network, that makes use of publicly available eye fundus datasets, for the classification of DME risk. As such, the influence of dataset domain bias is studied, simulating the deployment of the proposed method in a multi-center setting with different image acquisition and annotation protocols and equipment. Furthermore, the method is tested on a private dataset (ScreenDR), which more closely represents a realistic screening scenario [37]. Different architectures for DME risk classification are investigated focused on producing accurate results across the multiple datasets and that can generalize well to unseen data domains.

2   Methodology

2.1   Model Architectures

Given the goal of DME risk classification from an eye fundus image, multiple approaches have been proposed, namely exudate detection, segmentation and classification of images in varying degrees of DME risk. However, for the purpose of screening, the most important task is the separation of images/patients in two classes, those at risk of DME and those at no risk, so that those at risk can be referred to an ophtalmologist. As such, the aim of this study is the classification of eye fundus images in two classes: **M0)** images without hard exudates and **M1)** images with at least one hard exudate.

For this purpose, two different residual neural network architectures (ResNet) were tested: ResNet50 and ResNet512. ResNets are very deep convolutional neural networks which use residual learning through the use of skip-connections [12]. Residual networks have shown good results in image classification tasks in general and have already been used for the classification of patches of eye fundus images regarding the presence/absence of hard exudates [1,24].

A schematic of the ResNet architecture used is shown in Fig. 1. Both ResNets receive as input a 3-channel RGB image, resized and cropped to size $3 \times 512 \times 512$, and output two nodes with the probability of belonging to each of the classes M0 and M1. The convolutional block of the network is composed of residual blocks, each composed of repetitions of convolutional layers of varying kernel size and number of output channels, followed by batch normalization (BatchNorm) [15] and a rectified linear unit activation (ReLu) [10]. Each residual block repetition is connected by a skip connection which, for the first repetition of each residual block includes a convolutional layer followed by BatchNorm. This convolutional block is followed by a classification block which is composed

of two dense fully-connected layers with 256 and 2 neurons respectively and
separated by ReLu and a dropout layer [35] (0.2 probability). Softmax [4] is
then applied to the output of the last dense layer to obtain the probability of
each class.

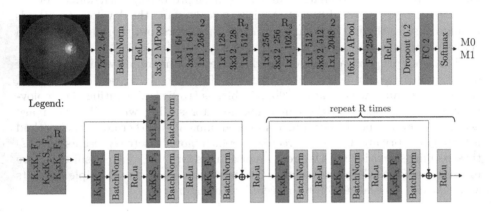

Fig. 1. ResNet architecture. Blue blocks ($K \times KS, F$) represent convolutional layers
with a $K \times K$ kernel, stride S and F output channels (if S is not given a stride of 1 was
used). Yellow MPool blocks ($K \times KS$ MPool) represent max-pooling operations with
a $K \times K$ kernel and stride S. Yellow APool blocks ($K \times K$ APool) represent average
pooling operations with a $K \times K$ kernel. Orange FC blocks (FCN) correspond to fully-
connected layers with N output neurons. Green blocks ($K_1 \times K_1, F_1, ..., K_n \times K_n, F_n, R$)
correspond to the residual blocks used where $K_l \times K_l, F_l$ corresponds to the kernels,
stride and output channels of convolutional layer l and R is the number of repetitions
of the convolutional layers with direct skip-connection. The number of repetitions of
residual blocks 2 and 3 (R_2 and R_3) is 3 and 5 for ResNet50 and 7 and 35 for ResNet152.
(Color figure online)

2.2   Model Training

The loss function used during training was weighted binary cross-entropy:

$$L_{CE} = \frac{\sum_{n=1}^{N} \sum_{c=0}^{C-1} -w_c y_{n,c} log(p_{n,c})}{\sum_{n=1}^{N} \sum_{c=0}^{C-1} w_c y_{n,c}}, \tag{1}$$

where N is the number of images per batch, C is the number of classes, w_c is
the class weight for class c, $y_{n,c}$ is the ground-truth label of class c for image n
and $p_{n,c}$ is the predicted probability of image n belonging to class c. The class
weights w_c are the number of images per class in the full training set and are
used to address the issue of class imbalance. Model optimization was performed
using Adam [18] with a learning rate of 0.0001 and a batch size of 32. Models
were trained for a maximum of 150 epochs with a patience of 20 epochs as an
early stopping criterion.

Table 1. Description of datasets used. N_A is the number of independent annotations per image available in the dataset and FOV is the image field of view.

Dataset	Annotation	N_A	FOV	Image size (px)	Number of images		
					Total	M0	M1
STARE [13]	Image	1	35°	700 × 605	372	275	97
DIARETDB0 [17]	Image	1	50°	1500 × 1152	130	59	71
DIARETDB1 [16]	Pixel	4	50°	1500 × 1152	89	48	41
e-ophta-EX [7]	Pixel	1	40°	2030 × 1352	82	35	47
DRiDB [29]	Pixel	5	45°	720 × 576	50	18	32
DR2 [25]	Image	1	45°	857 × 569	445	366	79
Messidor [8]	Image	1	45°	1859 × 1238	1200	974	226
IDRiD [26–28]	Image	1	50°	4288 × 2848	516	222	294
ScreenDR [37]	Image	4	45°	1975 × 1834	542	391	151
Total					3426	2388	1038

3   Experiments

All experiments were conducted on PyTorch 1.1.0 on Nvidia GTX 1080 and Nvidia GTX TITAN Xp graphic cards.

3.1   Datasets

A total of eight public datasets and one private dataset were used in this study as shown in Table 1. While the public datasets represent more controlled environments, the private dataset, ScreenDR, is a more accurate representation of a realistic screening scenario [37]. Because the annotation protocols vary between datasets, a conversion to M0 (no hard exudates) and M1 (at least one hard exudate) image-level classes was performed. For datasets having multiple independent annotations for the same image, a final image class was obtained by majority vote among all experts where, in the case of a tie, class M1 was attributed. For DIARETDB1, pixel-level annotations included a degree of confidence and lesions annotated with low confidence were excluded.

3.2   Multi-dataset DME Risk Classification

In order to test the multi-dataset performance of the proposed methodologies for the classification of DME risk, each architecture was trained on the training set of IDRiD-*Segmentation and grading challenge - part B*, and tested on all public datasets as well as on the IDRiD-part B test set. The IDRiD-part B dataset was chosen for this purpose as it is the most recent public dataset, it has a sufficiently large number of images to allow for a reasonable training of the network and it presents a predefined division into train and test set, making comparison to other studies straightforward.

Model training was performed starting from pretrained models from ImageNet [32] for a more efficient convergence of the model. Two training scenarios were tested for each architecture (ResNet50 and ResNet152). In the first scenario, the full network was trained, whereas in the second scenario only the final classification layers were trained and the rest of the network weights were frozen. The rationale for selective training (i.e. freezing the initial layers of the network during training) is that the initial layers represent generic features, while the final layers focus on the specific task for which the model was trained, and thus only the final layers need to be retrained.

Models were then tested on all images of each public dataset, except for IDRiD where only the test set was used. Model performance was computed at image level with receiver operating characteristic (ROC) curves [5] and area under the curve (AUC) is reported. For datasets with multiple independent annotations, interobserver variability was also computed. For comparison with previous studies, accuracy is also reported.

The efficiency of each of the proposed models was also investigated by computing the number of parameters (both total and trainable) of each of the ResNets as well as the number of floating-point operations (FPO) needed for inference. The number of total parameters correlates to the amount of memory consumed by the model, being thus a good measure of model efficiency whereas the number of trainable parameters is related to the amount of work needed to train the model. The number of FPOs directly computes the amount of work done in a single inference of the model, being strongly correlated to the running time of the model [33].

3.3   ScreenDR DME Risk Classification

To assess the performance of the proposed architecture in a dataset which more realistically represents a screening environment, the ScreenDR dataset was used for testing under two different training scenarios. First, using the model trained on the IDRiD training set only. Secondly, using a model trained on all public datasets (except the IDRiD test set). Given the heterogeneity of the data in a realistic screening scenario, it is expected that the extra data (in both quantity and diversity) provided by the datasets beyond IDRiD can lead to improved performance. As in Sect. 3.2, performance is assessed through AUC and interobserver variability is also reported.

To further validate the proposed architecture, Grad-CAM [34] was used to visualize the location of the regions responsible for the network predictions. Grad-CAM performs a weighted combination of the feature maps of the last convolutional layers for a given input image, generating a heatmap that explains which regions of that input image were responsible for the activation of a label. In this case M1 activations were generated to assess which structures were responsible for M1 classification.

Table 2. Number of parameters (total and trainable) and FPOs of each architecture and AUC obtained on all datasets.

Model	Parameters (M)		FPOs (G)	AUC all datasets
	Total	Trainable		(Mean ± 95% CI)
ResNet152 Frozen	58.67	0.53	60.38	0.773 ± 0.009
ResNet152 Unfrozen	58.67	58.67	60.38	0.895 ± 0.019
ResNet50 Frozen	24.03	0.53	21.47	0.766 ± 0.006
ResNet50 Unfrozen	24.03	24.03	21.47	0.891 ± 0.013

4 Results

4.1 Multi-dataset DME Risk Classification

Figure 2 shows the ROC curves for each dataset with each of the proposed architectures and training approaches. It can be seen that the unfrozen approach, for both models, obtain the highest performances in any of the datasets. However, performance varies between datasets, with particularly poor performance in STARE, while good performance is shown for other datasets such as DR2.

Table 2 shows the number of parameters and FPOs of each architecture and the AUC obtained in the combination of all public datasets. It can be seen that the ResNet50 architecture is much more efficient, with half the total parameters and one third of the FPOs per inference, achieving results comparable to the ResNet152. Comparing the frozen and unfrozen models, while the training is much more efficient in the frozen models, with only 530k trainable parameters, the performance is significantly degraded in comparison to the unfrozen models.

Figure 3 shows a comparison of the obtained Acc and AUC for each dataset with other state-of-the-art approaches. It can be seen that the proposed approach has competitive performance in comparison to previously reported methods though below the best results reported for each dataset.

4.2 ScreenDR DME Risk Classification

Given that the ResNet50 unfrozen showed better performance and greater efficiency, experiments on ScreenDR were limited to this architecture.

Figure 4 shows the ROC curves for the Resnet50 unfrozen approach when the model was trained on the IDRiD train set or trained on the combination of all publicly available dataset (AllData). Both models were tested on the IDRiD test set and ScreenDR dataset. It is shown that the model trained on all datasets maintains the performance on IDRiD while significantly improving the performance on ScreenDR (AUC 0.911 ± 0.009).

Figure 5 shows examples of Grad-CAM activations for ScreenDR images using the ResNet50 unfrozen trained on all public datasets. The first two rows show original images and respective Grad-CAM maps where the model successfully classifies the image, exhibiting high activation on the exudate regions (left

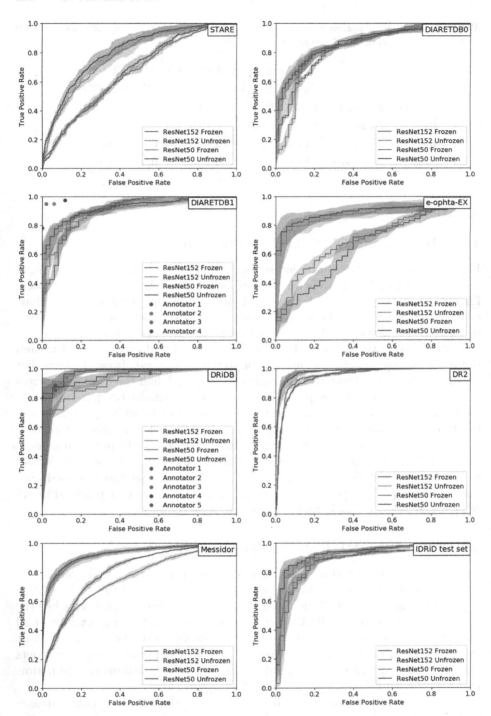

Fig. 2. ROC curves of the ResNet152 and ResNet50 trained on the IDRiD trainset for all datasets. Shaded region corresponds to the 95% confidence interval. Interobserver agreement is shown for DIARETDB1 and DRiDB.

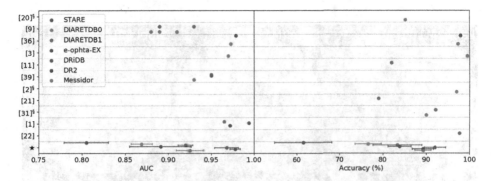

Fig. 3. Comparison of the proposed ResNet50 trained on the IDRiD train set (indicated as ★) to state-of-the-art methods tested on public datasets in terms of AUC and Acc (%). § indicates that only a subset of the data was used.

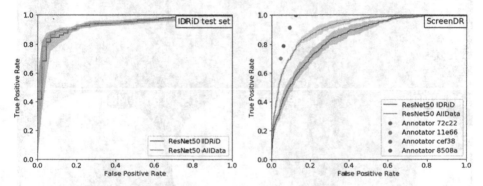

Fig. 4. ROC curves of the ResNet50 trained on the IDRiD train set and trained on all public datasets (except IDRiD test set). Shaded region corresponds to the 95% confidence interval. Interobserver agreement is shown for ScreenDR.

and middle columns). The right column shows that in spite of the activation on regions with acquisition artifacts, these are not sufficiently relevant for M1 classification. The third and forth rows show examples of failed classifications of M0 images, showing activations on acquisition artifacts (left column) and exudate-like structures (right column) and a failed classification for M1, in spite of a correct activation on the exudate region.

5   Discussion

In this study, the automatic classification of eye fundus images according to DME risk is proposed. The evaluation of the proposed method in a multi-dataset scenario allows a more in-depth evaluation of the method, less susceptible to overfitting to a single database and assuring the generalization of the model.

Comparing the performance obtained across datasets, a good performance was obtained for most datasets, though a significant degradation of performance

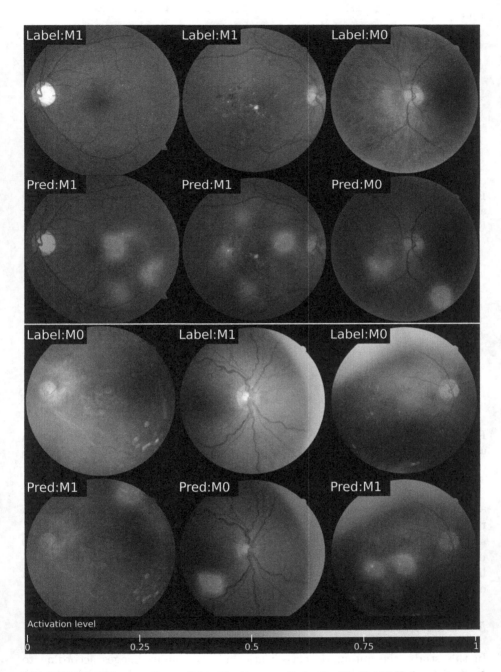

Fig. 5. ScreenDR examples and classification results by the ResNet50 Unfrozen trained on all public datasets. The first and third rows show the original eye fundus images, and the second and forth rows show their respective Grad-CAM activations normalized to the maximum activation present in the image.

is observed for STARE and DIARETDB0. These are direct examples of dataset bias. For STARE, the acquisition conditions (field-of-view, image size and quality of acquired images) differ within the dataset and significantly from IDRiD and the fact that images are significantly downsized and JPEG-compressed might be responsible for the lower performance. For DIARETDB0, also the acquisition conditions differ from IDRiD, as DIARETDB0 represents "practical situations", thus more representative of a clinical/screening scenario [17]. On the other hand, the proposed model presents the best transfer of knowledge when applied to DR2 due to the fact that these images are structurally similar to those of IDRiD - with extremely good image quality and virtually free of acquisition artifacts.

Comparing the different training approaches, the training of the whole network proved the most beneficial, allowing for a more refined estimation of the features most significant for the detection of hard exudates. As for the two ResNet implementation, results indicate that deeper networks do not always correlate to improved performance especially for relatively small training sets.

Regarding the comparison to state-of-the-art methods, it is shown that the developed approach cannot outperform the previous best results in a single dataset. However, this work is the first, to the authors' knowledge, in which the full range of available public datasets is evaluated and previous results have likely benefited from in-dataset training and validation, avoiding the issue of dataset domain bias. Nevertheless, the differences in performance accross datasets are indicative that the dataset domain bias is significant, strengthening the case for multi-dataset validation in future studies and also for multi-dataset training strategies that can breach the dataset domain issue.

Such joint learning strategy was implemented by training the ResNet50 with the combination of all public datasets. It can be seen that, for ScreenDR, a joint learning strategy produced significantly better results. Using only the IDRiD dataset for training reveals the difficulties in translation to a realistic screening scenario, where typically lower image quality, image artifacts and lesions/anatomical features not previously seen by the model are present. However, increasing the quantity and diversity of data by pooling images from multiple sources increases the generalization capabilities of the model and improves performance significantly, indicating a reduction of dataset domain bias between train set and test set. The marginal improvement in performance on the IDRiD test set further complements this conclusion, as the knowledge gained from other datasets was not beneficial to performance as improved performance could already be achieved with in-dataset training and validation.

Finally, Grad-CAM results obtained for ScreenDR show that the model learned to detect exudates correctly (Fig. 5 top row, left and center columns). As for classification errors, it can be seen that the model struggles to differentiate between exudates and some acquisition artifacts (Fig. 5 bottom row, left column) as these were probably not well represented in the training set, even after joining multiple datasets as these are typically of higher image quality than in a screening scenario. It can also be seen that the model is able to detect other structures which are extremely similar to exudates but not labeled as such, which could indicate possible mislabeling by the ophtalmologists (Fig. 5 bottom row, right column). On the other hand, on most true-negative classifications, regions containing exudates were activated by the model but with low probability. This indicates that the convolutional section of the model is not failing to detect the suspicious regions of the image (Fig. 5 bottom row, middle column), but that the probability at which they are detected is not sufficient to grant the M1 class.

6   Conclusions

In conclusion, the present work proposes a method for classification of eye fundus images in terms of DME risk by using residual neural networks in a multi-center and screening scenarios. By using multiple public datasets, a broad validation is performed, obtaining competitive results with state of the art (AUC 0.891 ± 0.013), while revealing the difficulties in translating artificial intelligence applications across datasets. This is further explored by validation in an external dataset that represents a realistic screening scenario with lower image quality and greater diversity in terms of image features, where it is shown that using data from multiple datasets can help mitigate the effects of dataset domain bias.

Acknowledgments. This work is financed by the ERDF - European Regional Development Fund through the Operational Programme for Competitiveness and Internationalisation - COMPETE 2020 Programme and by National Funds through the FCT Fundação para a Ciência e a Tecnologia within project CMUP-ERI/TIC/0028/2014.

The Messidor database was kindly provided by the Messidor program partners (see http://www.adcis.net/en/third-party/messidor/).

References

1. Abbasi-Sureshjani, S., Dashtbozorg, B., ter Haar Romeny, B.M., Fleuret, F.: Boosted exudate segmentation in retinal images using residual nets. In: Cardoso, M., et al. (eds.) Fetal, Infant and Ophthalmic Medical Image Analysis. Lecture Notes in Computer Science, vol. 10554, pp. 210–218. Springer, Cham (2017). https://doi.org/10.1007/978-3-319-67561-9_24
2. Acharya, U.R., Mookiah, M.R.K., Koh, J.E., Tan, J.H., Bhandary, S.V., Rao, A.K., et al.: Automated diabetic macular edema (DME) grading system using DWT, DCT features and maculopathy index. Comput. Biol. Med. **84**, 59–68 (2017)
3. Akram, M.U., Tariq, A., Khan, S.A., Javed, M.Y.: Automated detection of exudates and macula for grading of diabetic macular edema. Comput. Methods Programs Biomed. **114**(2), 141–152 (2014)
4. Bishop, C.M.: Pattern Recognition and Machine Learning. Springer, Cham (2006). https://doi.org/10.1007/978-1-4615-7566-5
5. Bradley, A.P.: The use of the area under the ROC curve in the evaluation of machine learning algorithms. Pattern Recogn. **30**(7), 1145–1159 (1997)
6. Bresnick, G.H., Mukamel, D.B., Dickinson, J.C., Cole, D.R.: A screening approach to the surveillance of patients with diabetes for the presence of vision-threatening retinopathy. Ophthalmology **107**(1), 19–24 (2000)

7. Decencière, E., Cazuguel, G., Zhang, X., Thibault, G., Klein, J.C., Meyer, F., et al.: TeleOphta: machine learning and image processing methods for teleophthalmology. Irbm **34**(2), 196–203 (2013)
8. Decencière, E., Zhang, X., Cazuguel, G., Lay, B., Cochener, B., Trone, C., et al.: Feedback on a publicly distributed image database: the Messidor database. Image Anal. Stereol. **33**(3), 231–234 (2014)
9. Giancardo, L., Meriaudeau, F., Karnowski, T.P., Li, Y., Garg, S., Tobin Jr., K.W., et al.: Exudate-based diabetic macular edema detection in fundus images using publicly available datasets. Med. Image Anal. **16**(1), 216–226 (2012)
10. Glorot, X., Bordes, A., Bengio, Y.: Deep sparse rectifier neural networks. In: 14th International Conference on Artificial Intelligence and Statistics, pp. 315–323 (2011)
11. Harangi, B., Hajdu, A.: Automatic exudate detection by fusing multiple active contours and regionwise classification. Comput. Biol. Med. **54**, 156–171 (2014)
12. He, K., Zhang, X., Ren, S., Sun, J.: Deep residual learning for image recognition. In: IEEE Conference on Computer Vision and Pattern Recognition, pp. 770–778 (2016)
13. Hoover, A.: Structured Analysis of the Retina. https://www.cecas.clemson.edu/ahoover/stare
14. Imani, E., Pourreza, H.R.: A novel method for retinal exudate segmentation using signal separation algorithm. Comput. Methods Programs Biomed. **133**, 195–205 (2016)
15. Ioffe, S., Szegedy, C.: Batch normalization: accelerating deep network training by reducing internal covariate shift. arXiv preprint arXiv:1502.03167 (2015)
16. Kälviäinen, R., Uusitalo, H.: DIARETDB1 diabetic retinopathy database and evaluation protocol. In: Medical Image Understanding and Analysis, vol. 2007, p. 61. Citeseer (2007)
17. Kauppi, T., Kalesnykiene, V., Kamarainen, J.K., Lensu, L., Sorri, I., Uusitalo, H., et al.: DIARETDB0: evaluation database and methodology for diabetic retinopathy algorithms. Mach. Vis. Pattern Recogn. Res. Group Lappeenranta Univ. Technol. Finl. **73**, 1–17 (2006)
18. Kingma, D.P., Ba, J.: Adam: a method for stochastic optimization. arXiv preprint arXiv:1412.6980 (2014)
19. Lam, C., Yu, C., Huang, L., Rubin, D.: Retinal lesion detection with deep learning using image patches. Investig. Ophthalmol. Vis. Sci. **59**(1), 590–596 (2018)
20. Lim, S., Zaki, W.M.D.W., Hussain, A., Lim, S., Kusalavan, S.: Automatic classification of diabetic macular edema in digital fundus images. In: 2011 IEEE Colloquium on Humanities, Science and Engineering, pp. 265–269. IEEE (2011)
21. Liu, Q., Zou, B., Chen, J., Ke, W., Yue, K., Chen, Z., et al.: A location-to-segmentation strategy for automatic exudate segmentation in colour retinal fundus images. Comput. Med. Imaging Graph. **55**, 78–86 (2017)
22. Long, S., Huang, X., Chen, Z., Pardhan, S., Zheng, D.: Automatic detection of hard exudates in color retinal images using dynamic threshold and SVM classification: algorithm development and evaluation. BioMed Res. Int. **2019**, 13 (2019)
23. Medeiros, M.D., Mesquita, E., Papoila, A.L., Genro, V., Raposo, J.F.: First diabetic retinopathy prevalence study in Portugal: RETINODIAB study - evaluation of the screening programme for Lisbon and Tagus Valley region. Br. J. Ophthalmol. **99**(10), 1328–1333 (2015)
24. Mo, J., Zhang, L., Feng, Y.: Exudate-based diabetic macular edema recognition in retinal images using cascaded deep residual networks. Neurocomputing **290**, 161–171 (2018)

25. Pires, R., Jelinek, H.F., Wainer, J., Valle, E., Rocha, A.: Advancing bag-of-visual-words representations for lesion classification in retinal images. PloS One **9**(6), e96814 (2014)
26. Porwal, P., Pachade, S., Kamble, R., Kokare, M., Deshmukh, G., Sahasrabuddhe, V., et al.: Indian diabetic retinopathy image dataset (IDRiD). IEEE Dataport (2018)
27. Porwal, P., Pachade, S., Kamble, R., Kokare, M., Deshmukh, G., Sahasrabuddhe, V., et al.: Indian diabetic retinopathy image dataset (IDRiD): a database for diabetic retinopathy screening research. Data **3**(3), 25 (2018)
28. Porwal, P., Pachade, S., Kokare, M., Deshmukh, G., Son, J., Bae, W., et al.: IDRiD: diabetic retinopathy-segmentation and grading challenge. Med. Image Anal. **59**, 101561 (2020)
29. Prentašić, P., Lončarić, S., Vatavuk, Z., Benčić, G., Subašić, M., Petković, T., et al.: Diabetic retinopathy image database (DRiDB): a new database for diabetic retinopathy screening programs research. In: 2013 8th International Symposium on Image and Signal Processing and Analysis (ISPA), pp. 711–716. IEEE (2013)
30. Rahim, S.S., Palade, V., Almakky, I., Holzinger, A.: Detection of diabetic retinopathy and maculopathy in eye fundus images using deep learning and image augmentation. In: Holzinger, A., Kieseberg, P., Tjoa, A.M., Weippl, E. (eds.) CD-MAKE 2019. LNCS, vol. 11713, pp. 114–127. Springer, Cham (2019). https://doi.org/10.1007/978-3-030-29726-8_8
31. Rekhi, R.S., Issac, A., Dutta, M.K., Travieso, C.M.: Automated classification of exudates from digital fundus images. In: 2017 International Conference and Workshop on Bioinspired Intelligence (IWOBI), pp. 1–6. IEEE (2017)
32. Russakovsky, O., et al.: ImageNet large scale visual recognition challenge. Int. J. Comput. Vis. **115**(3), 211–252 (2015). https://doi.org/10.1007/s11263-015-0816-y
33. Schwartz, R., Dodge, J., Smith, N.A., Etzioni, O.: Green AI. arXiv preprint arXiv:1907.10597 (2019)
34. Selvaraju, R.R., Cogswell, M., Das, A., Vedantam, R., Parikh, D., Batra, D.: Gradcam: visual explanations from deep networks via gradient-based localization. In: IEEE International Conference on Computer Vision, pp. 618–626 (2017)
35. Srivastava, N., Hinton, G., Krizhevsky, A., Sutskever, I., Salakhutdinov, R.: Dropout: a simple way to prevent neural networks from overfitting. J. Mach. Learn. Res/ **15**(1), 1929–1958 (2014)
36. Tariq, A., Akram, M.U., Shaukat, A., Khan, S.A.: Automated detection and grading of diabetic maculopathy in digital retinal images. J. Digit. Imaging **26**(4), 803–812 (2013)
37. Wanderley, D.S., Araújo, T., Carvalho, C.B., Maia, C., Penas, S., Carneiro, Â., et al.: Analysis of the performance of specialists and an automatic algorithm in retinal image quality assessment. In: 2019 IEEE 6th Portuguese Meeting on Bioengineering (ENBENG), pp. 1–4. IEEE (2019)
38. Zander, E., Herfurth, S., Bohl, B., Heinke, P., Herrmann, U., Kohnert, K.D., et al.: Maculopathy in patients with diabetes mellitus type 1 and type 2: associations with risk factors. Br. J. Ophthalmol. **84**(8), 871–876 (2000)
39. Zhang, X., Thibault, G., Decencière, E., Marcotegui, B., Laÿ, B., Danno, R., et al.: Exudate detection in color retinal images for mass screening of diabetic retinopathy. Med. Image Anal. **18**(7), 1026–1043 (2014)
40. Zhang, Y., Wu, H., Liu, H., Tong, L., Wang, M.D.: Improve model generalization and robustness to dataset bias with bias-regularized learning and domain-guided augmentation. arXiv preprint arXiv:1910.06745 (2019)

Enhancement of Retinal Fundus Images via Pixel Color Amplification

Alex Gaudio[1]([✉])[iD], Asim Smailagic[1][iD], and Aurélio Campilho[2,3][iD]

[1] Carnegie Mellon University, Pittsburgh, PA 15213, USA
agaudio@andrew.cmu.edu, asim@cs.cmu.edu
[2] INESC TEC, Porto, Portugal
[3] Faculty of Engineering, University of Porto, Porto, Portugal
campilho@fe.up.pt

Abstract. We propose a pixel color amplification theory and family of enhancement methods to facilitate segmentation tasks on retinal images. Our novel re-interpretation of the image distortion model underlying dehazing theory shows how three existing priors commonly used by the dehazing community and a novel fourth prior are related. We utilize the theory to develop a family of enhancement methods for retinal images, including novel methods for whole image brightening and darkening. We show a novel derivation of the Unsharp Masking algorithm. We evaluate the enhancement methods as a pre-processing step to a challenging multi-task segmentation problem and show large increases in performance on all tasks, with Dice score increases over a no-enhancement baseline by as much as 0.491. We provide evidence that our enhancement preprocessing is useful for unbalanced and difficult data. We show that the enhancements can perform class balancing by composing them together.

Keywords: Image enhancement · Medical image analysis · Dehazing · Segmentation · Multi-task learning

1 Introduction

Image enhancement is a process of removing noise from images in order to improve performance on a future image processing task. We consider image-to-image pre-processing methods intended to facilitate a downstream image processing task such as Diabetic Retinopathy lesion segmentation, where the goal is to identify which pixels in an image of a human retina are pathological. In this setting, image enhancement does not in itself perform segmentation, but rather it elucidates relevant features. Figure 1 shows an example enhancement with our method, which transforms the color of individual pixels and enhances fine detail.

Our main contributions are to re-interpret the distortion model underlying dehazing theory as a theory of pixel color amplification. Building on the widely known Dark Channel Prior method [5], we show a novel relationship between three previously known priors and a fourth novel prior. We then use these four

© Springer Nature Switzerland AG 2020
A. Campilho et al. (Eds.): ICIAR 2020, LNCS 12132, pp. 299–312, 2020.
https://doi.org/10.1007/978-3-030-50516-5_26

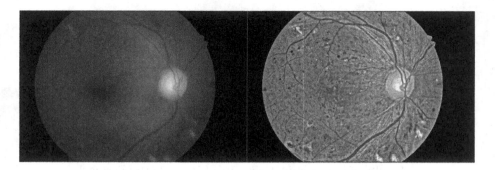

Fig. 1. Comparing unmodified image (left) to our enhancement of it (right).

priors to develop a family of brightening and darkening methods. Next, we show how the theory can derive the Unsharp Masking method for image sharpening. Finally, show that the pre-processing enhancement methods improve performance of a deep network on five retinal fundus segmentation tasks. We also open source our code for complete reproducibility [4].

2   Related Work

Natural images are distorted by refraction of light as it travels through the transmission medium (such as air), causing modified pixel intensities in the color channels of the image. A widely used physical theory for this distortion has traditionally been used for single image dehazing [2,5,8,14]:

$$\mathbf{I}(\mathbf{x}) = \mathbf{J}(\mathbf{x})t(\mathbf{x}) + \mathbf{A}(1 - t(\mathbf{x})), \tag{1}$$

where each pixel location, $\mathbf{x}$, in the distorted RGB image, $\mathbf{I}$, can be constructed as a function of the distortion-free radiance image $\mathbf{J}$, a grayscale transmission map image t quantifying the relative portion of the light ray coming from the observed surface in $\mathbf{I}(\mathbf{x})$ that was not scattered (and where $t(\mathbf{x}) \in [0,1] \; \forall \; \mathbf{x}$), and an atmosphere term, $\mathbf{A}$, which is typically a RGB vector that approximates the color of the uniform scattering of light. Distortion is simply a non-negative airlight term $\mathbf{A}(1 - t(\mathbf{x}))$. We refer to [2] for a deeper treatment of the physics behind the theory in a dehazing context. Obtaining a distortion free image $\mathbf{J}$ via this theory is typically a three step process. Given $\mathbf{I}$, define an atmosphere term $\mathbf{A}$, solve for the transmission map t, and then solve for $\mathbf{J}$. We develop new insights into this theory by demonstrating ways in which it can behave as a pixel amplifier when t and $\mathbf{A}$ are allowed to be three channel images.

The well known Dark Channel Prior (DCP) method [5,7] addresses the dehazing task for Eq. (1) by imposing a prior assumption on RGB images. The assumption differentiates the noisy (hazy) image, $\mathbf{I}$, from its noise free (dehazed) image, $\mathbf{J}$. That is, in any haze-free multi-channel region of a RGB image, at least one pixel has zero intensity in at least one channel ($\{(0, g, b), (r, 0, b), (r, g, 0)\}$), while

Dark Channel Prior (Dehazing)

Inverted DCP (Illumination Correction)

$$A = (r, g, b). \tag{2}$$

$$\tilde{t}(\mathbf{x}) = 1 - \min_c \min_{\mathbf{y} \in \Omega_{I(\mathbf{x})}} \frac{I^c(\mathbf{y})}{A^c} \tag{3}$$

$$t(\mathbf{x}) = \texttt{guidedFilter}(\mathbf{I}, \tilde{t}(\mathbf{x})). \tag{4}$$

$$\mathbf{J}(\mathbf{x}) = \frac{\mathbf{I}(\mathbf{x}) - \mathbf{A}}{\max(t(\mathbf{x}), \epsilon)} + \mathbf{A}. \tag{5}$$

$$\implies \mathbf{J} = f_{\text{DCP}}(\mathbf{I}, \mathbf{A}) \tag{6}$$

$$\mathbf{A} = (1, 1, 1). \tag{7}$$

$$\mathbf{J} = 1 - f_{\text{DCP}}(1 - \mathbf{I}, \mathbf{A}) \tag{8}$$

Bright Channel Prior (Exposure Correction)

$$\tilde{t}(x) = 1 - \max_c \max_{\mathbf{y} \in \Omega_{I(\mathbf{x})}} \frac{I^c(\mathbf{y})}{A^c} \tag{9}$$

$$t(\mathbf{x}) = \texttt{guidedFilter}(\mathbf{I}, \tilde{t}(\mathbf{x})). \tag{10}$$

$$\implies \mathbf{J} = f_{\text{BCP}}(\mathbf{I}, \mathbf{A}) \tag{11}$$

Fig. 2. Left: Dark Channel Prior (DCP) method for dehazing. Given an (inverted) image **I** and atmosphere **A**, obtain transmission map **t** and then recover **J**, the undistorted image. **Top and Bottom Right:** Two priors based on inversion of the Dark Channel Prior.

a hazy region will have no pixels with zero intensity ($r > 0, g > 0, b > 0$). The assumption is invalid if any channel of a distorted image is sparse or if all channels of the undistorted image are not sparse. To quantify distortion in an image, the assumption justifies creating a fourth channel, known as the dark channel, by applying a min operator convolutionally to each region of the images **I** and **J**. Specifically, $\tilde{I}^{\text{dark}}(\mathbf{x}) = \min_c \min_{\mathbf{y} \in \Omega_I(\mathbf{x})} \frac{I^{(c)}(\mathbf{y})}{A^c}$, where c denotes the color channel (red, green or blue) and $\Omega_I(\mathbf{x})$ is a set of pixels in **I** neighboring pixel **x**. The min operator causes $\tilde{I}^{\text{dark}}$ to lose fine detail, but an edge-preserving filter known as the guided filter [6] restores detail $\mathbf{I}^{\text{dark}} = g(\tilde{\mathbf{I}}^{\text{dark}}, \mathbf{I})$. While $\mathbf{J}^{\text{dark}}(\mathbf{x})$ always equals zero and therefore cancels out of the equations, $\mathbf{I}^{\text{dark}}(\mathbf{x})$ is non-zero in hazy regions. By observing that the distortion free image $\mathbf{J}^{\text{dark}}$ is entirely zero while $\mathbf{I}^{\text{dark}}$ is not entirely zero, solving Eq. (1) for **t** leads to Eq. (4) and then Eq. (5) in Fig. 2. In practice, the denominator of (5) is $\max(t(\mathbf{x}), \epsilon)$ to avoid numerical instability or division by zero; this amounts to preserving a small amount of distortion in heavily distorted pixels. Figure 2 summarizes the mathematics.

The DCP method permits various kinds of inversions. The bright channel prior [15] solves for a transmission map by swapping the min operator for a max operator in Eq. (4). This prior was shown useful for exposure correction. Figure 2 shows our variation of the bright channel prior based more directly on DCP mathematics and with an incorporated guided filter. Another simple modification of the DCP method is to invert the input image **I** to perform illumination correction [3,12,13]. The central idea is to invert the image, apply the dehazing equations, and then invert the dehazed result. We demonstrate the mathematics of this inverted DCP method in Fig. 2. Color illumination literature requires the assumption that $\mathbf{A} = (1, 1, 1)$, meaning the image is white-balanced.

In the dehazing context, this assumption would mean the distorted pixels are too bright, but in the color illumination context, distorted pixels are too dark. In the Methods section, we expand on the concept of brightness and darkness as pixel color amplification, show the theory supports other values of $\mathbf{A}$, and we also expand on the concept of inversion of Eqs. (4) and (5) for a wider variety of image enhancements.

3   Methods

The distortion theory Eq. (1) is useful for image enhancement. In Sect. 3.1, we show how the theory is a pixel color amplifier. In Sect. 3.2, we show ways in which the theory is invertible. We apply these properties to derive a novel prior and present a unified view of amplification under four distinct priors. Sections 3.2 and 3.2 apply the amplification theory to three specific enhancement methods: whole image brightening, whole image darkening and sharpening.

3.1   The Distortion Theory Amplifies Pixel Intensities

We assume that $\mathbf{A}$, $\mathbf{I}$ and $\mathbf{J}$ share the same space of pixel intensities, so that in any given channel c and pixel location $\mathbf{x}$, the intensities A^c, $I^c(\mathbf{x})$ and $J^c(\mathbf{x})$ can all have the same maximum or minimum value. We can derive the simple equation $t(\mathbf{x}) = \frac{I^{(c)}(\mathbf{x})-A^{(c)}}{J^{(c)}(\mathbf{x})-A^{(c)}} \in [0,1]$ from Eq. (1) by noting that the distortion theory presents a linear system containing three channels. The range of $\mathbf{t}$ implies the numerator and denominator must have the same sign. For example, if $A^{(c)} \geq I^{(c)}(\mathbf{x})$, then the numerator and denominator are non-positive and $J^{(c)}(\mathbf{x}) \leq I^{(c)}(\mathbf{x}) \leq A^{(c)}$. Likewise, when $A^{(c)} \leq I^{(c)}(\mathbf{x})$, the order is reversed $J^{(c)}(\mathbf{x}) \geq I^{(c)}(\mathbf{x}) \geq A^{(c)}$. These two ordering properties show the distortion theory amplifies pixel intensities. The key insight is that the choice of $\mathbf{A}$ determines how the color of each pixel in the recovered image $\mathbf{J}$ changes. Models that recover $\mathbf{J}$ using Eq. (1) will simply amplify color values for each pixel $\mathbf{x}$ in the direction $\mathbf{I}(\mathbf{x}) - \mathbf{A}$.

Atmosphere Controls the Direction of Amplification in Color Space. The atmosphere term $\mathbf{A}$ is traditionally a single RGB color vector with three scalar values, $A = (r, g, b)$, but it can also be an RGB image matrix. As a RGB vector, $\mathbf{A}$ does not provide precise pixel level control over the amplification direction. For instance, two pixels with the same intensity are guaranteed to change color in the same direction, even though it may be desirable for these pixels to change color in opposite directions. Fortunately, considering $\mathbf{A}$ as a three channel RGB image enables precise pixel level control of the amplification direction. It is physically valid to consider $\mathbf{A}$ as an image since the atmospheric light may shift color across the image, for instance due to a change in light source. As an image, $\mathbf{A}$ can be chosen to define the direction of color amplification $I^c(\mathbf{x}) - A^c(\mathbf{x})$ for each pixel and each color channel independently.

Transmission Map and Atmosphere Both Control the Rate of Amplification. Both the transmission map t and the magnitude of the atmosphere term **A** determine the amount or rate of pixel color amplification. The effect on amplification is shown in the equation $\mathbf{J} = \frac{\mathbf{I}-\mathbf{A}}{\mathbf{t}} + \mathbf{A}$, where the difference $\mathbf{I} - \mathbf{A}$ controls the direction and magnitude of amplification and t affects the amount of difference to amplify. The transmission map itself is typically a grayscale image matrix, but it can also be a scalar constant or a three channel color image. Each value $t(\mathbf{x}) \in [0, 1]$ is a mixing coefficient specifying what proportion of the signal is not distorted. When $t(\mathbf{x}) = 1$, there is no distortion; the distorted pixel $\mathbf{I}(\mathbf{x})$ and corresponding undistorted pixel $\mathbf{J}(\mathbf{x})$ are the same since $\mathbf{I}(\mathbf{x}) = \mathbf{J}(\mathbf{x}) + 0$. As $t(\mathbf{x})$ approaches zero, the distortion caused by the difference between the distorted image $\mathbf{I}$ and the atmosphere increases.

3.2 Amplification Under Inversion

The distortion theory supports several kinds of inversion. The Eqs. (4) and (5) are invertible. The input image $\mathbf{I}$ can also undergo invertible transformations. We prove these inversion properties and show why they are useful.

Inverting Eq. (4) Results in a Novel DCP-Based Prior. We discussed in Related Work three distinct priors that provide a transmission map: the traditional DCP approach with Eq. (4); the bright channel prior in Eq. (10); and color illumination via Eq. (8). Bright channel prior and color illumination respectively perform two types of inversion; the former changes the min operator to a max operator while the latter inverts the image $1 - \mathbf{I}$. Combining these two inversion techniques results in a novel fourth prior. In Table 1, we show the four transmission maps. We show that each prior has a solution using either the min or max operator, which is apparent by the following two identities:

$$\texttt{solve\_t}(\mathbf{I}, \mathbf{A}) = 1 - \min_{c} \min_{y \in \Omega_{I(\mathbf{x})}} \frac{I^c(\mathbf{y})}{A^c} \equiv \max_{c} \max_{y \in \Omega_{I(\mathbf{x})}} \frac{1 - I^c(\mathbf{y})}{A^c} \quad (12)$$

$$\texttt{solve\_t}(\mathbf{I}, \mathbf{A}) = 1 - \max_{c} \max_{y \in \Omega_{I(\mathbf{x})}} \frac{I^c(\mathbf{y})}{A^c} \equiv \min_{c} \min_{y \in \Omega_{I(\mathbf{x})}} \frac{1 - I^c(\mathbf{y})}{A^c} \quad (13)$$

The unified view of these four priors in Table 1 provides a novel insight into how they are related. In particular, the table provides proof that the Color Illumination Prior and Bright Channel Prior are inversely equivalent and utilize statistics of the maximum pixel values across channels. Similarly, DCP and our prior are also inversely equivalent and utilize statistics of the minimum pixel values across channels. This unified view distinguishes between weak and strong amplification, and amplification of bright and dark pixel neighborhoods.

In Fig. 3, we visualize these four transmission maps to demonstrate how they collectively perform strong or weak amplification of bright or dark regions of the input image. In this paper, we set $\mathbf{A} = 1$ when solving for t. Any choice of $A^c \in (0, 1]$ is valid, and when all A^c are equal, smaller values of $\mathbf{A}$ are guaranteed to amplify the differences between these properties further.

Table 1. Four transmission maps derived from variations of Eq. (4). For clear notation, we used the vectorized functions $\mathbf{t} = \mathtt{solveMin\_t}(\mathbf{I}, \mathbf{A}) = 1 - \min_c \min_{\mathbf{y} \in \Omega_I(\mathbf{x})} \frac{I^c(\mathbf{y})}{A^c}$ and $\mathbf{t} = \mathtt{solveMax\_t}(\mathbf{I}, \mathbf{A}) = 1 - \max_c \max_{\mathbf{y} \in \Omega_I(\mathbf{x})} \frac{I^c(\mathbf{y})}{A^c}$.

	Amplify dark areas	Amplify bright areas
Weak amplification	$\mathtt{solveMin\_t}(1 - \mathbf{I}, \mathbf{A} = 1)$ $1 - \mathtt{solveMax\_t}(\mathbf{I}, \mathbf{A} = 1)$ Color Illumination Prior	$\mathtt{solveMin\_t}(\mathbf{I}, \mathbf{A} = 1)$ $1 - \mathtt{solveMax\_t}(1 - \mathbf{I}, \mathbf{A} = 1)$ Standard Dark Channel Prior
Strong amplification	$1 - \mathtt{solveMin\_t}(\mathbf{I}, \mathbf{A} = 1)$ $\mathtt{solveMax\_t}(1 - \mathbf{I}, \mathbf{A} = 1)$ Our novel prior	$1 - \mathtt{solveMin\_t}(1 - \mathbf{I}, \mathbf{A} = 1)$ $\mathtt{solveMax\_t}(\mathbf{I}, \mathbf{A} = 1)$ Bright Channel Prior

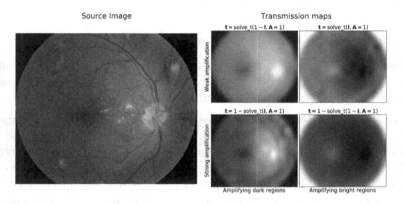

Fig. 3. The transmission maps (right) obtained from source image (left) selectively amplify bright or dark regions. Dark pixels correspond to a larger amount of amplification. We randomly sample a retinal fundus image from the IDRiD dataset (see Sect. 4.1). We set the blue channel to all ones when computing the transmission map for the top right and bottom left maps because the min values of the blue channel in retinal fundus images are noisy. (Color figure online)

Inverting Eq. (5) Motivates Brightening and Darkening. Given an image $\mathbf{I}$, transmission map $\mathbf{t}$ and an atmosphere $\mathbf{A}$, solving for the recovered image $\mathbf{J}$ with Eq. (5) can be computed two equivalent ways, as we demonstrate by the following identity:

$$\mathbf{J} = \mathtt{solve\_J}(\mathbf{I}, \mathbf{t}, \mathbf{A}) \equiv 1 - \mathtt{solve\_J}(1 - \mathbf{I}, \mathbf{t}, 1 - \mathbf{A}) \tag{14}$$

The proof is by simplification of $\frac{\mathbf{I} - \mathbf{A}}{\mathbf{t}} + \mathbf{A} = 1 - \left(\frac{(1 - \mathbf{I}) - (1 - \mathbf{A})}{\mathbf{t}} + (1 - \mathbf{A}) \right)$. It implies the space of possible atmospheric light values, which is bounded in $[0, 1]$, is symmetric under inversion.

We next prove that solving for $\mathbf{J}$ via the color illumination method [3,12,13] is equivalent to direct attenuation $\mathbf{J} = \frac{\mathbf{I}}{\mathbf{t}}$, a fact that was not clear in prior work. As we noted in Eq. (8), color illumination solves $\mathbf{J} = 1 - \frac{(1 - \mathbf{I}) - \mathbf{A}}{\mathbf{t}} + \mathbf{A}$ under the required assumption that $\mathbf{A} = 1$. We can also write the atmosphere as $\mathbf{A} = 1 - \mathbf{0}$.

Then, the right hand side of (14) leads to $\mathbf{J} = 1 - \texttt{solve\_J}(1 - \mathbf{I}, \mathbf{t}, \mathbf{A} = 1 - 0) = \frac{1-0}{t} + \mathbf{0}$. Therefore, color illumination actually performs whole image brightening with the atmosphere $\mathbf{A} = (0, 0, 0)$ even though the transmission map uses a white-balanced image assumption that $\mathbf{A} = (1, 1, 1)$. Both this proof and the invertibility property Eq. (14) motivate Sect. 3.2 where we perform brightening and darkening with all priors in Table 1.

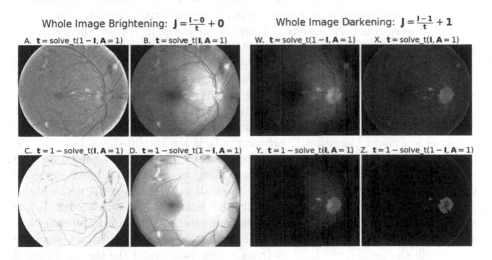

Fig. 4. Whole image brightening (left) and darkening (right) using the corresponding four transmission maps in Fig. 3. Note that $\mathbf{A} = 1$ when solving for $\mathbf{t}$, but $\mathbf{A} = 0$ or $\mathbf{A} = 1$, respectively, for brightening or darkening. (Color figure online)

Application to Whole Image Brightening and Darkening. Brightening versus darkening of colors is a matter of choosing an amplification direction. Extremal choices of the atmosphere term $\mathbf{A}$ result in brightening or darkening of all pixels in the image. For instance, $\mathbf{A} = (1, 1, 1)$ guarantees for each pixel $\mathbf{x}$ that the recovered color $\mathbf{J}(\mathbf{x})$ is darker than the distorted color $\mathbf{I}(\mathbf{x})$ since $\mathbf{J} \leq \mathbf{I} \leq \mathbf{A}$, while $\mathbf{A} = (0, 0, 0)$ guarantees image brightening $\mathbf{J} \geq \mathbf{I} \geq \mathbf{A}$. More generally, any $\mathbf{A}$ satisfying $1 \geq A^c \geq \max_{\mathbf{x}} I^c(\mathbf{x})$ performs whole image brightening and any $\mathbf{A}$ satisfying $0 \leq A^c \leq \min_{\mathbf{x}} I^c(\mathbf{x})$ performs whole image darkening. We utilize the four distinct transmission maps from Table 1 to perform brightening $\mathbf{A} = 0$ or darkening $\mathbf{A} = 1$, resulting in eight kinds of amplification. We visualize these maps and corresponding brightening and darkening techniques applied to retinal fundus images in Fig. 4. Our application of the Bright Channel Prior and Color Illumination Prior for whole image darkening is novel. Utilizing our prior for brightening and darkening is also novel.

Application to Image Sharpening. We show a novel connection between dehazing theory and *unsharp masking*, a deblurring method and standard image sharpening technique that amplifies fine detail [9]. Consider $\mathbf{A}$ as a three channel image obtained by applying a non-linear blur operator to $\mathbf{I}$, $\mathbf{A} = \text{blurry}(\mathbf{I})$. Solving Eq. (1) for $\mathbf{J}$ gives $\mathbf{J} = \frac{1}{t}\mathbf{I} - \frac{(1-t)}{t}\mathbf{A}$. Since each scalar value $t(\mathbf{x})$ is in $[0, 1]$, we can represent the fraction $t(\mathbf{x}) = \frac{1}{u(x)}$. Substituting, we have the simplified matrix form $\mathbf{J} = \mathbf{u} \circ \mathbf{I} - (\mathbf{u} - 1) \circ \text{blurry}(\mathbf{I})$ where the $\circ$ operator denotes element-wise multiplication with broadcasting across channels. This form is precisely *unsharp masking*, where $\mathbf{u}$ is either a constant, or $\mathbf{u}$ is a 1-channel image matrix determining how much to sharpen each pixel. The matrix form of $\mathbf{u}$ is known as locally adaptive unsharp masking. Thus, we show the distortion theory in Eq. (1) is equivalent to image sharpening by choosing $\mathbf{A}$ to be a blurred version of the original input image.

We present two sharpening algorithms, Algorithm 1 and 2, and show their respective outputs in Fig. 5. Sharpening amplifies differences between an image and a blurry version of itself. In unevenly illuminated images, the dark or bright regions may saturate to zero or one respectively. Therefore, the use of a scalar transmission map (Algorithm 1), where all pixels are amplified, implies that the input image should ideally have even illumination. The optional guided filter in the last step provides edge preserving smoothing and helps to minimize speckle noise, but can cause too much blurring on small images, hence the if condition.

Algorithm 2 selectively amplifies only the regions that have an edge. Edges are found by deriving a three channel transmission map from a Laplacian filter applied to a morphologically smoothed fundus image. We enhance edges by recursively sharpening the Laplace transmission map under the theory. Figure 6 shows the results of sharpening each image in Fig. 4 with Algorithm 2.

4   Experiments

Our primary hypothesis is that enhancement facilitates a model's ability to learn retinal image segmentation tasks. We introduce a multi-task dataset and describe our deep network implementation.

4.1   Datasets

The **Indian Diabetic Retinopathy Dataset (IDRiD)** [11] contains 81 retinal fundus images for segmentation, with a train-test split of 54:27 images. Each image is 4288×2848 pixels. Each pixel has five binary labels for presence of: Microaneurysms (MA), Hemorrhages (HE), Hard Exudates (EX), Soft Exudates (SE) and Optic Disc (OD). Only 53:27 and 26:14 images present HE and SE, respectively. Table 2 shows the fraction of positive pixels per category is unbalanced both across categories (left columns) and within categories (right columns).

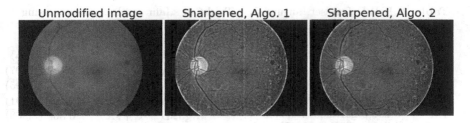

Fig. 5. Sharpening a retinal fundus image with Algorithm 1 (middle) and 2 (right). Image randomly sampled from IDRiD training dataset (described in Sec. 4.1). (Color figure online)

Algorithm 1: Image Sharpening, simple	**Algorithm 2:** Image Sharpening, complex
Input: I (input fundus image)	**Input: I** (input fundus image)
Result: J (sharpened image)	**Result: J** (sharpened image)
$\mathbf{A} = \texttt{blur}(\mathbf{I}, \texttt{blur\_radius})$;	$\tilde{\mathbf{t}} = \texttt{Algo\_1}($
$\mathbf{t} = 0.15$;	$\quad\quad \texttt{morphological\_laplace}(\mathbf{I}, (2,2,1)))$;
if $\min(img\_width, img\_height) >$	
1500 **then**	$\tilde{\mathbf{t}} = 1 - \frac{\tilde{\mathbf{t}} - \min(\tilde{\mathbf{t}})}{\max(\tilde{\mathbf{t}}) - \min(\tilde{\mathbf{t}})}$;
$\quad$ $\mathbf{J} = \texttt{guidedFilter}(\texttt{guide} =$	$\epsilon = \max(10^{-8}, \frac{\min(\tilde{\mathbf{t}})}{2})$;
$\quad$ $\mathbf{I}, \texttt{src} = \frac{\mathbf{I}-\mathbf{A}}{\mathbf{t}} + \mathbf{A})$;	$\mathbf{t} = \texttt{elementwise\_max}(\tilde{\mathbf{t}}, \epsilon)$;
else	$\mathbf{J} = \texttt{Algo\_1}(\mathbf{I}, \mathbf{t} = \mathbf{t})$;
$\quad$ $\mathbf{J} = \frac{\mathbf{I}-\mathbf{A}}{\mathbf{t}} + \mathbf{A}$;	
end	

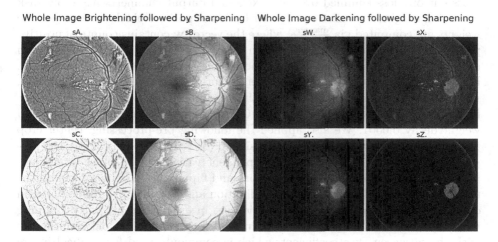

Fig. 6. The result of sharpening each image in Fig. 4 using Algorithm 2. (Color figure online)

Table 2. IDRiD Dataset, an unbalanced class distribution.

Category	Pos/∑Pos		Pos/(Pos + Neg)	
	Train	Test	Train	Test
MA	0.027	0.024	0.0007	0.0003
HE	0.253	0.256	0.0066	0.0036
EX	0.207	0.261	0.0054	0.0036
SE	0.049	0.043	0.0013	0.0006
OD	0.464	0.416	0.0120	0.0058

Table 4. Main results, pre-processing yields large improvements.

Task	Method	Dice (delta)
EX	avg4:sA + sC + sX + sZ	0.728 (0.407)
	avg2:sA + sZ	0.615 (0.295)
HE	avg3:sA + sC + sX	0.491 (0.491)
	avg3:sB + sC + sX	0.368 (0.368)
MA	avg4:A + B + C + X	0.251 (0.251)
	avg2:sB + sX	0.219 (0.219)
OD	avg4:sA + sC + sX + sZ	0.876 (0.359)
	avg2:sA + sZ	0.860 (0.343)
SE	avg4:sA + sC + sX + sZ	0.491 (0.332)
	avg3:B + C + X	0.481 (0.322)

Table 3. Competing method results, best per category of A, B, D, or X.

Task	Method	Dice (delta)
EX	A	0.496 (0.175)
HE	A	0.102 (0.102)
MA	A	0.122 (0.122)
OD	A	0.849 (0.332)
SE	A	0.423 (0.264)

Blackbox Evaluation: Does an Enhancement Method Improve Performance? We implement and train a standard U-Net model [10] and evaluate change in performance via the Dice coefficient. We apply this model simultaneously to five segmentation tasks (MA, HE, SE, EX, OD) on the IDRiD dataset; the model has five corresponding output channels. We use a binary cross entropy loss summed over all pixels and output channels. We apply task balancing weights to ensure equal contribution of positive pixels to the loss. The weights are computed via $\frac{\max_i w_i}{\mathbf{w}}$, where the vector $\mathbf{w}$ contains counts of positive pixels across all training images for each of the 5 task categories (see left column of Table 2). Without the weighting, the model did not learn to segment MA, HE, and EX even with our enhancements. For the purpose of our experiment, we show in the results that this weighting is suboptimal as it does not balance bright and dark categories. The Adam Optimizer has a learning rate 0.001 and weight decay 0.0001. We also applied the following pre-processing: center crop the fundus to minimize background, resize to (512 × 512) pixels, apply the pre-processing enhancement method (independent variable), clip pixel values into [0, 1], randomly rotate and flip. Rotations and flipping were only applied on the training set; we excluded them from validation and test sets. We randomly hold out two training images as the validation set in order to apply early stopping with a patience of 30 epochs. We evaluate test set segmentation performance with the Sørensen-Dice coefficient, which is commonly used for medical image segmentation.

4.2   Pre-processing Enhancement Methods for Retinal Fundus Images

We combine the brightening, darkening and sharpening methods together and perform an ablation study. We assign the eight methods in Fig. 4 a letter. The brightening methods, from top left to bottom right are A, B, C, D. The corresponding darkening methods are W, X, Y, Z. We also apply sharpening via Algorithm 2. Combined methods assume the following notation: $A + X$ is the average of A and X, which is then sharpened; $sA + sX$ is the average of sharpened A with sharpened X. A standalone letter X is a sharpened X. All methods have the same hyperparameters, which we chose using a subset of IDRiD training images. When solving for t, the size of the neighborhood Ω is (5×5); the guided filter for t has radius $= 100$, $\epsilon = 1e^{-8}$. When solving for J, the max operator in the denominator is $\max\langle\min(t)/2, 1e-8\rangle$. For sharpening, we blur using a guided filter (radius $= 30$, $\epsilon = 1e^{-8}$), and we do not use a guided filter to denoise as the images are previously resized to (512×512).

5   Results

5.1   Our Pre-processing Enhancement Methods Significantly Improve Performance on All Tasks

We show the top two models with highest test performance in each category in Table 4. The delta values show the pre-processing enhancement methods significantly improve performance over the no-enhancement (identity function) baseline for all tasks, underscoring the value of our theory and methods.

Enhancement Improves Detection of Rare Classes. The smallest delta improvement in Table 4 is 0.219 for MA, the rarest category across and within categories (as shown in Table 2). Our smallest improvement is a large increase considering the largest possible Dice score is one.

Enhancement can be Class Balancing. The IDRiD results support the primary hypothesis that enhancement makes the segmentation task easier. The delta values show the baseline identity model did not learn to segment MA or HE. Indeed, during the implementation, we initially found that the model learned to segment only the optic disc (OD). Of the categories, OD has the most extremal intensities (brightest) and is typically the largest feature by pixel count in a fundus image. In our multi-task setting, the gradients from other tasks were therefore overshadowed by OD gradients. After we implemented a category balancing weight, the no-enhancement baseline model was still unable to learn MA and HE. As an explanation for this phenomenon, we observe that EX, SE and OD are bright features while MA and HE are dark features. Considering the class balancing weights, the bright features outnumber the dark features three to two. This need to carefully weigh the loss function suggests differences in color intensity values *cause* differences in performance. It is therefore particularly interesting that the enhancement methods were able to learn

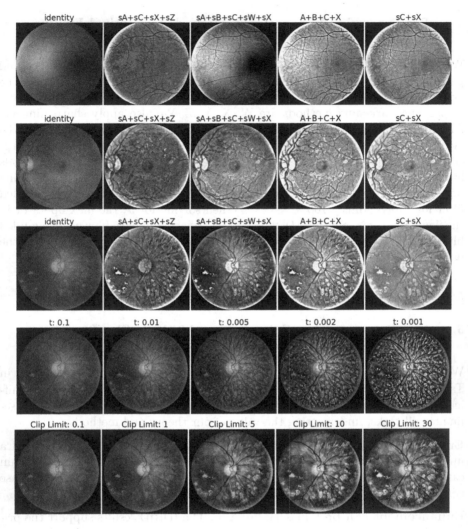

Fig. 7. Visualization of our enhancement methods. Each row is an image. Each column is an enhancement method. Last two rows compare our Algorithm 1 with CLAHE. (Color figure online)

despite also being subject to these issues. In fact, we can observe that the best enhancements in the table incorporate the Z method, which performs a strong darkening of bright regions. We interpret this result as strong evidence that our enhancement methods make the segmentation task easier, and in fact, that they can be used as a form of class balancing by image color augmentation.

5.2 Comparison to Existing Work

The methods A, D, X and arguably B correspond to existing work and were visualized (with sharpening) in Fig. 6. The A method outperforms B, D and X on all tasks. Its values, reported in Table 3, are substantially lower than the values in Table 4. We attribute the low scores to our intentional category imbalance.

Contrast Limited Adaptive Histogram Equalization (CLAHE) applied to the luminance channel of LAB colorspace is useful for retinal fundus enhancement [1]. We compare it to Algorithm 1 in bottom rows of Fig. 7, using the LAB conversion for both methods. We observe that CLAHE preserves less detail, and both methods overemphasize uneven illumination. CLAHE is faster to compute and could serve as a simple drop-in replacement, with clip limit as a proxy for the scalar t.

5.3 Qualitative Analysis

We visualize a subset of our image enhancement methods in the top three rows of Fig. 7. Each row presents a different fundus image from a private collection. We observe that the input images are difficult to see and appear to have little detail, while the enhanced images are quite colorful and very detailed. The halo effect around the fundus is caused by the guided filter (with $\epsilon = 1e^{-8}$) rather than the theory. The differences in bright and dark regions across each row provide intuitive sense of how averaging the models (Fig. 4 and 6) can yield a variety of different colorings.

6 Conclusion

In this paper, we re-interpret a theory of image distortion as pixel color amplification and utilize the theory to develop a family of enhancement methods for retinal fundus images. We expose a relationship between three existing priors commonly used for image dehazing with a fourth novel prior. We apply our theory to whole image brightening and darkening, resulting in eight enhancement methods, five of which are also novel (methods B, C, W, Y, and Z). We also show a derivation of the Unsharp Masking algorithm for image sharpening and develop a sharpening algorithm for retinal fundus images. Finally, we evaluate our enhancement methods as pre-processing steps for multi-task deep network retinal fundus image segmentation. We show the enhancement methods give strong improvements and can perform class balancing. Our pixel color amplification theory applied to retinal fundus images yields a variety of rich and colorful enhancements, as shown by our compositions of methods A-D and W-Z, and the theory shows great promise for wider adoption by the community.

Acknowledgements. We thank Dr. Alexander R. Gaudio, a retinal specialist and expert in degenerative retinal diseases, for his positive feedback and education of fundus images.

Supported in part by the National Funds through the Fundação para a Ciência e a Tecnologia within under Project CMUPERI/TIC/0028/2014.

References

1. Cao, L., Li, H., Zhang, Y.: Retinal image enhancement using low-pass filtering and alpha-rooting. Signal Process. **170**, 107445 (2020). https://doi.org/10.1016/j.sigpro.2019.107445. http://www.sciencedirect.com/science/article/pii/S0165168419304967
2. Fattal, R.: Single image dehazing. ACM Trans. Graph. **27**(3), 72:1–72:9 (2008). https://doi.org/10.1145/1360612.1360671
3. Galdran, A., Bria, A., Alvarez-Gila, A., Vazquez-Corral, J., Bertalmio, M.: On the duality between retinex and image dehazing. In: 2018 IEEE/CVF Conference on Computer Vision and Pattern Recognition (June 2018). https://doi.org/10.1109/cvpr.2018.00857
4. Gaudio, A.: Open source code (2020). https://github.com/adgaudio/ietk-ret
5. He, K., Sun, J., Tang, X.: Single image haze removal using dark channel prior. IEEE Trans. Pattern Anal. Mach. Intell. **33**(12), 2341–2353 (2011). https://doi.org/10.1109/TPAMI.2010.168
6. He, K., Sun, J., Tang, X.: Guided image filtering. IEEE Trans. Pattern Anal. Mach. Intell. **35**(6), 1397–1409 (2013). https://doi.org/10.1109/TPAMI.2012.213
7. Lee, S., Yun, S., Nam, J.H., Won, C.S., Jung, S.W.: A review on dark channel prior based image dehazing algorithms. EURASIP J. Image Video Process. **2016**(1), 4 (2016). https://doi.org/10.1186/s13640-016-0104-y
8. Narasimhan, S.G., Nayar, S.K.: Vision and the atmosphere. IJCV **48**(3), 233–254 (2002)
9. Petrou, M., Petrou, C.: Image Processing: The Fundamentals, pp. 357–360. Wiley, Chichester (2011)
10. Ronneberger, O., Fischer, P., Brox, T.: U-Net: convolutional networks for biomedical image segmentation. In: Navab, N., Hornegger, J., Wells, W.M., Frangi, A.F. (eds.) MICCAI 2015. LNCS, vol. 9351, pp. 234–241. Springer, Cham (2015). https://doi.org/10.1007/978-3-319-24574-4_28
11. Sahasrabuddhe, V., et al.: Indian diabetic retinopathy image dataset (IDRID) (2018). https://doi.org/10.21227/H25W98
12. Savelli, B., et al.: Illumination correction by dehazing for retinal vessel segmentation. In: 2017 IEEE 30th International Symposium on Computer-Based Medical Systems (CBMS), pp. 219–224 (June 2017). https://doi.org/10.1109/CBMS.2017.28
13. Smailagic, A., Sharan, A., Costa, P., Galdran, A., Gaudio, A., Campilho, A.: Learned pre-processing for automatic diabetic retinopathy detection on eye fundus images. In: Karray, F., Campilho, A., Yu, A. (eds.) ICIAR 2019. LNCS, vol. 11663, pp. 362–368. Springer, Cham (2019). https://doi.org/10.1007/978-3-030-27272-2_32
14. Tan, R.T.: Visibility in bad weather from a single image. In: 2008 IEEE Conference on Computer Vision and Pattern Recognition, pp. 1–8 (June 2008). https://doi.org/10.1109/CVPR.2008.4587643
15. Wang, Y., Zhuo, S., Tao, D., Bu, J., Li, N.: Automatic local exposure correction using bright channel prior for under-exposed images. Signal Process. **93**(11), 3227–3238 (2013). https://doi.org/10.1016/j.sigpro.2013.04.025. http://www.sciencedirect.com/science/article/pii/S0165168413001680

Wavelet-Based Retinal Image Enhancement

Safinaz ElMahmoudy[1](✉), Lamiaa Abdel-Hamid[1], Ahmed El-Rafei[2],
and Salwa El-Ramly[3]

[1] Department of Electronics and Communication, Faculty of Engineering,
Misr International University, Cairo, Egypt
safinaz.mohamed@miuegypt.edu.eg
[2] Department of Mathematics and Physics, Faculty of Engineering, Ain Shams University,
Cairo, Egypt
[3] Department of Electronics and Communication, Faculty of Engineering,
Ain Shams University, Cairo, Egypt

Abstract. Retinal images provide a simple non-invasive method for the detection
of several eye diseases. However, many factors can result in the degradation of the
images' quality, thus affecting the reliability of the performed diagnosis. Enhancement of retinal images is thus essential to increase the overall image quality. In
this work, a wavelet-based retinal image enhancement algorithm is proposed that
considers four different common quality issues within retinal images (1) noise
removal, (2) sharpening, (3) contrast enhancement and (4) illumination enhancement. Noise removal and sharpening are performed by processing the wavelet
detail subbands, such that the upper detail coefficients are eliminated, whereas
bilinear mapping is used to enhance the lower detail coefficients based on their relevance. Contrast and illumination enhancement involve applying contrast limited
adaptive histogram equalization (CLAHE) and the proposed luminance boosting
method to the approximation subband, respectively. Four different retinal image
quality measures are computed to assess the proposed algorithm and to compare its
performance against four other methods from literature. The comparison showed
that the introduced method resulted in the highest overall image improvement followed by spatial CLAHE for all the considered quality measures; thus, indicating
the superiority of the proposed wavelet-based enhancement method.

Keywords: Retinal image enhancement · Wavelet transform · Bilinear
mapping · CLAHE · Luminance boosting

1 Introduction

The World Health Organization (WHO) estimates that there is more than 2.2 billion
people worldwide suffering from vision impairments or blindness, from which at least
one billion could have been prevented by early diagnosis and proper treatment [1].
Retinal images have been widely used by ophthalmologists as well as in computer aided
diagnosis (CAD) for detection and diagnosis of retinal diseases such as age-related
macular degeneration (AMD), glaucoma and diabetic retinopathy. Retinal imaging has

© Springer Nature Switzerland AG 2020
A. Campilho et al. (Eds.): ICIAR 2020, LNCS 12132, pp. 313–324, 2020.
https://doi.org/10.1007/978-3-030-50516-5_27

the advantages of being non-invasive, cost effective, and requiring no patient preparation. In order to assure reliable diagnosis, retinal images should be of adequate quality such that retinal structures and disease lesions are sufficiently clear and separable within the images [2]. However, several factors can cause degradation of the quality of retinal images such as camera settings, experience of the operating person, patient movement or blinking, pupil dilation, retina's curved structure, ocular media opacity, and presence of diseases [3]. Bad quality retinal images can thus be characterized by blurriness, low contrast, appearance of noise and/or non-uniform illumination. Studies have shown that practical retinal image datasets commonly include a large number of bad quality retinal images, that can be as high as 60% of the total dataset images [4, 5]. Accordingly, retinal image enhancement is essential to improve the overall image quality in order to assure that the processed images are of sufficient quality for reliable diagnosis.

2 Related Work

Retinal image enhancement methods can be generally divided into two categories: *spatial* and *transform*. Spatial methods consider the direct manipulation of image pixels, most commonly by relying on histogram based enhancement techniques. Contrast limited adaptive histogram equalization (CLAHE) is the most widely implemented technique for retinal image contrast enhancement in the spatial domain. CLAHE works by dividing an image into small local tiles then applying adaptive histogram equalization to the clipped histogram of each tile. It has the advantage of improving the overall image contrast enhancement with minimal introduction of artifacts. Setiawan et al. [6] showed that for color retinal image enhancement, applying CLAHE to only the green channel resulted in significantly better results than its application to each of the RGB channels. Ramasubramanian et al. [7] compared several different methods for enhancement of the green channel of retinal images reaching the conclusion that the most robust results were achieved by applying median filter followed by CLAHE. Jintasuttisak et al. [8] applied CLAHE to the improved nonlinear hue-saturation-intensity (iNHSI) channels in order to enhance the overall appearance of color retinal images. Jin et al. [9] applied CLAHE to each of the normalized CIELAB channels for overall color retinal image enhancement. Zhou et al. [10] proposed a color retinal image enhancement algorithm that makes use of both the HSV and CIELAB color channels. Specifically, gamma correction of the (V) channel was used for luminosity enhancement, whereas CLAHE was applied to the (L) channel for contrast enhancement. Sonali et al. [11] introduced a weighted median filter, shown to outperform typical median filtering for image denoising, which they applied to each of the RGB channels alongside CLAHE for contrast enhancement.

On the other hand, transform retinal image enhancement methods benefit from transforming the image into another domain, most commonly the discrete wavelet transform (DWT). DWT is a multi-resolution technique characterized by being localized in both time and frequency, in addition to having the advantage of being consistent with the human visual system [12]. DWT decomposes an image by applying low pass (L) and high pass (H) filters to the image rows and columns resulting in four subbands: LL, LH, HL and HH. The approximation subband (LL) represents the illumination of the image, whereas the detail subbands (LH, HL and HH) convey the image high frequency information (edges and/or noise) in three orientations (horizontal, vertical and diagonal directions, respectively). Palanisamy et al. [13] considered channels from both HSV and CIELAB for enhancing the visual perception of color retinal images. Their method combines singular value decomposition (SVD) of the gamma corrected (V) channel's approximation subband, followed by applying CLAHE to the spatial L-channel. More commonly though, wavelet-based retinal image enhancement methods consider only the image's green channel which, owing to its high contrast, is the most commonly considered channel for disease diagnosis [6, 14]. Soomro et al. [14] combined the Retinex algorithm with wavelet thresholding for contrast enhancement and noise removal, respectively. In a later work by the authors [15], they enhanced the retinal image's green channel by applying bottom-top hat morphological operation for illumination enhancement, wavelet thresholding for image denoising and CLAHE for contrast enhancement. Li et al. [16, 17] proposed an enhancement method based on Dual-Tree Complex Wavelet Transform (DTCWT). For contrast enhancement, they applied morphological top hat transform to the approximation subband, whereas an improved Bayes Shrink thresholding of the detail subbands was considered for image denoising.

Most spatial retinal image enhancement algorithms rely on applying CLAHE either separately or alongside other methods, to one or more of the image channels for improving the contrast of the retinal color images. However, CLAHE has the limitation of being subject to contrast overstretching and noise enhancement problems [18]. On the other side, wavelet-based retinal image enhancement algorithms are usually applied to the green channel of retinal images which are most commonly considered for retinal image diagnosis due to their high contrast [6, 14]. So far, wavelet-based retinal image enhancement algorithms have been mainly considered for the purpose of noise removal by applying wavelet thresholding methods, whereas contrast or illumination enhancement were usually performed within the spatial domain.

In this work, a wavelet-based retinal image enhancement algorithm is introduced that considers four major quality issues: (1) noise removal, (2) sharpening, (3) contrast enhancement and (4) illumination enhancement. Generally, wavelet transform has the advantage of being consistent with the human visual system by separating the retinal images' edge and luminance information in its detail and approximation subbands, respectively. Hence, the retinal images' detail subbands were considered for noise removal and sharpness enhancement, whereas contrast and illumination improvement were applied by processing the images' approximation subbands. Noise removal was performed within the upper detail wavelet subbands, while lower detail subbands consisting of retinal structures related information were used for the image sharpening. On the other hand, contrast and illumination enhancement were applied to the level 1 approximation subband using DWT-CLAHE [18] and the introduced DWT luminance boosting, respectively.

3   Proposed Wavelet-Based Retinal Image Enhancement Algorithm

In retinal images, the green channel has the highest contrast, whereas the red and blue channels are characterized by being over and under illuminated, respectively, scarcely containing any valuable information [19]. The green channel of retinal images is hence the most commonly used for disease diagnosis [7, 16, 17], and in turn will be considered in this work. For the proposed algorithm, the green channel of each retinal image is decomposed into three wavelet levels using Daubechies 4 (db4) mother wavelet. Noise removal and sharpening are then applied to the detail (high frequency) subbands, whereas contrast enhancement and luminance boosting are applied to the approximation (low frequency) subbands. Finally, inverse wavelet transform is used to attain the enhanced retinal image. The details of the proposed wavelet-based retinal image enhancement algorithm are summarized in Fig. 1.

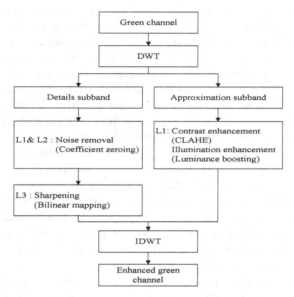

Fig. 1. Flowchart of the proposed wavelet-based retinal image enhancement algorithm.

3.1 Noise Removal

Generally, noise, characterized by its high frequency components [20], can appear in retinal images during the image acquisition process [11]. DWT has the advantage of separating an image's high frequency components in its detail subbands. However, high frequency components of an image can include both useful sharpness information as well as noise. Previous studies have shown that noise is more concentrated in the upper detail subbands (i.e. levels 1 & 2), and that the amplitude of the noise coefficients tend to rapidly decay across wavelet levels [21, 22]. On the other hand, relevant information related to the retinal structures was shown to be found within lower detail subbands (i.e. level 3) [23, 24]. Consequently for noise removal, the detail subband coefficients of both the first and second levels (L1 & L2) were zeroed, as previously performed in [13].

3.2   Sharpening

Wavelet decomposition separates the retinal image's high frequency content in the subsequent detail subbands. In the previous subsection, upper detail levels (L1 and L2) were eliminated in order to denoise the retinal images. However, lower detail subbands were previously shown to include information relevant to the main retinal structures [23, 24], where larger detail coefficients correspond to sharper edges.

Bilinear mapping [20] is a non-linear function that was used in the present study to enhance the sharpness of the retinal structures by multiplying Level 3 (L3) detail coefficients by a certain scaling factor depending on the priority of those coefficients as follows:

$$E(x) = \begin{cases} K_2x - K_1T, & x < -T \\ K_1x, & |x| \le T \\ K_2x + K_1T, & x > T \end{cases} \tag{1}$$

where x is the wavelet coefficient within the detail subband, K_1 & K_2 are the scaling factors and T is the threshold.

Since most of the significant edge details (larger coefficients) are typically within the upper half of the image histogram, the threshold value was empirically chosen to be the 80th percentile of the considered detail subband. Bilinear mapping hence divides the total range of coefficients into three regions based on their relevance as can be observed in Fig. 2. In order to avoid enhancement of noise represented by low amplitude coefficients, the middle region is linearly represented with $K_1 = 1$. However, for the other two regions including relevant information related to the retinal structures, the edges were enhanced by scaling the coefficients with gain value $K_2 = 2$ as well as adding the scaled threshold value (K_1T) while considering the coefficient's sign.

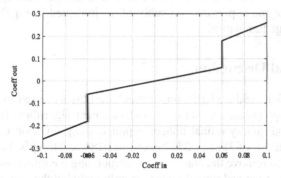

Fig. 2. Implemented bilinear mapping function.

3.3   Contrast Enhancement

CLAHE is widely used with retinal images in order to enhance the contrast between the retinal structures (i.e. vessels and disease lesions) and the retinal background to attain a clear image suitable for reliable disease diagnoses. CLAHE has been, however, typically applied to the spatial channels of the retinal images.

Wavelet decomposition allows separation of an image into illumination (low frequency) and edge or noise (high frequency) information in its approximation and detail subbands, respectively. Lidong et al. [18] have shown that for natural images, applying DWT-CLAHE to the image's L1 approximation subband has the advantage of improving the original image while effectively limiting enhancement of the noise (separated in the L1 detail subbands). To the best of the authors' knowledge, DWT-CLAHE has never been implemented for retinal image enhancement. In the introduced algorithm, CLAHE was applied to the 5 × 5 median filtered L1 approximation subband in order to improve the retinal green channel's contrast without excessively amplifying noise.

3.4 Illumination Enhancement

Under-illumination is a common issue in retinal images due to the retina's naturally curved structure. Dark images can affect the visibility of some early disease lesions, which in turn can lead to misdiagnosis. Luminance boosting is introduced in the present study in order to enhance the overall illumination of retinal images as given by the following equation:

$$I_{enh} = I_{CLAHE} + (k \times I_{BLUR}) \tag{2}$$

where (I_{BLUR}) is the blurred version of the CLAHE enhanced L1 approximation subband representing the image luminance information, whereas k is the scaling factor multiplied to (I_{BLUR}) prior to its addition to (I_{CLAHE}) contrast enhanced image. In the introduced DWT *luminance boosting* algorithm, a Gaussian blurring filter was used to extract the illumination component, whereas a scaling factor of 0.8 was empirically chosen. Since the approximation subband represents the image luminance information [2], adding the blurred version of the approximation subband, as indicated in Eq. 2, is equivalent to boosting the image luminance.

4   Results and Discussion

The high-resolution fundus (HRF) image database [25] was used to evaluate the performance of the proposed wavelet-based enhancement algorithm. The HRF database consists of 18 bad quality retinal images captured using a Canon CR-1 fundus camera having resolutions of 3888 × 2592 pixels or 5184 × 3456 pixels. All images were cropped in order to remove the black region surrounding the retina using the method in [2], such that the processed retinal images mainly included the diagnostically relevant macular region (Fig. 3). The final processed images thus had resolutions of 2116 × 1235 pixels or 2812 × 1636 pixels. All experiments were executed using MATLAB R2015a.

Two approaches are commonly used to evaluate retinal image enhancement algorithms: *subjective* and *objective* [16]. Subjective evaluation involves visual inspection of the enhanced retinal images for their quality assessment [26]. Its results may vary from one person to another depending on the expertise of the human inspector. Objective evaluation, on the other hand, rely on quality related measures computed from the enhanced

Fig. 3. Bad quality retinal image's green channel before (left) and after (right) cropping.

images, that are intended to quantify the improvement related to specific image quality issues, i.e. sharpness, contrast or illumination. Objective measures hence facilitate comparison between different enhancement approaches. In this work, both subjective and objective evaluations are utilized.

The introduced algorithm is compared to four methods from literature: spatial CLAHE (RGB) [6], in addition to the methods of Jin et al. [9] (CIELAB: CLAHE), Zhou et al. [10] (HSV: gamma correction & CIELAB: CLAHE) and Palanisamy et al. [13] (DWT-HSV: gamma correction – SVD & CIELAB: CLAHE). CLAHE is seen to be common among all methods since it is by far the most widely used enhancement method for retinal images due to its efficiency [6]. However, all methods process the images in their spatial domain, except that of Palanisamy et al. which combines wavelet enhancement with spatial CLAHE.

4.1 Subjective Evaluation

Two examples of bad quality images from the HRF dataset along with their enhanced versions from all four methods in addition to the introduced method are demonstrated in Fig. 4. The first column (Fig. 4a) shows the original bad quality retinal images characterized by blurry blood vessels, low contrast and under illumination. The second column (Fig. 4b) includes the spatial domain CLAHE [6] enhanced images showing improved contrast between the vessels and background, yet with the low sharpness issue still existing. The third column (Fig. 4c) illustrates images enhanced using Jin et al.'s algorithm [9], which despite their improved contrast, are relatively dark. The fourth column (Fig. 4d) was enhanced by Zhou et al.'s method [10]. In this case, the illumination appears better than Jin et al.'s method, but the images are still not very sharp. The fifth column (Fig. 4e) includes images enhanced using Palanisamy et al.'s algorithm [13] which is shown to introduce dark artifacts in the considered images. In comparison to the previous methods, the proposed wavelet-based enhancement algorithm shown in the last column (Fig. 4f), results in images with sharp retinal structures, clear contrast, and overall improved illumination.

4.2 Objective Evaluation

Objective evaluation provides measurable metrics to assess the different aspects of image quality. In this work, four different objective quality measures were considered (1) retinal image quality index, (2) mean, (3) standard deviation and (4) entropy. All four measures

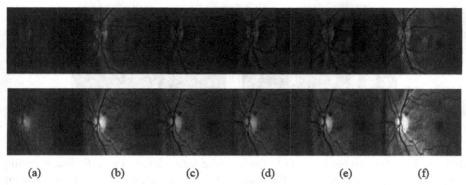

(a)   (b)   (c)   (d)   (e)   (f)

Fig. 4. Retinal images' [25] green channel for, from left to right: (a) original images, (b) CLAHE, (c) Jin et al. [9], (d) Zhou et al. [10], (e) Palanisamy et al. [13] and (f) proposed algorithm.

utilized in this work have been previously implemented in literature for the specific task of retinal image quality assessment [16, 27]. The retinal image quality index is a wavelet based quality measure for retinal images sharpness assessment, whereas the mean, standard deviation and entropy measures are spatial measures based on the properties of the human visual system [16].

Retina Image Quality Index (Qr) is a wavelet-based quality measure that has been specifically devised for evaluation of the sharpness of retinal images [27]. Qr is computed as the ratio between the detail and approximation subbands' entropies under the intuition that better quality retinal images have sharper structures, resulting in more information content within their detail subbands with respect to the approximation subband. Higher Qr measures are thus associated with sharper retinal image structures. The retinal image quality index is computed using the following equations:

$$Qr = \frac{|En_H| + |En_V|}{|En_A|} \tag{3}$$

$$En_{Sb} = -\sum_{i=1}^{N} |C_i|^2 \log|C_i|^2 \tag{4}$$

where En_H, En_V and En_A are the wavelet Shannon entropies of the third level wavelet detail horizontal, vertical and approximation subbands, respectively, N is the number of coefficients in the wavelet subband and C_i is the wavelet coefficient having an index i within the respective subband.

Mean (E_{MV}) measures the average value of image pixels in order to evaluate the overall image illumination [16]. Hence, the brighter the retinal image illumination, the higher its mean. The retinal image mean is defined as:

$$E_{MV} = \frac{1}{M \times N} \sum_{x=1}^{M} \sum_{y=1}^{N} I_{enh}(x, y) \tag{5}$$

where I_{enh} is the enhanced image, (x, y) are the spatial coordinates of the image and M & N represent the width and height of the image, respectively.

Standard Deviation (E_{SD}) measures the dispersion of the image pixel values around the image mean (E_{MV}), thus measuring the contrast of the retinal image [16]. A high E_{SD} value is equivalent to better gray scale distribution within the image, thus indicating an image with higher contrast. E_{SD} is defined as follows:

$$E_{SD} = \sqrt{\frac{1}{M \times N} \sum_{x=1}^{M} \sum_{y=1}^{N} (I_{enh}(x, y) - E_{MV})^2} \tag{6}$$

where I_{enh} is the enhanced image, (x, y) are the spatial coordinates of the image and M & N represent the width and height of the image, respectively.

Shannon Information Entropy (E_{IIE}) measures the information content within an image as defined by the following equations [16]:

$$E_{IIE} = -\sum_{x=1}^{M} \sum_{y=1}^{N} p_{x,y} log_2(p_{x,y}) \tag{7}$$

$$p_{x,y} = \frac{I_{enh}(x, y)}{M \times N} \tag{8}$$

where $p_{x,y}$ represents the probability of a certain intensity occurring in the image whose coordinates are (x, y), under the condition $\sum_{x=1}^{M} \sum_{y=1}^{N} p_{x,y} = 1$, I_{enh} is the enhanced image and M & N represent the width and height of the image, respectively. Generally, larger E_{IIE} corresponds to more information content, and in turn better enhanced retinal images [16].

Table 1 summarizes the values of the indicated evaluation metrics for the 18 bad quality HRF retinal images processed using the proposed wavelet-based retinal image enhancement algorithm. Table 2 compares the average values of each measure for the original and enhanced images, considering the four different enhancement algorithms along with the proposed method. Highest average values for the four different quality measures were achieved by the proposed wavelet-based retinal image enhancement algorithm, followed by spatial CLAHE. Specifically, the introduced method results in average Q_r, E_{MV}, E_{SD} and E_{IIE} of 10.091, 0.32, 0.15 and 6.989, respectively whereas CLAHE [6] gives 6.921, 0.228, 0.111 and 6.528, respectively. The average values summarized in Table 2 thus illustrate the superiority of the proposed method in enhancing the different common quality issues within retinal images in comparison to different enhancement methods from literature.

Table 1. Objective measures for original (org.) and enhanced (enh.) bad quality HRF images.

Image no.	Q_r		E_{MV}		E_{SD}		E_{IIE}	
	Org.	Enh.	Org.	Enh.	Org.	Enh.	Org.	Enh.
1	0.746	**9.392**	0.295	**0.494**	0.097	**0.161**	6.392	**7.352**
2	0.369	**4.865**	0.147	**0.258**	0.044	**0.077**	5.483	**6.323**
3	0.872	**6.636**	0.213	**0.457**	0.094	**0.205**	6.202	**7.474**

(continued)

Table 1. (*continued*)

Image no.	Q_r		E_{MV}		E_{SD}		E_{IIE}	
	Org.	Enh.	Org.	Enh.	Org.	Enh.	Org.	Enh.
4	1.401	**11.414**	0.215	**0.494**	0.094	**0.222**	6.370	**7.635**
5	1.228	**8.823**	0.107	**0.277**	0.058	**0.146**	5.185	**6.945**
6	0.504	**4.747**	0.152	**0.247**	0.035	**0.069**	5.087	**6.086**
7	1.262	**14.009**	0.171	**0.389**	0.076	**0.171**	5.878	**7.233**
8	0.949	**9.899**	0.100	**0.258**	0.047	**0.132**	5.313	**6.893**
9	0.692	**7.223**	0.102	**0.251**	0.047	**0.128**	5.371	**6.868**
10	1.589	**13.988**	0.218	**0.453**	0.104	**0.229**	6.278	**7.534**
11	1.903	**17.699**	0.104	**0.288**	0.057	**0.157**	5.539	**7.117**
12	2.274	**18.482**	0.115	**0.332**	0.071	**0.190**	5.699	**7.272**
13	1.149	**10.552**	0.105	**0.294**	0.074	**0.177**	5.583	**7.115**
14	1.091	**9.534**	0.097	**0.282**	0.058	**0.161**	5.477	**7.103**
15	0.559	**6.265**	0.092	**0.213**	0.035	**0.089**	4.986	**6.421**
16	0.537	**5.730**	0.095	**0.212**	0.038	**0.093**	5.069	**6.432**
17	1.105	**11.278**	0.113	**0.280**	0.055	**0.141**	5.487	**6.986**
18	1.031	**11.102**	0.114	**0.279**	0.058	**0.148**	5.622	**7.010**
Avg.	1.070	**10.091**	0.142	**0.320**	0.063	**0.150**	5.612	**6.989**

Table 2. Averages of objective measures for the different enhancement methods.

	Q_r	E_{MV}	E_{SD}	E_{IIE}
Original	1.070	0.142	0.063	5.612
CLAHE [6]	6.921	0.228	0.111	6.528
Jin et al. [9]	3.716	0.178	0.074	6.042
Zhou et al. [10]	4.436	0.236	0.079	6.100
Palanisamy et al. [13]	5.828	0.223	0.097	6.376
Proposed	**10.091**	**0.320**	**0.150**	**6.989**

5   Conclusion

Retinal images are widely used in medical diagnosis as they include valuable information while being non-invasive and requiring no patient prep. Retinal image enhancement is an essential step used to increase the reliability of performed diagnosis, both by ophthalmologists or automated systems. A wavelet-based retinal image enhancement algorithm was introduced that considers four different quality issues within the retinal images: noise removal, sharpening, contrast enhancement and illumination enhancement. Wavelet transform has the advantage of separating the image related information in different subbands. Therefore, the proposed algorithm operates by manipulating the wavelet coefficients in the subbands corresponding to the addressed quality issue. This allows to efficiently focus on each issue separately providing the most adequate enhancement for it without affecting the other quality issues. Subjective and objective image quality assessment of the enhanced images indicated the ability of the introduced method to provide sharp retinal images with high contrast and enhanced illumination. Furthermore, comparison of the introduced wavelet-based enhancement algorithm to four retinal image enhancement methods showed the superiority of the devised method.

References

1. WHO: World Report on vision: executive summary (2019)
2. Abdel-Hamid, L., El-Rafei, A., El-Ramly, S., Michelson, G., Hornegger, J.: Retinal image quality assessment based on image clarity and content. J. Biomed. Opt. **21**(9), 096007 (2016). https://doi.org/10.1117/1.JBO.21.9.096007
3. Youssif, A.A., Ghalwash, A.Z., Ghoneim, A.S.: Comparative study of contrast enhancement and illumination equalization methods for retinal vasculature segmentation. In: Cairo International Biomedical Engineering Conference, pp. 1–5 (2006)
4. Yao, Z., Zhang, Z., Xu, L.Q., Fan, Q., Xu, L.: Generic features for fundus image quality evaluation. In: IEEE 18th International Conference e-Health Networking, Applications and Services (Healthcom), pp. 1–6 (2016). https://doi.org/10.1109/HealthCom.2016.7749522
5. Yu, H., Agurto, C., Barriga, S., Nemeth, S.C., Soliz, P., Zamora, G.: Automated image quality evaluation of retinal fundus photographs in diabetic retinopathy screening. In: IEEE Southwest Symposium on Image Analysis Interpretation, pp. 125–128 (2012)
6. Setiawan, A.W., Mengko, T.R., Santoso, O.S., Suksmono, A.B.: Color retinal image enhancement using CLAHE. In: International Conference on ICT for Smart Society ICISS, pp. 215–217 (2013). https://doi.org/10.1109/ICTSS.2013.6588092
7. Ramasubramanian, B., Selvaperumal, S.: A comprehensive review on various preprocessing methods in detecting diabetic retinopathy. In: International Conference on Communication Signal Processing ICCSP, pp. 642–646 (2016). https://doi.org/10.1109/ICCSP.2016.7754220
8. Jintasuttisak, T., Intajag, S.: Color retinal image enhancement by Rayleigh contrast-limited adaptive histogram equalization. In: IEEE 14th International Conference on Control Automation and Systems (ICCAS), pp. 692–697 (2014)
9. Jin, K., et al.: Computer-aided diagnosis based on enhancement of degraded fundus photographs. Acta Ophthalmol. **96**(3), e320–e326 (2018). https://doi.org/10.1111/aos.13573
10. Zhou, M., Jin, K., Wang, S., Ye, J., Qian, D.: Color retinal image enhancement based on luminosity and contrast adjustment. IEEE Trans. Biomed. Eng. **65**(3), 521–527 (2018). https://doi.org/10.1109/TBME.2017.2700627
11. Sonali, Sahu, S., Singh, A.K., Ghrera, S.P., Elhoseny, M.: An approach for de-noising and contrast enhancement of retinal fundus image using CLAHE. Opt. Laser Technol. **110**(3), 87–98 (2019). https://doi.org/10.1016/j.optlastec.2018.06.061
12. Ninassi, A., Le Meur, O., Le Callet, P., Barba, D.: On the performance of human visual system based image quality assessment metric using wavelet domain. In: SPIE Human Vision Electronic Imaging XIII, vol. 6806, pp. 680610–680611 (2008). https://doi.org/10.1117/12.766536

13. Palanisamy, G., Ponnusamy, P., Gopi, V.P.: An improved luminosity and contrast enhancement framework for feature preservation in color fundus images. Signal Image Video Process. **13**(4), 719–726 (2019). https://doi.org/10.1007/s11760-018-1401-y

14. Soomro, T.A., Gao, J.: Non-invasive contrast normalisation and denosing technique for the retinal fundus image. Ann. Data Sci. **3**(3), 265–279 (2016). https://doi.org/10.1007/s40745-016-0079-7

15. Soomro, T.A., Gao, J., Khan, M.A.U., Khan, T.M., Paul, M.: Role of image contrast enhancement technique for ophthalmologist as diagnostic tool for diabetic retinopathy. In: International Conference on Digital Image Computing: Technical Applications DICTA, pp. 1–8 (2016). https://doi.org/10.1109/DICTA.2016.7797078

16. Li, D., Zhang, L., Sun, C., Yin, T., Liu, C., Yang, J.: Robust retinal image enhancement via dual-tree complex wavelet transform and morphology-based method. IEEE Access **7**, 47303–47316 (2019). https://doi.org/10.1109/ACCESS.2019.2909788

17. Li, D.M., Zhang, L.J., Yang, J.H., Su, W.: Research on wavelet-based contourlet transform algorithm for adaptive optics image denoising. Optik (Stuttg) **127**(12), 5029–5034 (2016). https://doi.org/10.1016/j.ijleo.2016.02.042

18. Lidong, H., Wei, Z., Jun, W., Zebin, S.: Combination of contrast limited adaptive histogram equalisation and discrete wavelet transform for image enhancement. IET Image Process. **9**(10), 908–915 (2015). https://doi.org/10.1049/iet-ipr.2015.0150

19. Walter, T., Klein, J.C.: Automatic analysis of color fundus photographs and its application to the diagnosis of diabetic retinopathy. In: Suri, J.S., Wilson, D.L., Laxminarayan, S. (eds.) Handbook of Biomedical Image Analysis. ITBE, pp. 315–368. Springer, Boston (2007). https://doi.org/10.1007/0-306-48606-7_7

20. Stefanou, H., Kakouros, S., Cavouras, D., Wallace, M.: Wavelet-based mammographic enhancement. In: 5th International Networking Conference INC, pp. 553–560 (2005)

21. Jia, C., Nie, S., Zhai, Y., Sun, Y.: Edge detection based on scale multiplication in wavelet domain. Jixie Gongcheng Xuebao/Chin. J. Mech. Eng. **42**(1), 191–195 (2006). https://doi.org/10.3901/JME.2006.01.191

22. Mallat, S., Hwang, W.L.: Singularity detection and processing with wavelets. IEEE Trans. Info. Theory **38**(2), 617–643 (1992). https://doi.org/10.1109/18.119727

23. Abdel Hamid, L.S., El-Rafei, A., El-Ramly, S., Michelson, G., Hornegger, J.: No-reference wavelet based retinal image quality assessment. In: 5th Eccomas Thematic Conference on Computational Vision and Medical Image Processing VipIMAGE, pp. 123–130 (2016)

24. Nirmala, S.R., Dandapat, S., Bora, P.K.: Wavelet weighted distortion measure for retinal images. Signal Image Video Process. **7**(5), 1005–1014 (2010). https://doi.org/10.1007/s11760-012-0290-8

25. Köhler, T., Budai, A., Kraus, M.F., Odstrčilik, J., Michelson, G., Hornegger, J.: Automatic no-reference quality assessment for retinal fundus images using vessel segmentation. In: Proceedings of the IEEE 26th Symposium Computer-Based Medical Systems CBMS, pp. 95–100 (2013). https://doi.org/10.1109/CBMS.2013.6627771

26. Dai, P., Sheng, H., Zhang, J., Li, L., Wu, J., Fan, M.: Retinal fundus image enhancement using the normalized convolution and noise removing. Int. J. Biomed. Imaging (2016). https://doi.org/10.1155/2016/5075612

27. Abdel-Hamid, L., El-Rafei, A., Michelson, G.: No-reference quality index for color retinal images. Comput. Biol. Med. **90**, 68–75 (2017). https://doi.org/10.1016/j.compbiomed.2017.09.012

An Interpretable Data-Driven Score for the Assessment of Fundus Images Quality

Youri Peskine[1]([⊠]), Marie-Carole Boucher[2], and Farida Cheriet[1]

[1] Polytechnique Montréal, Montréal, QC H3T 1J4, Canada
youri.peskine@gmail.com
[2] Hôpital Maisonneuve-Rosemont, Montréal, QC H1T 2M4, Canada

Abstract. Fundus images are usually used for the diagnosis of ocular pathologies such as diabetic retinopathy. Image quality need however to be sufficient in order to enable grading of the severity of the condition. In this paper, we propose a new method to evaluate the quality of retinal images by computing a score for each image. Images are classified as gradable or ungradable based on this score. First, we use two different U-Net models to segment the macula and the vessels in the original image. We then extract a patch around the macula in the image containing the vessels. Finally, we compute a quality score based on the presence of small vessels in this patch. The score is interpretable as the method is heavily inspired by the way clinicians assess image quality, according to the Scottish Diabetic Retinopathy Grading Scheme. The performances are evaluated on a validation database labeled by a clinician. This method presented a sensitivity of 95% and a specificity of 100% on this database.

Keywords: Diabetic retinopathy · Image quality · Deep learning · Structure-based · Data-driven

1 Introduction

Diabetic retinopathy is the leading cause of visual impairment in the working age population as this condition can appear without any symptom. Regular eye examinations are required to enable its detection and treatment. Technicians acquire retinal images of the patient's eyes and retina specialists assess them. Technicians most often take multiple images for each eye of the patient. However, some images may not be used by specialists due to their lack of quality. In the worst case, none of the images meet the quality requirements and specialists cannot grade the images. This can lead to a significant waste of time and resources for technicians, clinicians, and patients. Also, the task of evaluating the quality of a retinal image may have a subjective component as different clinicians or technicians provide different evaluations, based on their experience.

A. Campilho et al. (Eds.): ICIAR 2020, LNCS 12132, pp. 325–331, 2020.
https://doi.org/10.1007/978-3-030-50516-5_28

A detailed survey of the image quality assessment methods have been made by Raj et al. [9], dividing methods into three different categories: similarity-based, segmentation-based and machine learning based. Similarity-based methods rely on comparing fundus image features with those a selected set of good quality images. Segmentation-based methods segment precise structures in fundus images to assess its quality. Blood vessels are the main structure used in segmentation-based methods. For example, Hunter et al. [8] use blood vessels contrast to assess the fundus image quality. Machine learning based methods are data-driven. They learn to classify the data into different categories (e.g. "good quality", "poor quality"). Among these machine learning based methods, we can distinguish two other types of techniques. First, techniques that are based on hand-crafted features and techniques that are based on deep learning features. For example, in [4–6] the quality of the images are evaluated based on a set of hand-crafted features such as colour, texture or sharpness. These features are not interpretable for clinical purposes, they are typically used to assess the quality of natural images and not fundus images. Deep learning is used in [7,11] to assess the fundus image quality by extracting features. The features extracted by deep learning based methods are most often not interpretable and described as black boxes.

Our method uses deep learning to segment structures that are relevant to clinicians, resulting to an interpretable score. In this paper, we propose a segmentation-based deep learning method to evaluate the quality of each individual image to help technicians to better assess their quality and retake images when necessary. This new method is based on macula and vessel segmentations, which are regions of interest on retinal images. These regions are extracted using deep learning, but the resulting score is interpretable. This work is inspired by the way clinicians assess image quality in the Scottish Diabetic Retinopathy Grading Scheme [3]. Here, we are focused on evaluating the quality of images used to detect diabetic retinopathy but this work can also be applied to other diseases or condition detection that uses fundus images, such as age-macular degeneration or glaucoma.

2   Method

In this paper, we propose a new method inspired by the quality evaluation in the Scottish Diabetic Retinopathy Grading Scheme. According to this grading scheme, fundus images can be classified as gradable or ungradable based on quality. A gradable image contains the optic disk, the macula and "the third generation vessels radiating around the fovea". Also, the fovea needs to be more than 2 times the diameter of the optic disk from the edges of the image. Here, we will only consider the presence of the macula and the third generation vessels, as they are the most crucial regions of interest to assess the quality of the image. The goal of the method will focus on segmenting these two regions as well as giving an interpretable score to assess the quality of the image. We first segment the macula and the vessels independently in the original image. Then, we compute a score based on a patch around the macula in the vessel segmentation to evaluate the quality of the images. A threshold can be selected to classify the image as gradable or ungradable. Two different U-Net models [10] are used to segment the macula and the vessels. This model has proven to be very efficient in biomedical image segmentation. The two models are trained independently and have different architectures.

Figure 1 shows the pipeline of the method on a good quality example. Algorithm 1 presents the entire process.

2.1 Preprocessing

Both U-Net models are using the same preprocessing of the retinal images. This preprocessing consist of applying a Contrast-Limited Adaptive Histogram Equalization (CLAHE) on the LAB color space. This preprocessing is commonly used to enhance images for Diabetic Retinopathy examination.

2.2 Macula Segmentation

A U-Net model is first used to segment the macula on the preprocessed image. The architecture and the training strategy are similar to those used by Ronneberger et al. in [10]. The model is then trained on 200 retinal images annotated by retina specialists. This segmentation is used to locate the center of the macula. We compute the mean of the detected pixels to obtain the coordinates of the center of the macula. If the macula cannot be segmented, the image is automatically classified as ungradable. The result corresponds to Fig. 1 (b).

2.3 Vessel Segmentation

The vessel segmentation model is also a U-Net model but with a different architecture than the original one. Compared to the model used in [10], this model has twice the number of layers and half the number of features by layers. This prevents overfitting and widens the receptive field. This U-Net model is trained on 20 images from the DRIVE dataset [1] and 100 images from the MESSIDOR dataset [2]. The model is trained for 100 epoch with an ADAM optimizer. We also performed the following data augmentations during the training: rotation, flip and elastic deformations. The model is used to segment the vessels on the original image. The results corresponds to Fig. 1 (c).

2.4 Evaluation of the Quality of the Image

The coordinates of the center of the macula are used to extract a patch from the image containing the vessels. This patch is centered on the macula and contains the small vessels around the fovea. Note that the macula should not be visible in this patch because only the vessels are segmented. The size of the patch should cover approximately 1 diameter of the optic disk from the macula. We decided to set the size of the patch to a constant value of 25% of the original image height

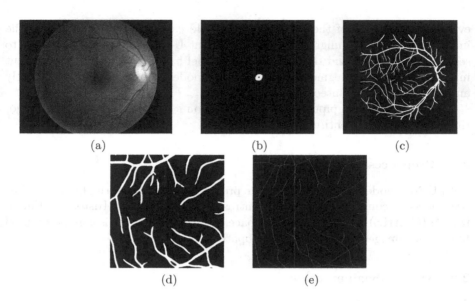

(a)   (b)   (c)

(d)   (e)

Fig. 1. pipeline of the method on a good quality example. (a) is the original fundus image. (b) is the macula segmented in the original image. The center of the macula is marked in red. (c) is the vessels segmented in the original image. The center of the macula found in (b) is marked in red. (d) is the patch extracted from (c) centered on the macula. (e) is the skeleton of image (d). (Color figure online)

Algorithm 1: the entire process of the method

Result: Score of the image
image = Image.open(fundusImage);
preprocessed_image = preprocess(image);
unet_macula = load_unet(model_macula);
unet_vessels = load_unet(model_vessels);
macula_segmented = unet_macula(preprocessed_image);
x, y = get_center_of_macula(macula_segmented);
vessels_segmented = unet_vessels(preprocessed_image);
patch = extract_patch(vessels_segmented, x, y);
skeleton = skeletonize(patch);
score = sum(skeleton);
return score;

---

and 25% of the original image width because the size of the optic disk is almost constant. Also, by not introducing another model that segments the optic disk, we reduce the propagation of errors, because an error in the segmentation of the optic disk leads to an error in the global evaluation of the quality of the image. The results corresponds to Fig. 1 (d).

Then, we compute a skeletonization on the patch of segmented vessels. This reduces the impact of large vessels and puts more weight on the number of visible vessels and their length. This is also explained by Hunter in [8]. This helps reducing the number of false positive as artifacts can sometimes be detected as vessels. With the skeletonization, their weight on the score is significantly reduced. We then simply count the number of remaining white pixels to obtain the final quality score. The results corresponds to Fig. 1 (e).

This score is an interpretable indication of the quality of the image. A null score means that the macula was not found on the image or that the vessels in the patch around the macula were not segmented, resulting in an ungradable image. A good score means that enough lengthy vessels were segmented, resulting in a gradable image. A low score means that only a few vessels were segmented around the macula. Here, we decided to set a threshold to separate gradable and ungradable images based on their score.

We set the threshold value to match the decisions of a clinician on a training set. The amount of vessels that needs to appear on the image is not addressed in the Scottish Diabetic Retinopathy Grading Scheme., the value is somewhat subjective. The threshold also helps reducing the number of false positives, at the cost of false negatives. In our case, precision is the most important measure because we often have multiple images of a patient's eye, and we only need one gradable image to evaluate the severity of the disease. Detecting all the gradable images is not as much important as making sure that the detected images are indeed gradable.

3   Results and Discussion

The validation set used to evaluate the performance of our method is constructed from the dataset used by Fasih in [6]. We used 88 images annotated by a clinician as gradable or ungradable. The dataset contains 44 images of each class. The proposed method obtained a sensitivity of 95% and a specificity of 100% on this validation set.

Only two classification errors were made; the macula on only two gradable images were not detected. Figure 2 (c) shows one of the images where the macula could not be found by our algorithm, resulting to a misclassification.

The interpretability of our score allows to better understand the output of our method. When an image is classified as ungradable, we can know if this classification is due to the lack of vessels in the patch or if the macula has not been segmented. This gives us important information to better assess the quality of an image. For example, clinicians are using multiple images of the same patient's eye to establish the diagnosis of the diabetic retinopathy. Images with low score are ungradable alone, but they still give some information that can be used by clinicians for the grading of the overall disease, paired with another image.

Figure 2 (a) and (b) show an example of a good quality retinal image having a low score with our method. This image is classified as gradable by our method, but its score is low and just above our threshold. Figure 2 (b) shows the vessel segmentation around the macula. Compared to Fig. 1 (e), very few vessels have

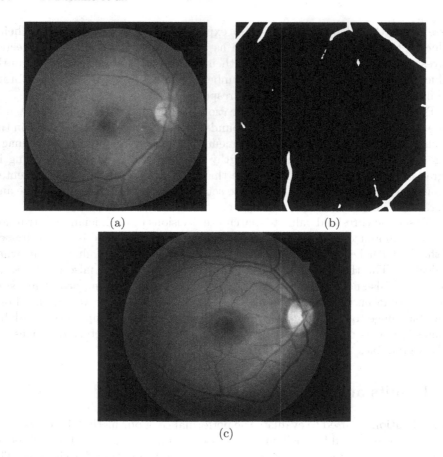

Fig. 2. A low score example, and a false negative example. (a) good quality image where the macula is successfully detected. (b) is the insufficient vessel segmentation around the macula of (a). (c) is a good quality image where the macula has not been detected (Color figure online)

been segmented in this example. This means that the image can be gradable, but its quality may not be optimal. If other images of the same patient's eye are available with a better score, they should be prioritized for the diagnostic.

The importance of the macula segmentation is also highlighted in the results of our method. We noted that 80% of the ungradable images were successfully classified due to the macula detection method. This shows how crucial this detection is. The macula is the main criterion to filter out bad quality images.

In this database, all the good quality images were centered on the macula. This is not the case for real life examples. Multiple images are taken for each of the patient's eyes and they are centered on the optic disk as well as on the macula. Our work can be further improved by generalizing this methodology on databases representative of real-life fundus image acquisitions.

4 Conclusion

In this paper, we proposed a new method for evaluating the gradability of fundus images based on the Scottish Diabetic Retinopathy Grading Scheme. This method uses two different U-Net models for the macula and vessel segmentation. The score computed is interpretable and helps understanding the evaluation of quality detection. We achieved a sensitivity of 95% and specificity of 100% while showing the importance of macula segmentation in this methodology. In future works, this method can be used to filter out ungradable images to improve the reliability of deep learning algorithms for diabetic retinopathy grading. Introducing optic disk segmentation is required to further generalize this method by taking into account other relevant information used by clinicians to assess fundus image quality.

Acknowledgements. The authors wish to acknowledge the financial support from the CIHR SPOR Network in Diabetes and its Related Complications (DAC) and the department of ophthalmology at the university of Montreal, Quebec, Canada.

References

1. Drive Database. https://drive.grand-challenge.org/
2. Messidor Database. http://www.adcis.net/fr/logiciels-tiers/messidor-fr/
3. Scottish Diabetic Retinopathy Grading Scheme (2007). https://www.ndrs.scot.nhs.uk/
4. Abdel-Hamid, L., El-Rafei, A., El-Ramly, S., Michelson, G., Hornegger, J.: Retinal image quality assessment based on image clarity and content. J. Biomed. Opt. **21**(9), 1–17 (2016). https://doi.org/10.1117/1.JBO.21.9.096007
5. Dias, J.M.P., Oliveira, C.M., da Silva Cruz, L.A.: Retinal image quality assessment using generic image quality indicators. Inf. Fusion **19**, 73–90 (2014). https://doi.org/10.1016/j.inffus.2012.08.001. http://www.sciencedirect.com/science/article/pii/S1566253512000656. Special Issue on Information Fusion in Medical Image Computing and Systems
6. Fasih, M.: Retinal image quality assessment using supervised classification. Masters thesis, École Polytechnique de Montréal (2014)
7. Fu, H., et al.: Evaluation of retinal image quality assessment networks in different color-spaces. arXiv e-prints arXiv:1907.05345 (July 2019)
8. Hunter, A., Lowell, J.A., Habib, M., Ryder, B., Basu, A., Steel, D.: An automated retinal image quality grading algorithm. In: 2011 Annual International Conference of the IEEE Engineering in Medicine and Biology Society, pp. 5955–5958 (August 2011). https://doi.org/10.1109/IEMBS.2011.6091472
9. Raj, A., Tiwari, A.K., Martini, M.G.: Fundus image quality assessment: survey, challenges, and future scope. IET Image Process. **13**(8), 1211–1224 (2019). https://doi.org/10.1049/iet-ipr.2018.6212
10. Ronneberger, O., Fischer, P., Brox, T.: U-Net: convolutional networks for biomedical image segmentation. arXiv e-prints arXiv:1505.04597 (May 2015)
11. Zago, G.T., Andreão, R.V., Dorizzi, B., Salles, E.O.T.: Retinal image quality assessment using deep learning. Comput. Biol. Med. **103**, 64–70 (2018). https://doi.org/10.1016/j.compbiomed.2018.10.004. http://www.sciencedirect.com/science/article/pii/S001048251830297X

Optic Disc and Fovea Detection in Color Eye Fundus Images

Ana Maria Mendonça[1,2], Tânia Melo[1,2(✉)], Teresa Araújo[1,2],
and Aurélio Campilho[1,2]

[1] INESC TEC - Institute for Systems and Computer Engineering,
Technology and Science, Porto, Portugal
{tania.f.melo,tfaraujo}@inesctec.pt
[2] Faculty of Engineering of the University of Porto, Porto, Portugal
{amendon,campilho}@fe.up.pt

Abstract. The optic disc (OD) and the fovea are relevant landmarks in fundus images. Their localization and segmentation can facilitate the detection of some retinal lesions and the assessment of their importance to the severity and progression of several eye disorders. Distinct methodologies have been developed for detecting these structures, mainly based on color and vascular information. The methodology herein described combines the entropy of the vessel directions with the image intensities for finding the OD center and uses a sliding band filter for segmenting the OD. The fovea center corresponds to the darkest point inside a region defined from the OD position and radius. Both the Messidor and the IDRiD datasets are used for evaluating the performance of the developed methods. In the first one, a success rate of 99.56% and 100.00% are achieved for OD and fovea localization. Regarding the OD segmentation, the mean Jaccard index and Dice's coefficient obtained are 0.87 and 0.94, respectively. The proposed methods are also amongst the top-3 performing solutions submitted to the IDRiD online challenge.

Keywords: Fundus image analysis · Optic disc localization · Optic disc segmentation · Fovea localization · Sliding band filter

1 Introduction

As shown in Fig. 1, the retinal vasculature, the optic disc (OD) and the fovea are the major landmarks of the retina that can be identified and examined through the analysis of eye fundus images. The segmentation of retinal vessels can be useful for locating the other two anatomical structures.

The OD is the area where the nervous fibers and blood vessels converge in the retina and it appears as a yellowish oval region in fundus images [1]. A correct localization and segmentation of the OD allows not only to detect appearance changes (characteristic of some eye diseases, such as the glaucoma) but also to reduce the number of bright lesions incorrectly detected in this region due to color similarities [2].

© Springer Nature Switzerland AG 2020
A. Campilho et al. (Eds.): ICIAR 2020, LNCS 12132, pp. 332–343, 2020.
https://doi.org/10.1007/978-3-030-50516-5_29

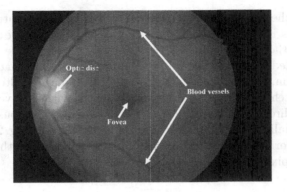

Fig. 1. An eye fundus image with the main retinal landmarks identified

Although some methods developed for OD localization are only based on the OD appearance (color, shape and size) [3], several approaches use the vascular network as the starting point for OD detection [1,4].

Several methods have also been proposed for OD segmentation, which can be divided into four categories: template-based, deformable models-based, morphology-based and pixel classification-based methods [5].

The fovea is a small depression located in the center of the macula, corresponding to the portion of the retina responsible for the central vision. Since the severity of several eye diseases depends on the distance between the lesions and the fovea center, a correct localization of this anatomical structure is crucial for a good diagnosis [6]. Furthermore, the color characterization of the region around the fovea can be relevant for the detection of red lesions (microaneurysms and hemorrhages) within the macular region, and consequently improve the performance of the red lesion detectors.

Although many approaches have already been proposed for OD and fovea detection, the results are still not satisfactory when the images have poor quality or there are lesions superimposed on these anatomical structures. This paper presents a new method for the detection of the fovea in color eye fundus images. This new approach requires the previous localization of the OD center and the estimation of its radius to define the region of the image where the fovea center is going to be looked for. The rest of this paper contains four sections. The methods for OD detection and segmentation, as well as the new proposal for fovea detection are described in Sect. 2. Sections 3 and 4 present the datasets used for methods' assessment and the obtained results, respectively. Finally, the conclusions of this work are stated in Sect. 5.

2   Methods

2.1   Optic Disc Localization

The methodology herein proposed for locating the OD in fundus images is an extended version of the method described in [1]. In the first stage, an initial estimation of the OD center is obtained by computing the entropy of the vascular directions in a low resolution version of the input image (resize factor is dependent on the field-of-view, FOV, diameter). For that, the retinal vasculature is segmented using the method proposed in [7] (Fig. 2b). In order to find the vessels' direction, the green component of the RGB image is filtered with twelve directional matched filters. The direction associated with each vessel pixel corresponds to the direction of the filter for which a maximum response is obtained at that point. Finally, the entropy of the vascular directions (Fig. 2c) is computed based on the normalized histogram of the vessel directions in the neighborhood of each image pixel (Eq. 1).

$$H = -\sum_{i=1}^{n} p_i \cdot log(p_i) \tag{1}$$

where H is the entropy of the vascular directions, n is the number of directions and p_i is the probability of occurrence of vessels in direction i.

The first estimation of the OD center corresponds to the pixel of maximum entropy. Although the OD center corresponds to the point of maximum entropy in most fundus images, there are cases for which this does not verify. For instance, when the images have poor quality or there are lesions in the OD region, the vessels may not be well segmented and, consequently, the entropy may present several local maxima, which may not correspond to the OD center. Since the OD corresponds to a bright region of the fundus images, the intensity information is used for improving the results. Taking into account that the red and the green channels present a higher contrast between the OD and the background, they are combined into a new intensity image (Fig. 2d).

In order to reduce the effect of the vessels on the selection of the OD candidates, a sequence of morphological operations is performed. Afterwards, the illumination equalization process described in [1] is applied and an initial set of OD candidates is obtained by thresholding (Fig. 2e). Since the OD is a rounded structure, a shape criterion is also used for removing the most elongated candidates.

Taking into account that the point of maximum entropy does not always correspond to the OD center and the image brightest regions are sometimes bright lesions, a validation step is included. After locating the maximum value of the entropy (ME) in the OD candidates and the maximum absolute value of the entropy (MAE) in the image, three different situations are considered:

1. ME is less than $0.4 \times$ MAE: in this case it is assumed that the ME value is not reliable and the OD center is established as the location of the MAE.
2. ME is higher than $0.6 \times$ MAE: the local entropy maxima in OD candidates are considered as potential OD centers. If there is just one candidate, the position of the maximum is defined as the initial OD center. When several local maxima are available, the entropy is calculated using the full resolution image and the location of its absolute maximum is selected as OD center.
3. For the other cases, the decision is taken between two candidates: the largest one and the one with the highest entropy. Although most of the times these

two candidates are the same, when this condition is not verified, the largest candidate is selected if its intensity is greater and its maximal entropy is still high (higher than $0.7 \times ME$); otherwise the candidate with the maximum entropy is retained.

The position of the highest entropy in the selected candidate is considered as the initial OD center. Afterwards, this position is refined through the calculation of the weighted centroid of all candidates that are inside a circular area around the initial position having a radius identical to the expected OD radius, and considering the geometric mean of the intensity and entropy values at each pixel position.

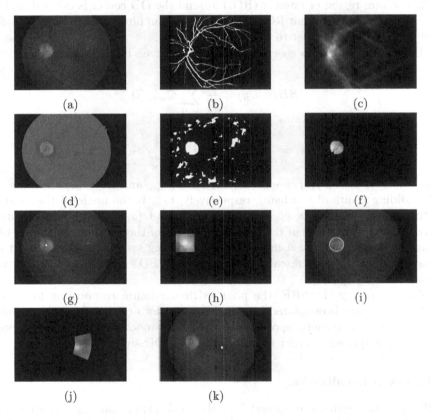

(a)   (b)   (c)

(d)   (e)   (f)

(g)   (h)   (i)

(j)   (k)

Fig. 2. Illustrative example: (a) input image; (b) map of the vasculature; (c) entropy map of the vessel directions; (d) combination of the red and green components of (a), after brightness equalization; (e) intensity-based OD candidates; (f) OD candidates used in the refinement stage multiplied by (c) and (d); (g) final OD center (green point); (h) SBF response in a region centered on the OD; (i) OD boundary (green line); (j) mask of the search region for fovea detection multiplied by the complement of the grayscale image obtained from (a) after histogram equalization and mean filtering; (k) final fovea center (blue point) (Color figure online)

2.2   Optic Disc Segmentation

The methodology used for segmenting the OD region derives from the one proposed in [5]. It requires as input an RGB eye fundus image, the map of the retinal vasculature and the initial coordinates of the OD center. The other two inputs are obtained as mentioned in Subsect. 2.1.

Initially, the input image is down-sampled (by a factor which is half the scale factor used for OD localization) and a grey-scale image is created from its red and green components. In order to reduce the effect of the vessels on the following steps, the vessel pixels are replaced in the new image by an estimate of their neighboring background. The OD boundary is then obtained through the application of a sliding band filter (SBF) [8,9]. Since this is a time-consuming task, a square region of interest (ROI) around the OD center is defined and the SBF is only applied to that ROI (Fig. 2h). Gaussian filtering is performed before applying the SBF in order to remove image noise.

The SBF output at a specific point (x, y) is given by

$$SBF(x,y) = \frac{1}{N} \sum_{i=0}^{N-1} C_{max}(i) \tag{2}$$

$$C_{max}(i) = \max_{R_{min} \leq n \leq R_{max}} \left(\frac{1}{d} \sum_{m=n}^{n+d} cos(\theta_{i,m}) \right) \tag{3}$$

where N is the number of support region lines, R_{min} and R_{max} are the inner and outer sliding limits of the band, respectively, $\theta_{i,m}$ is the angle of the gradient vector at the point m *pixels* away from the point (x, y) in direction i and d corresponds to the width of the band. The values of the parameters used in filter definition were estimated using the IDRiD training set. In the algorithm, these values are automatically rescaled according to the FOV diameter and FOV angle of the input image.

After applying the SBF, the point with maximum response is found and the OD boundary is estimated based on the positions of the sliding band points which contribute to that response. In the end, the smoothing algorithm proposed in [5] is also applied in order to obtain the final OD shape (Fig. 2i).

2.3   Fovea Localization

Although the methods proposed for OD localization and segmentation have derived from previous works, a novel methodology is herein proposed for fovea localization.

Taking into account that there is a relation between the OD-fovea distance and the OD diameter [10] and the fovea center is located temporal and slightly below the OD center [11], a search region is defined. As shown in Fig. 2j, this region corresponds to a circular sector defined by two circumferences centered on the OD and two radial lines with origin in the OD center. The radii of the circumferences are chosen based on the mean and the standard deviation of the ratios between the OD-fovea distances and the OD diameters computed in [10] (2.65 and 0.30, respectively), whereas the slope of the lines is selected based on the mean OD-fovea angle presented in [11] ($-5.6°$).

Once the search region is defined, a search for the fovea center is performed. First, the original RGB image is converted into a grey-scale image and its complement is obtained. Histogram equalization is then applied to the image in order to enhance the contrast. Since the fovea can be confused with the red lesions due to color similarities, the pixels which are potential lesion candidates are identified and their intensity is set to 0 in the enhanced image. These lesion candidates are obtained after subtracting the result of a morphological closing from the enhanced image.

Afterwards, for each pixel of the search region, the average intensity within a 15×15 window centered on that pixel is computed (Fig. 2j). The final fovea candidates correspond to the image regions where the average intensity is higher than 90% of the highest value found. If only one candidate is obtained, the fovea center corresponds to the pixel whose average intensity in its neighborhood is maximum. Otherwise, the distance between the candidates' centroid and the centroid of the search region is computed. If the maximum intensity of the candidate closest to the centroid of the search region is higher or equal to 95% of the absolute maximum intensity, its centroid is considered the fovea center. If this does not happen, the centroid of the largest candidate or the centroid of the candidate with maximum intensity is selected as the fovea center, depending on the maximum intensity of the largest candidate.

3   Datasets

The methods developed for OD and fovea detection were evaluated using two public datasets: the Messidor database [12] and the IDRiD test set [13].

The Messidor database is composed by 1200 retinal photographs with 45° FOV and three different resolutions (1440×960, 2240×1488 and 2304×1536 *pixels*). In this work, only the 1136 images for which there are the OD masks and the fovea centers were used. Since the coordinates of the OD centers are not provided for these images, we assume that they correspond to the centroids of the OD masks.

Regarding the IDRiD test set, the images provided for evaluating the OD and fovea localization methods are different from those used for comparing the performance of OD segmentation approaches. While one set contains 103 images with information about the OD and fovea centers, the other one includes 27 images with the corresponding OD masks. All images were acquired using a camera with 50° FOV and have a resolution of 2848×4288 *pixels*.

In order to define the parameters of the SBF (applied for OD segmentation), the 54 images of the IDRiD training set for which the OD masks are available were also used.

4   Results and Discussion

In order to compare the performance of the methods proposed for OD and fovea localization with other works, the mean Euclidean distance (ED) between the coordinates of the detected centers and the *ground-truth* (GT), as well as the percentage of images for which that distance is less than the OD radius (S_{1R}) are computed. For evaluating the OD segmentation performance, the mean Jaccard index (J) and Dice's coefficient are used instead. Tables 1 and 2 shows the results for the three different tasks (OD localization, OD segmentation and fovea localization) in the Messidor and IDRiD datasets respectively.

For the two datasets, the developed methods are amongst the top-performing solutions in both localization and segmentation tasks. Since the datasets contain images with different resolutions and FOV, this proves that the proposed methods are robust to inter-image variability.

Although the mean ED between the detected OD centers and the GT is much higher in the IDRiD test set, the mean ratio of the ED to the OD radius

Table 1. Performance of different methods in the Messidor dataset

		OD localization		OD segmentation		Fovea localization	
	Approach	ED	S_{1R}	J	Dice	ED	S_{1R}
Our work	SBF + Handcrafted features	9.77	99.56%	0.87	0.92	9.68	100%
[14]	UOLO	9.40	99.74%	0.88	0.93	10.44	99.38%
[14]	YOLOv2	6.86	100%	-	-	9.01	100%
[14]	U-Net	-	-	0.88	0.93	-	-
[15]	Intensity profile analysis	23.17	99.75%	-	-	34.88	99.40%
[16]	Circular Hough Transform	-	98.83%	0.86	-	-	-
[17]	Sequential CNNs	-	97%	-	-	-	96.6%
[18]	Handcrafted features	-	-	-	-	16.09	98.24%
[19]	Handcrafted features	-	-	-	-	-	96.92%
[20]	Variational model	-	-	0.91	-	-	-
[21]	Supervised segmentation	-	100%	0.84	-	-	-
[22]	Mathematical morphology	-	-	0.82	-	-	-

Table 2. Performance of the top-performing methods in the IDRiD test set [13]

		OD localization	OD segmentation	Fovea localization
	Approach	ED	J	ED
Our team	SBF + Handcrafted features	29.18	0.89	59.75
Team A	ResNet + VGG	21.07	-	64.49
Team B	U-Net	33.54	0.93	68.47
Team C	Mask R-CNN	33.88	0.93	570.13
Team D	Mask R-CNN	36.22	0.79	85.40
Team E	SegNet	-	0.86	-

is very similar in both datasets (0.114 in the Messidor and 0.122 in the IDRiD). The same does not happen for fovea localization. While in the Messidor the ED between the detected centers and the GT is, on average, approximately 6% of the OD radius, in the IDRiD test set it corresponds to approximately 25% of the OD radius. Nevertheless, the method herein proposed for fovea localization outperforms all the other solutions submitted to the IDRiD challenge for this task. Figures 3 and 4 show some examples of the results obtained in the Messidor and IDRiD datasets, respectively.

Since the OD localization and segmentation methods use mainly the vascular and intensity information for estimating the OD center and boundary, they tend to fail in images which present large bright artifacts or lesions near the OD (Fig. 3b). The localization of fovea centers is also affected by the presence of lesions in the macular region (Fig. 4c). Furthermore, there are images where the contrast between the fovea center and its neighborhood is very low and it is very difficult to find the precise coordinates of the fovea center (Fig. 3a).

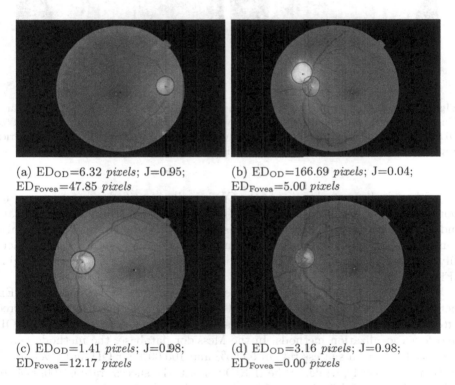

(a) ED_{OD}=6.32 *pixels*; J=0.95; ED_{Fovea}=47.85 *pixels*

(b) ED_{OD}=166.69 *pixels*; J=0.04; ED_{Fovea}=5.00 *pixels*

(c) ED_{OD}=1.41 *pixels*; J=0.98; ED_{Fovea}=12.17 *pixels*

(d) ED_{OD}=3.16 *pixels*; J=0.98; ED_{Fovea}=0.00 *pixels*

Fig. 3. Examples of images of the Messidor database where the developed methods fail in one of the three tasks (a-b) or succeed in all of them (c-d). Blue line: OD boundary; blue and green dots: detected OD and fovea centers, respectively; black line: GT of OD segmentation; black dots: GT of OD and fovea locations. The values of the metrics appear below each image. (Color figure online)

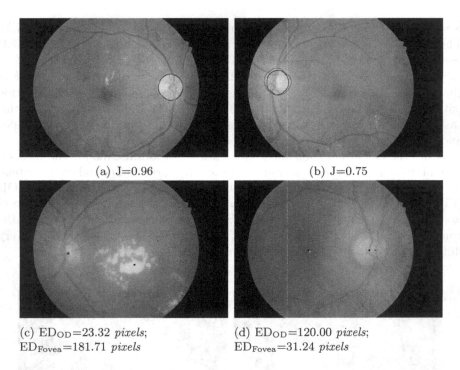

(a) J=0.96   (b) J=0.75

(c) ED$_{OD}$=23.32 *pixels*;   (d) ED$_{OD}$=120.00 *pixels*;
ED$_{Fovea}$=181.71 *pixels*   ED$_{Fovea}$=31.24 *pixels*

Fig. 4. Examples of results obtained in the IDRiD test set. Blue line: OD boundary; blue and green dots: detected OD and fovea centers, respectively; black line: GT of OD segmentation; black dots: GT of OD and fovea locations. The values of the metrics appear below each image. (Color figure online)

Although the central region of the OD is easily identified in most images, its boundary is not always well delimited (Fig. 4b). In such situations, the Jaccard index and the Dice's coefficient are penalized by that ambiguity. There are also some images where the precise position of the OD center is not easy to identify, because there is not an obvious point for which the vessels are converging (Fig. 4d).

Therefore, we consider that the percentage of images for which the ED between the detected centers and the GT is less than the OD radius (referred afterwards as success rate) is more relevant than the mean ED for evaluating OD and fovea localization methods. In the Messidor database, the methods herein proposed achieve a success rate of 99.56% and 100.00% for OD and fovea localization respectively, while, in the IDRiD test set, the success rate is 100.00% for OD localization and 98.06% for fovea localization.

Although the deep models proposed in [14] present slightly better results than the methods herein described (Table 1), they require larger amounts of annotated data and processing power for training. Note that, in our case, only few samples of the IDRiD training set were used for estimating the filters' parameters. Then, the same set of parameters was applied to compute the results both in the

Messidor and IDRiD test set. This ability of generalization (without depending on large training sets) is a great advantage of the developed methods.

5   Conclusion

In the research herein presented, three inter-dependent modules were proposed for OD localization, OD segmentation and fovea localization. While the methods proposed for OD localization and segmentation are extended versions of previous works, the method used for finding the coordinates of fovea centers is new. It is based on the relation that exists between the fovea-OD distance and the OD diameter. Therefore, both the OD center coordinates and the OD radius (estimated from OD segmentation) are used as inputs for defining a search region. The fovea center is then assigned to the darkest point inside that region.

The performance of the three modules is evaluated in the Messidor and IDRiD datasets. In both datasets, the proposed methods are amongst the top-performing solutions. This proves that they are therefore robust to variations in image resolution, field-of-view and acquisition conditions.

Contrarily to some of the deep learning approaches proposed for OD and fovea detection, the methods herein described take advantage of medical knowledge to find the retinal structures of interest and do not rely on large amounts of data and processing power for training.

Acknowledgments. This work is financed by the ERDF – European Regional Development Fund through the Operational Programme for Competitiveness and Internationalisation – COMPETE 2020 Programme, and by National Funds through the FCT – Fundação para a Ciência e a Tecnologia within project. CMUP-ERI/TIC/0028/2014.
Tânia Melo is funded by the FCT grant SFRH/BD/145329/2019. Teresa Araújo is funded by the FCT grant SFRH/BD/122365/2016.

References

1. Mendonça, A.M., Sousa, A., Mendonça, L., Campilho, A.: Automatic localization of the optic disc by combining vascular and intensity information. Comput. Med. Imaging Graph. **37**(5–6), 409–417 (2013). https://doi.org/10.1016/j.compmedimag.2013.04.004
2. Jelinek, H., Cree, M.: Automated Image Detection of Retinal Pathology, 1st edn. CRC Press, Boca Raton (2009)
3. Aquino, A., Gegúndez-Arias, M., Marin, D.: Automated optic disc detection in retinal images of patients with diabetic retinopathy and risk of macular edema. Int. J. Biol. Life Sci. **8**(2), 87–92 (2012). https://doi.org/10.5281/zenodo.1085129
4. Youssif, A., Ghalwash, A., Ghoneim, A.: Optic disc detection from normalized digital fundus images by means of a vessels' direction matched filter. IEEE Trans. Med. Imaging **27**(1), 11–18 (2008). https://doi.org/10.1109/TMI.2007.900326
5. Dashtbozorg, B., Mendonça, A.M., Campilho, A.: Optic disc segmentation using the sliding band filter. Comput. Biol. Med. **56**, 1–12 (2015). https://doi.org/10.1016/j.compbiomed.2014.10.009

6. Medhi, J., Dandapat, S.: An effective fovea detection and automatic assessment of diabetic maculopathy in color fundus images. Comput. Biol. Med. **74**, 30–44 (2016). https://doi.org/10.1016/j.compbiomed.2016.04.007
7. Mendonça, A.M., Campilho, A.: Segmentation of retinal blood vessels by combining the detection of centerlines and morphological reconstruction. IEEE Trans. Med. Imaging **25**(9), 1200–1213 (2006). https://doi.org/10.1109/TMI.2006.879955
8. Pereira, C.S., Mendonça, A.M., Campilho, A.: Evaluation of contrast enhancement filters for lung nodule detection. In: Kamel, M., Campilho, A. (eds.) ICIAR 2007. LNCS, vol. 4633, pp. 878–888. Springer, Heidelberg (2007). https://doi.org/10.1007/978-3-540-74260-9_78
9. Esteves, T., Quelhas, P., Mendonça, A.M., Campilho, A.: Gradient convergence filters and a phase congruency approach for in vivo cell nuclei detection. Mach. Vis. Appl. **23**(4), 623–638 (2012). https://doi.org/10.1007/s00138-012-0407-7
10. Jonas, R., et al.: Optic disc - Fovea distance, axial length and parapapillary zones. The Beijing eye study 2011. PLoS ONE **10**(9), 1–14 (2015). https://doi.org/10.1371/journal.pone.0138701
11. Rohrschneider, K.: Determination of the location of the fovea on the fundus. Invest. Ophthalmol. Vis. Sci. **45**(9), 3257–3258 (2004). https://doi.org/10.1167/iovs.03-1157
12. Decencière, E., Zhang, X., Cazuguel, G., Lay, B., Cochener, B., Trone, C., et al.: Feedback on a publicly distributed image database: the Messidor database. Image Anal. Stereol. **33**(3), 231–234 (2014). https://doi.org/10.5566/ias.1155
13. Porwal, P., et al.: IDRiD: diabetic retinopathy - segmentation and grading challenge. Med. Image Anal. **59**, 101561 (2020). https://doi.org/10.1016/j.media.2019.101561
14. Araújo, T., Aresta, G., Galdran, A., Costa, P., Mendonça, A.M., Campilho, A.: UOLO - automatic object detection and segmentation in biomedical images. In: Stoyanov, D., et al. (eds.) DLMIA/ML-CDS -2018. LNCS, vol. 11045, pp. 165–173. Springer, Cham (2018). https://doi.org/10.1007/978-3-030-00889-5_19
15. Kamble, R., Kokare, M., Deshmukh, G., Hussin, F., Mériaudeau, F.: Localization of optic disc and fovea in retinal images using intensity based line scanning analysis. Comput. Biol. Med. **87**, 382–396 (2017). https://doi.org/10.1016/j.compbiomed.2017.04.016
16. Aquino, A., Gegúndez-Arias, M., Marin, D.: Detecting the optic disc boundary in digital fundus images using morphological, edge detection, and feature extraction techniques. IEEE Trans. Med. Imaging **29**(11), 1860–1869 (2010). https://doi.org/10.1109/TMI.2010.2053042
17. Al-Bander, B., Al-Nuaimy, W., Williams, B., Zheng, Y.: Multiscale sequential convolutional neural networks for simultaneous detection of fovea and optic disc. Biomed. Signal Process. Control **40**, 91–101 (2018). https://doi.org/10.1016/j.bspc.2017.09.008
18. Aquino, A.: Establishing the macular grading grid by means of fovea centre detection using anatomical-based and visual-based features. Comput. Biol. Med. **55**, 61–73 (2014). https://doi.org/10.1016/j.compbiomed.2014.10.007
19. Gegundez-Arias, M., Marin, D., Bravo, J., Suero, A.: Locating the fovea center position in digital fundus images using thresholding and feature extraction techniques. Comput. Med. Imaging Graph. **37**(5–6), 386–393 (2013). https://doi.org/10.1016/j.compmedimag.2013.06.002
20. Dai, B., Wu, X., Bu, W.: Optic disc segmentation based on variational model with multiple energies. Pattern Recogn. **64**, 226–235 (2017). https://doi.org/10.1016/j.patcog.2016.11.017

21. Roychowdhury, S., Koozekanani, D., Kuchinka, S., Parhi, K.: Optic disc boundary and vessel origin segmentation of fundus images. IEEE J. Biomed. Health Inf. **20**(6), 1562–1574 (2016). https://doi.org/10.1109/JBHI.2015.2473159
22. Morales, S., Naranjo, V., Angulo, J., Alcañiz, M.: Automatic detection of optic disc based on PCA and mathematical morphology. IEEE Trans. Med. Imaging **32**(4), 786–796 (2013). https://doi.org/10.1109/TMI.2013.2238244

The Effect of Menopause on the Sexual Dimorphism in the Human Retina – Texture Analysis of Optical Coherence Tomography Data

Ana Nunes[1], Pedro Serranho[1,2], Hugo Quental[1],
Miguel Castelo-Branco[1,3], and Rui Bernardes[1,3(✉)]

[1] Coimbra Institute for Biomedical Imaging and Translational Research (CIBIT), Institute of Nuclear Sciences Applied to Health (ICNAS), University Coimbra, Coimbra, Portugal
[2] Department of Sciences and Technology, Universidade Aberta, Lisboa, Portugal
[3] Coimbra Institute for Biomedical Imaging and Translational Research (CIBIT), Faculty of Medicine (FMUC), University Coimbra, Coimbra, Portugal
rmbernardes@fmed.uc.pt

Abstract. Sexual dimorphism in the human retina has recently been connected to gonadal hormones. In the study herein presented, texture analysis was applied to computed mean value fundus (MVF) images from optical coherence tomography data of female and male healthy adult controls. Two separate age-group analyses that excluded the probable perimenopause period of the women in the present study were performed, using a modified MVF image computation method that further highlights texture differences present in the retina. While distinct texture characteristics were found between premenopausal females and age-matched males, these differences almost disappeared in the older groups (postmenopausal women vs age-matched men), suggesting that sex-related texture differences in the retina can be correlated to the hormonal changes that women go through during the menopausal transition. These findings suggest that texture-based metrics may be used as biomarkers of physiology and pathophysiology of the retina and the central nervous system.

Keywords: Texture analysis · Optical coherence tomography · Retina · Healthy controls · Sexual dimorphism · Menopause

1 Introduction

Sexual dimorphism is a well-known phenomenon, manifested at different organ systems. In the visual system, sex-based differences exist both in the physiology

This study was supported by The Portuguese Foundation for Science and Technology (PEst-UID/NEU/04539/2019 and UID/04950/2017), by FEDER-COMPETE (POCI-01-0145-FEDER-007440 and POCI01-0145-FEDER-016428), and by Centro 2020 FEDER-COMPETE (BIGDATIMAGE, CENTRO-01-0145-FEDER-000016).

A. Campilho et al. (Eds.): ICIAR 2020, LNCS 12132, pp. 344–357, 2020.
https://doi.org/10.1007/978-3-030-50516-5_30

and in the pathology of the eye and, more specifically, the retina [26]. Sex-based differences in the human retina have been extensively reported for the full retinal thickness [2,9,17,19,32,35] and less often, for the different retinal layers [27]. Furthermore, there is a link between sex and retinal disorders, which has been attributed to gonadal hormones [26]. Importantly, these hormones have a significant role in neurological disorders, such as Alzheimer's disease (AD) [28], which are closely related to the physiological processes within the retina, given the well-established relationship between the retina and the central nervous system (CNS) [21,33].

Optical coherence tomography (OCT) studies addressing the sex-related differences in the retina rely mostly on layer thickness measurements. While most of these studies report thickness differences amongst the retinas of female and male healthy individuals [2,9,17,19,27,32,35], inconclusive results [6,13] have been reported as well. One limitation of these studies that may be influencing the sex-related differences found (or lack thereof) is the typical inclusion of study groups with a wide age range and an unbalanced female-to-male ratio. On the other hand, the use of thickness as the metric of choice in these studies is somewhat restrictive, as new methodologies, notably the ones using texture analysis [3,25], have shown to reveal information not conveyed by thickness.

Recently, our group has been applying texture analysis to OCT data to gain further insight on the differences in the retinal structure, between distinct healthy control groups [23] and between healthy individuals and patients diagnosed with several neurodegenerative disorders, namely Alzheimer's and Parkinson's diseases [25] and multiple sclerosis [24]. Furthermore, the same analysis methodology was applied to animal studies, in which OCT data from a mouse model of AD and control mice groups were examined [10,22]. These studies further confirmed the power of texture analysis to unveil information from OCT data, yet unexplored.

The present study constitutes a follow-up analysis of previous work [23], where texture features were computed from OCT data of female and male healthy adult controls, to identify differences in the retina associated with the subjects' sex. In the work herein presented, the same texture analysis methodology was applied to a larger population. Furthermore, this population was split into younger and older age-groups, leaving out females at the menopausal transition period. As the sexual dimorphism in both the healthy and the diseased retina is correlated to hormonal differences [26], the goal of this grouping of the subjects in the present study is to examine how the cessation of menstrual cycles manifests in the retina, namely by analysing its impact on the previously identified sex-based differences [23]. Also, this study aims to assess the impact of two distinct methods to compute fundus images for each retinal layer in spotting texture differences.

2 Materials and Methods

2.1 Participants

Data from 100 healthy controls, 50 females and 50 males, with no previous retinal pathologies, were gathered from the authors' institutional database. The data collection protocol used for gathering OCT data was approved by the Ethics Committee of the Faculty of Medicine of the University of Coimbra and performed according to the tenets stated in the Declaration of Helsinki [36].

Female subjects were split into two groups: one younger, premenopausal group (30 women) and one older, postmenopausal group (20 women). As no information regarding the phase of the menstrual cycle/onset of menopausal transition at the time of image acquisition was available for any of the females in the present study, the median age of menopause, 51 years old [12], was the reference used. To exclude the probable perimenopause period, and maximise the number of included subjects amongst the ones available in the authors' institutional database, women aged 42–54 years old were not considered for analysis.

Male subjects were selected to ensure an exact age-match to the female subjects, where possible. All females and males were perfectly age-matched for the younger groups, while for the older groups, 7 out of 20 female/male pairs (35%) were not perfectly age-matched.

Both eyes of each subject were analysed, except for one right eye from the younger female group, one right eye from the older female group, and one left eye from the older male group, which were excluded due to scan quality, yielding a total of 197 eye scans. Demographic data for the four groups in the present study are shown in Table 1.

Table 1. Demographic data of the groups at study.

Group	N	Age (years) Mean(STD)	Age (years) Min(Max)	Eyes Right(Left)	# Eye scans
Younger Females	30	30(6.9)	19(42)	29(30)	59
Younger Males	30	30(6.9)	19(42)	30(30)	60
Older Females	20	64(5.3)	54(74)	19(20)	39
Older Males	20	68(8.8)	54(79)	20(19)	39

2.2 OCT Imaging

The Cirrus SD-OCT 5000 (Carl Zeiss Meditec, Dublin, CA, USA) was used to gather all eye scans. The 512×128 Macular Cube protocol was used, centred on the macula.

2.3   Image Processing

OCT data were processed using the OCT Explorer software (Retinal Image Analysis Lab, Iowa Institute for Biomedical Imaging, Iowa City, IA, USA) [1, 11,20] to segment the six innermost layers of the retina: the retinal nerve fibre layer (RNFL), the ganglion cell layer (GCL), the inner plexiform layer (IPL), the inner nuclear layer (INL), the outer plexiform layer (OPL) and the outer nuclear layer (ONL).

For each of the six layers at study, a mean value fundus (MVF) image was computed where each pixel is the average of the A-scan zeroing all A-scan values outside the layer of interest (see Fig. 1). This MVF image computation method is a modified approach based on the method originally developed by our group [14], where each pixel in the image is computed as the average of the A-scan values within the respective layer. The rationale for the use of the alternative MVF method is described in Sect. 2.5.

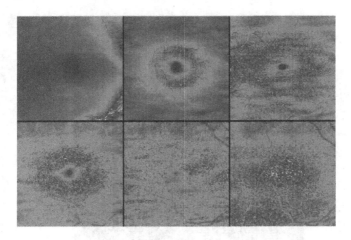

Fig. 1. Colour-coded MVF images from the right eye of a male healthy control, in which each pixel is the average of the corresponding A-scan zeroing all A-scan values outside the layer of interest. From left to right and top to bottom: RNFL, GCL, IPL, INL, OPL and ONL layer fundus images. For reference purposes only: images were intensity-corrected, pseudo-colour coded and downsampled to 128 × 128, for ease of visualisation. (Color figure online)

As both eyes from each subject were considered for analysis, all left-eye MVF images were horizontally flipped to match the temporal and nasal regions across all eye scans.

2.4   Texture Analysis

Texture features were computed for each MVF image, following an approach used previously [24,25]. Two different texture analysis approaches were considered: the grey-level co-occurrence matrix (GLCM) [16], used to highlight local intensity variations, and the dual-tree complex wavelet transform (DTCWT) [30], used to identify coarser texture properties [18].

For the GLCM analysis, each image was down-sampled to 128×128 pixels and split into 7 × 7 blocks (see Fig. 2), which were independently analysed, except for the blocks in the central (4th) row and column, which were not considered to exclude the foveal region. For each block, four GLCMs were computed for distinct pixel pair orientations (0°, 45°, 90°, and 135°), using the distance of one pixel and considering 180° apart angles to be the same. For each block and orientation, 20 features (defined in [7,8,15,16,31]) were computed. The maximum value across the four orientations was selected as the feature value for each block. Blocks were then aggregated into quadrants: the temporal-superior (Q1), nasal-superior (Q2), temporal-inferior (Q3) and nasal-inferior (Q4) quadrants, composed of 3×3 blocks each (Fig. 2). The average feature value, across the 3 × 3 blocks, was used as the final quadrant feature value.

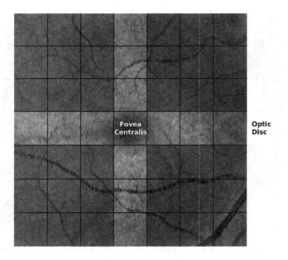

Fig. 2. Computed fundus image from the volumetric macular cube scan of the right eye of a healthy control subject [25]. Each of the 7×7 blocks show the individually analysed areas which results were later aggregated into larger regions (shaded areas). Image axes are: x-axis (horizontal) - temporal (left) to nasal (right) and y-axis (vertical) superior (top) to inferior (bottom).

The DTCWT method was applied to the full macular area covered by the MVF image. The variance of the magnitude of the DTCWT complex coefficients was computed for six image subbands (±15°, ±45° and ±75°), as described in [5,34]. These six variance values were used as the global features from each MVF image.

In total, 86 texture features were computed from each MVF of each of the six retinal layers being considered: 80 (20 × 4 quadrants) local GLCM-based features, and six global DTCWT-based features.

2.5 MVF Image Computation Methods

Let I_A and I_B be the average intensity values (of a particular layer of the retina) from two A-scans located at two neighbour locations A and B that will produce two pixels of the corresponding fundus image (the MVF [14]). For our demonstration, we assume that $I_B = I_A + \Delta I$. Also, let t_A and t_B be the number of samples within the layer (thickness) in the positions of the considered A-scans, with $t_A \approx t_B$ (because these are side-by-side A-scans). Let I_M e I_m, respectively, be the maximum and minimum intensity values in a block where the GLCM is to be computed. Finally, let us consider N the number of grey-levels considered to compute the GLCM, and K the total number of samples in an A-scan. We will illustrate that the alternative method to compute a MVF image considering the average over the entire A-scan has a more discriminative power to small differences in intensity (ΔI), except for the case where the thickness is constant across the GLCM block where both methods will produce the same results.

The Original MVF Computation Method
The intensity amplitude within the GLCM block, as given by

$$h_o = \frac{I_M - I_m}{N}.$$

Therefore, the level within the greyscale for I_A and I_B is given by

$$\alpha_o(I_A) = \frac{I_A - I_m}{h_o} \quad \text{and} \quad \alpha_o(I_B) = \alpha_o(I_A + \Delta I) = \alpha_o(A) + \frac{\Delta I}{h_o},$$

where the mapping of computed intensities to the individual greyscale levels (from 0 to N-1) is given by $\lfloor \alpha(.) \rfloor$, and $\lfloor . \rfloor$ is the floor operation. Therefore, the difference between the two intensity levels (I_A and I_B) is given by

$$\Delta \alpha_o = \frac{\Delta I}{h_o}.$$

The Alternative MVF Computation Method
In this case, using the same notation, the new intensity values are given by $(I_A t_A)/K$ and $(I_B t_B)/K$. Similarly, one defines the intensity amplitude within the GLCM block as

$$h_a = \frac{\max(I\,t) - \min(I\,t)}{NK} \leq \frac{I_M \max(t) - I_m \min(t)}{NK}. \tag{1}$$

Then one gets the level in greyscale for the computation of the GLCM given by

$$\alpha_a(I_A) = \frac{I_A t_A - I_m \min(t)}{K h_a}$$

and, by straightforward computations,

$$\alpha_a(I_B) = \frac{(I_A t_A - I_m \min(t)) + \Delta I t_A + \Delta t I_B}{K h_a} = \alpha_a(A) + \frac{\Delta I t_A + \Delta t I_B}{K h_a},$$

with $\Delta t = t_B - t_A$ and $\Delta I = I_B - I_A$, which means that the difference in greyscale is given by

$$\Delta \alpha_a = \frac{\Delta It_A + \Delta tI_B}{Kh_a}.$$

Comparison Between the Discriminant Powers

In the original method, I_A and I_B are assigned the levels of, respectively, $\alpha_o(I_A)$ and $\alpha_o(I_B)$, which differ by $\Delta I/h_o$. On the other hand, for the alternative method, the difference in assigned levels is $(\Delta It_A + \Delta tI_B)/(Kh_a)$, being the latter higher than the former. One can compute the ratio (γ) of the differences and show that this ratio is over the unit, meaning that the latter approach presents a better discriminative power.

$$\gamma = \frac{\frac{\Delta It_A + \Delta tI_B}{Kh_a}}{\frac{\Delta I}{h_o}} = \frac{h_o}{h_a K}\left(t_A + \frac{\Delta tI_B}{\Delta I}\right). \tag{2}$$

Therefore one has from (1)

$$\frac{h_o}{h_a K} = \frac{\frac{I_M - I_m}{N}}{K \frac{\max(I\,t) - \min(I\,t)}{NK}} \geq \frac{I_M - I_m}{I_M \max(t) - I_m \min(t)} \approx \frac{1}{\bar{t}} \tag{3}$$

with $\bar{t}$ the average thickness within the GLCM block.

From (2) and (3), one gets the approximate lower bound for γ as

$$\gamma \gtrsim \underbrace{\frac{t_A}{\bar{t}}}_{\approx 1} + \frac{\Delta tI_B}{\bar{t}\Delta I} \approx 1 + \frac{\Delta t}{\bar{t}} \cdot \frac{I_B}{\Delta I}.$$

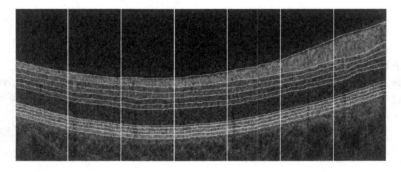

Fig. 3. Cropped OCT B-scan image from the right eye of male subject of the younger group. This B-scan crosses the third and fourth quadrants (Q3 and Q4) and shows the limits for the GLCM blocks in white solid vertical lines. The central block (4th) is not used in our analysis (see Fig. 2). Red lines show the segmentation of the different layers of the retina as computed by the method used [1,11,20]. Image axes are: x-axis (horizontal) - temporal (left) to nasal (right) and z-axis (vertical) anterior (vitreous – top) to posterior (choroid – bottom). (Color figure online)

The lower bound for γ, as defined above, is based on the assumption that the amplitude in thickness within a GLCM blocks is small – which verifies for the block divisions used (see Figs. 2 and 3), hence the ratio between t_A and the average thickness is close to 1 – and differences in image intensity across two neighbour pixels is small in comparison to their average value – which verifies in general except for lower intensity image regions, in which case both methods will map to the same greyscale level, therefore, producing the same result.

As shown, the discriminative power of the alternative method is superior to that of the original method for small differences between neighbour pixel intensities, which is the default scenario in healthy or close to the health condition retinas. Moreover, it is also clear that both approaches have similar results in the cases of small variations in thickness for a given layer.

2.6   Statistical Analysis

All texture features were tested for normality using the Kolmogorov-Smirnov test (10% significance level for a conservative normality decision). For each feature, when both the female and the male groups followed a normal distribution, the two-sample t-test was used to test mean differences. Otherwise, the Mann-Whitney U-test (non-parametric) was used. The obtained p-values were separated into three significance levels: $p \leq 0.05$, $p \leq 0.01$ and $p \leq 0.001$.

3   Results

The first analysis concerns the younger females and males in the present study. Table 2 shows the list of texture features presenting statistically significant differences between the two younger groups. Differences between the younger female and male groups are shown using the symbols ●, ■, and ✱, and can be found in every single layer. The highest statistical differences can be found in the IPL (local GLCM features) and the GCL, INL and ONL (global DTCWT features). In the INL, texture differences are spread the most, as they cover three consecutive quadrants. Q2, the nasal-superior quadrant, holds consistent differences at consecutive layers of the retina – from the IPL to the ONL, while the same occurs for Q3, the temporal-inferior quadrant, from the GCL to the INL.

The second analysis concerns the older females and males. Overall, a minimal number of statistically significant differences was found between the two older groups: only two local features in the GCL, and three local features the ONL, all at the 5% significance level (shown in square-brackets – Table 2).

4   Discussion

In order to reach a better understanding of the effect of gonadal hormones in different physiological and pathophysiological mechanisms, research, where the menopausal transition is taken into consideration, is particularly valuable, as it leverages the natural interruption of the production of the female hormones. Estrogen, for instance, is believed to have a major role in the sexual dimorphism observed not only in retinal disorders [26] but also in neurodegenerative diseases such as AD [28]. Since the study of retinal biomarkers for disorders like AD using the retina as a window to the CNS has been gaining momentum in recent years [21,33], understanding the interactions of estrogen and other gonadal hormones in the physiological mechanisms taking place in the retina, both pre- and postmenopause, is of utmost importance.

The main goal of the present work is a more in-depth exploration of the sex-based differences that were previously identified in the retinas of healthy adult subjects [23]. For this purpose, we studied a larger population than the one used before [23] and performed two separate age-group analyses that excluded the probable perimenopause period of the women in the present study, in order to have pre- and postmenopause sex group comparisons.

Our results reveal numerous texture differences in all retinal layers between the two younger groups, which almost disappear in the older groups. These results suggest that texture analysis is an adequate tool to identify sex-based differences in the retina since it highlights them before the menopause when sexual dimorphism is more pronounced. Furthermore, it reveals that the observed sex-related differences become negligible after women reach menopause when gonadal hormones eventually stabilise at their permanent postmenopausal levels [29].

These results add to those from [23] where the menopause transition period was not taken into consideration. The results from that [23] menopause-independent study showed numerous statistically significant differences between the female and male groups. In the present study, the differences across the two age group-analyses are very noticeable. The contrast in the number and significance of the sex-based differences across the younger and older groups supports the hypothesis that the differences across the younger and older groups can be correlated to menopause-based hormonal alterations.

Our group has been studying texture-based methods to extract new information from retinal fundus images computed from OCT data. Preliminary analyses suggested the MVF computation method herein presented may be more sensitive to differences present in the retina than the original method [14].

One limitation of the work herein reported is the fact that it is a retrospective study, which prevented the authors from collecting information regarding the phase of the menstrual cycle and the onset of menopausal transition for any of the females in the present study. The age of menopause has been defined as the age at the last menstrual period, which can only be determined retrospectively after a woman has stopped menstruating for 12 consecutive months [4]. The authors believe that the used reference mean age of menopause (51 years old) is reasonably accurate, as most studies focusing on the menopausal transition are longitudinal [12], and so there is little risk of distortion on the reported timing of the last menstruation.

Table 2. Texture features presenting statistically significant differences between the female and male subjects for the retinal nerve fibre layer (RNFL); ganglion cell layer (GCL); inner plexiform layer (IPL); inner nuclear layer (INL); outer plexiform layer (OPL), and; outer nuclear layer (ONL). Hyphens (–) represent a p-value > 0.05 (non-significant statistical difference), the green-coloured circles (●) represent a p-value ≤ 0.05, the orange-coloured squares (■) represent a p-value ≤ 0.01 and the red-coloured asterisks (✱) represent a p-value ≤ 0.001 for the younger groups. Square-brackets ([ ]) denote differences found between the older female and male groups (all at the 5% significance level). Q1 – Superior-temporal quadrant, Q2 – Superior-nasal quadrant, Q3 – Inferior-temporal quadrant and Q4 – Inferior-nasal quadrant. IDN stands for Inverse Difference Moment Normalized; in the features IMC1 and IMC2, IMC stands for Information Measure of Correlation; and INN stands for Inverse Difference Normalized.

Layer	Parameter	Q1	Q2	Q3	Q4	Layer	Parameter	Q1	Q2	Q3	Q4
RNFL	Autocorrelation	–	●	–	–	GCL	Autocorrelation	–	–	●	–
	Cluster Prominence	■	–	–	–		Cluster Prominence	–	●	●	–
	Cluster Shade	■	–	–	–		Correlation	–	●	■	–
	Correlation	■	■	–	–		Difference Entropy	–	●	■	–
	Difference Entropy	■	●	–	–		Difference Variance	–	–	■	–
	Difference Variance	■	■	–	–		Dissimilarity	–	–	■	–
	Dissimilarity	■	■	–	–		Entropy	–	●	●	–
	Entropy	■	●	–	–		Homogeneity	–	●	●	–
	Homogeneity	■	●	–	–		IDN	–	–	■	–
	IDN	■	■	–	–		IMC1	–	–	■	–
	IMC1	■	●	–	–		IMC2	–	–	●	–
	IMC2	■	■	–	–		Inertia	–	–	■	–
	Inertia	■	■	–	■		INN	–	–	■	–
	INN	■	●	–	–		Maximum Probability	–	●	■	–
	Maximum Probability	●	■	–	–		Sum Entropy	–	●	●	–
	Sum Average	■	–	–	–		Sum of Squares	–	–	[–]	●
	Sum Entropy	●	●	–	–		Sum Variance	–	[–]	●	–
	Sum Variance	●	–	●	–		Uniformity	–	●	■	–
	Uniformity	■	■	–	–		Variance 15$^+$ (Global)		✱		
	Variance 45$^+$ (Global)		●				Variance 15$^-$ (Global)		✱		
	Variance 45$^-$ (Global)		●				Variance 45$^+$ (Global)		✱		
	Variance 75$^+$ (Global)		■				Variance 45$^-$ (Global)		✱		
	Variance 75$^-$ (Global)		■				Variance 75$^+$ (Global)		✱		
							Variance 75$^-$ (Global)		✱		
IPL	Autocorrelation	●	–	✱	–	INL	Autocorrelation	–	–	●	–
	Cluster Prominence	–	●	✱	–		Cluster Prominence	■	–	●	●
	Correlation	–	●	✱	–		Cluster Shade	■	●	●	–
	Difference Entropy	–	●	■	–		Correlation	■	–	●	●
	Difference Variance	–	●	■	–		Difference Entropy	■	■	●	–
	Dissimilarity	–	●	■	–		Difference Variance	■	■	●	–
	Entropy	–	●	✱	–		Dissimilarity	■	■	●	–
	Homogeneity	–	●	✱	–		Entropy	■	–	●	–
	IDN	–	●	✱	–		Homogeneity	■	–	●	–
	IMC1	–	–	■	–		IDN	■	■	●	●
	IMC2	–	●	■	–		IMC1	●	■	●	●
	Inertia	–	●	■	–		IMC2	■	✱	●	–
	INN	–	●	■	–		Inertia	■	■	●	–
	Maximum Probability	–	●	■	–		INN	■	■	●	–
	Sum Entropy	–	–	■	–		Maximum Probability	●	■	●	–
	Sum Variance	●	–	✱	–		Sum Entropy	●	■	●	–
	Uniformity	●	●	✱	–		Sum Variance	■	■	●	–
	Variance 45$^+$ (Global)		●				Uniformity	■	■	●	–
	Variance 75$^-$ (Global)		●				Variance 15$^+$ (Global)		✱		
							Variance 15$^-$ (Global)		✱		
							Variance 45$^+$ (Global)		✱		
							Variance 45$^-$ (Global)		✱		
							Variance 75$^+$ (Global)		✱		
							Variance 75$^-$ (Global)		✱		
OPL	Autocorrelation	–	■	–	–	ONL	Cluster Shade	[●]	●	–	–
	Cluster Prominence	–	■	–	–		Correlation	–	■	–	–
	Difference Entropy	–	■	–	–		Difference Entropy	–	■	–	●
	Difference Variance	–	■	–	–		Difference Variance	–	■	–	–
	Dissimilarity	–	■	–	–		Dissimilarity	–	■	–	–
	Entropy	–	■	–	–		Entropy	–	●	–	–
	Homogeneity	–	■	–	–		Homogeneity	–	●	–	●
	IDN	–	■	–	–		IDN	–	●	–	●
	IMC1	–	■	–	–		IMC1	–	■	–	–
	IMC2	–	■	–	–		IMC2	–	■	–	–
	Inertia	–	■	–	–		Inertia	–	■	–	–
	INN	–	■	–	–		INN	–	●	–	●
	Maximum Probability	–	●	–	–		Sum Average	[–]	●	[■]	–
	Sum Entropy	–	●	–	–		Sum Entropy	[–]	●	–	–
	Sum of Squares	–	●	–	–		Sum of Squares	–	–	■	–
	Sum Variance	–	■	–	–		Variance 45$^+$ (Global)		■		
	Uniformity	–	■	–	–		Variance 45$^-$ (Global)		✱		
							Variance 75$^+$ (Global)		✱		
							Variance 75$^-$ (Global)		✱		

Nevertheless, in the present study, the older female group's lower age limit (54 years old) is reasonably close to the reference age of menopause (51 years old), so the youngest women in this group may be still menstruating. However, these women are most likely already going through the menopause transition period. Nevertheless, the limit of 54 years old was chosen to include the largest possible number of female/male pairs to populate both older groups, from all healthy control subjects in the authors' institutional database. Concerning the younger females' group (upper age limit of 42 years old), it is unlikely that the oldest women are already going through the menopausal transition, assuming that none of them had premature menopause (<40 years) [37]. For reference, Bromberger [4] used 47.5 years as the approximate age over which females would presumably be close to the time of menopause.

5   Conclusion

The present study confirms the existence of sex-based differences in the human retina, as measured by texture analysis of MVF images computed from OCT data. In this study, we first provide evidence suggesting that the neuroretina, the only part of the CNS directly accessible through optical means, presents distinct texture characteristics for premenopausal females and age-matched males and that differences almost disappear after women have gone through the menopausal period. Second, we apply a modified MVF image computation method that further highlights the retinal texture differences found.

Conflicts of Interest. The authors declare that they have no conflict of interest.

References

1. Abràmoff, M.D., Garvin, M.K., Sonka, M.: Retinal imaging and image analysis. IEEE Rev. Biomed. Eng. **3**, 169–208 (2010). https://doi.org/10.1109/RBME.2010.2084567
2. Adhi, M., Aziz, S., Muhammad, K., Adhi, M.I.: Macular thickness by age and gender in healthy eyes using spectral domain optical coherence tomography. PLoS ONE **7**(5), e37638 (2012). https://doi.org/10.1371/journal.pone.0037638
3. Anantrasirichai, N., Achim, A., Morgan, J.E., Erchova, I., Nicholson, L.: SVM-based texture classification in optical coherence tomography. In: IEEE 10th International Symposium on Biomedical Imaging: From Nano to Macro, pp. 1332–1335 (2013). https://doi.org/10.1109/ISBI.2013.6556778
4. Bromberger, J.T., Matthews, K.A., Kuller, L.H., Wing, R.R., Meilahn, E.N., Plantinga, P.: Prospective study of the determinants of age at menopause. Am. J. Epidemiol. **145**(2), 124–133 (1997). https://doi.org/10.1093/oxfordjournals.aje.a009083
5. Celik, T., Tjahjadi, T.: Multiscale texture classification using dual-tree complex wavelet transform. Pattern Recogn. Lett. **30**, 331–339 (2009). https://doi.org/10.1016/j.patrec.2008.10.006

6. Chan, A., Duker, J.S., Ko, T.H., Fujimoto, J.G., Schuman, J.S.: Normal macular thickness measurements in healthy eyes using stratus optical coherence tomography. Arch. Ophthalmol. **124**(2), 193–198 (2006). https://doi.org/10.1001/archopht.124.2.193

7. Clausi, D.A.: An analysis of co-occurrence texture statistics as a function of grey level quantization. Canadian J. Remote Sens. **28**(1), 45–62 (2002). https://doi.org/10.5589/m02-004

8. Conners, R.W., Trivedi, M.M., Harlow, C.A.: Segmentation of a high-resolution urban scene using texture operators (Sunnyvale, California). Comput. Vis. Graph. Image Process. **25**(3), 273–310 (1984). https://doi.org/10.1016/0734-189x(84)90197-x

9. Çubuk, M., Kasım, B., Koçluk, Y., Sukgen, E.A.: Effects of age and gender on macular thickness in healthy subjects using spectral optical coherence tomography/scanning laser ophthalmoscopy. Int. Ophthalmol. **38**(1), 127–131 (2017). https://doi.org/10.1007/s10792-016-0432-z

10. Ferreira, H., et al.: Characterization of the retinal changes of the 3xTg-AD mouse model of Alzheimer's disease. In: Henriques, J. (ed.) MEDICON 2019, IFMBE. vol. 76, pp. 1816–1821 (2020). https://doi.org/10.1007/978-3-030-31635-8

11. Garvin, M.K., Abràmoff, M.D., Wu, X., Russell, S.R., Burns, T.L., Sonka, M.: Automated 3-D intraretinal layer segmentation of macular spectral-domain optical coherence tomography Images. IEEE Trans. Med. Imag. **28**, 1436–1447 (2009). https://doi.org/10.1109/TMI.2009.2016958

12. Gold, E.B.: The timing of the age at which natural menopause occurs. Obstetr. Gynecol. Clin. North Am. **38**(3), 425–440 (2011). https://doi.org/10.1016/j.ogc.2011.05.002

13. Guedes, V., et al.: Optical coherence tomography measurement of macular and nerve fiber layer thickness in normal and glaucomatous human eyes. Ophthalmol. **110**(1), 177–189 (2003). https://doi.org/10.1016/s0161-6420(02)01564-6

14. Guimaraes, P., et al.: Ocular fundus reference images from optical coherence tomography. Comput. Med. Imag. Graph. **38**, 381–389 (2014). https://doi.org/10.1016/j.compmedimag.2014.02.003

15. Haralick, R.M.: Statistical and structural approaches to texture. Proc. IEEE **67**(5), 786–804 (1979). https://doi.org/10.1109/PROC.1979.11328

16. Haralick, R.M., Shanmugam, K., Dinstein, I.: Texture features for image classification. IEEE Trans. Syst. Man Cybern. **SMC–3**, 610–621 (1973). https://doi.org/10.1109/TSMC.1973.4309314

17. Kashani, A.H., et al.: Retinal thickness analysis by race, gender, and age using stratus OCT. Am. J. Ophthalmol. **149**, 496–502 (2010). https://doi.org/10.1016/j.ajo.2009.09.025

18. Kassner, A., Thornhill, R.E.: Texture analysis: a review of neurologic MR imaging applications. Am. J. Neuroradiol. **31**, 809–816 (2010). https://doi.org/10.3174/ajnr.A2061

19. Kelty, P.J., Payne, J.F., Trivedi, R.H., Kelty, J., Bowie, E.M., Burger, B.M.: Macular thickness assessment in healthy eyes based on ethnicity using stratus OCT optical coherence tomography. Investigative Ophthalmol. Vis. Sci. **49**(6), 2668–2672 (2008). https://doi.org/10.1167/iovs.07-1000

20. Li, K., Wu, X., Chen, D.Z., Sonka, M.: Optimal surface segmentation in volumetric images - a graph-theoretic approach. IEEE Trans. Pattern Anal. Mach. Intell. **28**, 119–134 (2006). https://doi.org/10.1109/TPAMI.2006.19

21. London, A., Benhar, I., Schwartz, M.: The retina as a window to the brain - from eye research to CNS disorders. Nat. Rev. Neurol. (2012). https://doi.org/10.1038/nrneurol.2012.227

22. Nunes, A., Ambrósio, A.F., Castelo-Branco, M., Bernardes, R.: Texture biomarkers of Alzheimer's disease and disease progression in the mouse retina. In: 2018 IEEE 18th International Conference on Bioinformatics and Bioengineering, pp. 41–46 (2018). https://doi.org/10.1109/BIBE.2018.00016

23. Nunes, A., Serranho, P., Quental, H., Ambrosio, A.F., Castelo-Branco, M., Bernardes, R.: Sexual dimorphism of the adult human retina assessed by optical coherence tomography. In: Henriques, J. (ed.) MEDICON 2019, IFMBE. vol. 76, pp. 1830–1834 (2020). https://doi.org/10.1007/978-3-030-31635-8

24. Nunes, A., et al.: Textural information from the retinal nerve fibre layer in multiple sclerosis. In: 2019 IEEE 6th Portuguese Meeting on Bioengineering (ENBENG) (2019). https://doi.org/10.1109/ENBENG.2019.8692454

25. Nunes, A., et al.: Retinal texture biomarkers may help to discriminate between Alzheimer's, Parkinson's, and healthy controls. PLoS ONE **14**(6), e0218826 (2019). https://doi.org/10.1371/journal.pone.0218826

26. Nuzzi, R., Scalabrin, S., Becco, A., Panzica, G.: Gonadal hormones and retinal disorders: a review. Front. Endocrinol. **9**, 66 (2018). https://doi.org/10.3389/fendo.2018.00066

27. Ooto, S., et al.: Effects of age, sex, and axial length on the three-dimensional profile of normal macular layer structures. Investigative Ophthalmol. Vis. Sci. **52**, 8769–8779 (2011). https://doi.org/10.1167/iovs.11-8388

28. Ratnakumar, A., Zimmerman, S.E., Jordan, B.A., Mar, J.C.: Estrogen activates Alzheimer's disease genes. Alzheimer's & Dementia: Translat. Res. Clin. Intervent. **5**, 906–917 (2019). https://doi.org/10.1016/j.trci.2019.09.004

29. Santoro, N.: Perimenopause: from research to practice. J. Women's Health **25**(4), 332–339 (2016). https://doi.org/10.1089/jwh.2015.5556

30. Selesnick, I.W.W., Baraniuk, R.G.G., Kingsbury, N.C.C.: The dual-tree complex wavelet transform. IEEE Signal Process. Mag. **22**, 123–151 (2005). https://doi.org/10.1109/MSP.2005.1550194

31. Soh, L., Tsatsoulis, C.: Texture analysis of SAR sea ice imagery using gray level co-occurrence matrices. IEEE Trans. Geosci. Remote Sens. **37**(2), 780–795 (1999). https://doi.org/10.1109/36.752194

32. Song, W.K., Lee, S.C., Lee, E.S., Kim, C.Y., Kim, S.S.: Macular thickness variations with sex, age, and axial length in healthy subjects: a spectral domain-optical coherence tomography study. Investigative Ophthalmol. Vis. Sci. **51**, 3913–3918 (2010). https://doi.org/10.1167/iovs.09-4189

33. Svetozarskiy, S.N., Kopishinskaya, S.V.: Retinal optical coherence tomography in neurodegenerative diseases (Review). Sovremennye Tehnologii v Medicine **7**(1), 116–123 (2015). https://doi.org/10.17691/stm2015.7.1.14

34. Wang, S., Lu, S., Dong, Z., Yang, J., Yang, M., Zhang, Y.: Dual-tree complex wavelet transform and twin support vector machine for pathological brain detection. Appl. Sci. **6**(169), 1–18 (2016). https://doi.org/10.3390/app6060169

35. Wong, A., Chan, C., Hui, S.: Relationship of gender, body mass index, and axial length with central retinal thickness using optical coherence tomography. Eye **19**, 292–297 (2005). https://doi.org/10.1038/sj.eye.6701466

36. World Medical Association: World Medical Association Declaration of Helsinki: ethical principles for medical research involving human subjects. J. Am. Med. Assoc. **310**(20), 2191–2194 (2013). https://doi.org/10.1001/jama.2013.281053

37. Zhu, D., et al.: Age at natural menopause and risk of incident cardiovascular disease: a pooled analysis of individual patient data. Lancet Public Health **4**, 553–564 (2019). https://doi.org/10.1016/S2468-2667(19)30155-0

Deep Retinal Diseases Detection and Explainability Using OCT Images

Mohamed Chetoui[ID] and Moulay A. Akhloufi[✉][ID]

Perception, Robotics, and Intelligent Machines Research Group (PRIME),
Department of Computer Science, Université de Moncton, Moncton,
NB E1A 3E9, Canada
{emc7409,moulay.akhloufi}@umoncton.ca

Abstract. Retinal disease classification is an important challenge in computer aided diagnosis (CAD) for medical applications. Eye diseases can cause different symptoms from mild vision problems to complete blindness if it is not timely treated. The early diagnosis is crucial to prevent blindness. In this work, we use deep Convolutional Neural Networks (CNN) on a 4-class classification problem to automatically detect choroidal neovascularization (CNV), diabetic macular edema (DME), drusen, and normal cases using Optical Coherence Tomography (OCT) images. The obtained results achieve state-of-the-art performance and show that the proposed network leads to higher classification rates with an accuracy of 98.46%, and an Area Under Curve (AUC) of 0.998. An explainability algorithm was also developed and shows the efficiency of the proposed approach in detecting retinal disease signs.

Keywords: Diabetic retinopathy · Diabetic Macular Edema (DME) · Convolutional Neural Networks · Optical Coherence Tomography (OCT)

1 Introduction

Retinal diseases can be related to aging, diabetes or other diseases, trauma to the eye, or family history. These diseases may lead to vision loss, whether temporarily or permanently. The symptoms may differ depending on the disease and its causes. If not timely treated, the patient may suffer from complete blindness. The Optical Coherence Tomography (OCT) imaging is extensively used to capture 3D cross sectional views of the human retina for the detection of many eye diseases such as CNV, DME, and drusen. As shown in Fig. 1, OCT is able to probe through the retina depth and image all the retinal layers at high resolution. This technique is now considered as a standard in ophthalmology for the examination and assessment of the response to retinal treatments. In daily clinical practice, these lesions are manually diagnosed by physicians using OCT images. However,

This work was supported in part by the New Brunswick Health Research Foundation (NBHRF). The NVIDIA Quadro P6000 was donated by the NVIDIA Corporation.

this task is tedious, time consuming, and requires an intensive effort due to the small size of the lesions and their lack of contrast. Thus, it is beneficial to develop an automatic diagnosis system to support physicians in their diagnosis work.

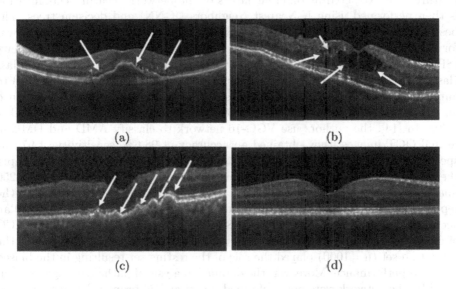

Fig. 1. Examples of OCT images. The arrows indicated the lesion sites. (a) Choroidal neovascularization (CNV); (b) Diabetic macular edema (DME); (c) Drusen; (d) Normal.

In the past, feature engineering and machine learning techniques have been proposed for retinal diseases detection [4,16]. Deep learning has proven to be the best approach in recent years and is now considered the state of the art for retinal diseases detection. In [2], the authors propose combining features from three pre-trained Convolutional Neural Networks: AlexNet [9], VggNet [21] and GoogleNet [22] and performing feature space reduction using Principal Component Analysis. They use a majority voting scheme based on a plurality rule between classifications from the three CNNs. Experiments were conducted using OCT datasets. The system achieves an accuracy of 93.75%. While the performance is interesting, feature extraction by the fusion of 3 CNNs is time-consuming. In addition, feature space reduction can lead to losing important information in DME detection. Chan et al. [3] propose the use of transfer learning For Diabetic Macular Edema (DME) detection on OCT images denoised using Block-Matching with 3-Dimension (BM3D) filtering and cropped through image boundary extraction. After these preprocessing, features are extracted using the first layers of AlexNet [9]. Classification is performed using a support vector machine (SVM) classifier. They show interesting results with an accuracy of 96%. Perdomo et al. [18] propose an end-to-end CNN called OCT-NET. This network consists of 16 layers including 2 FC layers for a binary classification of 2 classes DME and normal. The system is evaluated using a 32-fold Leave One Patient Out Cross

Validation (LOPO-CV). The proposed system achieves an accuracy (ACC) of 93.75%. The work of Awais *et al.* [1] focuses on the classification of abnormal and normal OCT image volumes using a pre-trained VGG16 CNN network [21]. Features are extracted at different layers of the network. Feature classification is then performed using K-Nearest Neighbors (K-NN) and decision trees. The best Classification results with noise removal and without image cropping was for K-NN (k = 3) achieving an accuracy (ACC), sensitivity (SE) and specificity (SP) of 90.6%, 100%, 81.25%, respectively. Li *et al.* [13] focused on a 4-class classification problem, they use an ensemble of four CNN models (ResNet50) to automatically classify CNV, DME, drusen, and normal. Their system obtained an accuracy of 97.9%, sensitivity of 96.8%, specificity of 99.4% and an AUC of 0.998. In [14], the authors use VGG-16 network to classify AMD and DME in retinal OCT images, they obtained an accuracy of 98.6%, sensitivity of 97.8%, specificity of 99.4%, and an AUC of 100%. Authors cite in their recent paper the validation set was used as a test set with insufficient quantity of data (250 images for each category), which makes the model sensitive to overfitting, so the reported scores could be biased. In Kermany *et al.* [11], the authors reported an accuracy of 96.6%, a sensitivity of 97.8%, a specificity of 97.4%, and an AUC of 0.999 for detection of CNV, DME, drusen and normal exams. As in [14], the validation set (n = 1000) played the role of the testing set resulting in the biased reported performance. Moreover, the detailed analysis of pathology regions identified by the network were not performed. Lu *et al.* [15] proposed a system based on deep learning, they use a CNN model Resnet-101 [6] to detect multiple diseases from OCT images. The obtained accuracy was 95.9%, sensitivity = 94.0%, specificity = 97.3% and AUC = 0.984. The authors apply a random split for OCT images into training, validation, and testing sets. The specific partition might include multiple images from the same eye so that the performance of the model may be biased.

2 Proposed Approach

In this work we propose the use of a recent convolutional neural network called EfficientNet [24]. This architecture will be fine-tuned to detect choroidal neovascularization (CNV), diabetic macular edema (DME), drusen, and normal using Optical Coherence Tomography (OCT) images. An explainability algorithm is also developed to visually show the pathology regions identified by the network. The proposed approach is illustrated in Fig. 2.

2.1 Dataset

The images for training and testing are from Optical Coherence Tomography (OCT) dataset [10]. Images were collected from retrospective cohorts of adult patients from the Shiley Eye Institute of the University of California San Diego, the California Retinal Research Foundation, Medical Center Ophthalmology Associates, the Shanghai First People's Hospital, and Beijing Tongren Eye

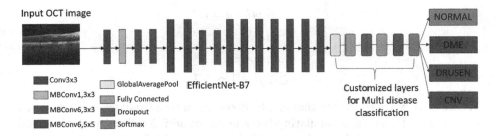

Fig. 2. Proposed deep CNN architecture for CNV, DME, drusen, and normal detection

Center between July 1, 2013 and March 1, 2017. The dataset contain 84,484 OCT high resolution images JPEG format (CNV = 37,455, DME = 11,598, drusen = 8,866, normal = 26,565). The OCT images were analyzed and labeled by experts as (disease)-(randomized patient ID)-(image number of this patient) and split into 4 directories: CNV, DME, drusen, and normal. Every image has a single label. For evaluation of the proposed approach, we randomly selected 3,512 images from CNV, 1,500 from DME category, 900 from drusen, and 2,494 from normal images, resulting in a total of 8,406 as test set (1,340 patients). After random split and to avoid bias, we verified that the patients in the test set are not included in the training set (using Patiend-ID). The 76,078 kept for model validation, contain 1,093 images (321 CNV, 249 DME, 225 drusen, 298 normal) to optimize the hyperparameters of our model. The remaining 74,982 images are used for training (15,216 patient). The OCT images were resized (keeping the aspect ratio) to a dimension of 224×224 pixels for our deep CNN architecture input.

2.2 Deep Learning Model

To detect CNV, DME, drusen, and normal using Optical Coherence Tomography images, we fine-tuned a CNN called EfficientNet-B7, based on the architecture proposed by Tan *et al.* [24]. The authors propose to balance all dimensions of the network: width, resolution and depth. Unlike standard CNN scaling approaches use one dimension scaling, EfficientNets are the first to quantify the relationship between all three dimensions empirically.

The Authors use MnasNet network [23] to improve their basic architecture. They use a multi-objective neural architecture search that optimizes accuracy and FLOPS. They create a powerful network called EfficientNet-B0 similar to MnasNet [23] but much bigger because of the larger FLOPS target. Mobile inverted bottleneck MBConv [19] block with squeeze and excitation optimization [7] are integrated in the architecture. Based on EfficientNet-B0, The authors apply a compound scaling approach using a compound coefficient ϕ to scale the network width, depth and resolution by the following equation:

$$depth : d = \alpha^{\phi}$$
$$width : w = \beta^{\phi}$$
$$resolution : r = \gamma^{\phi} \tag{1}$$
$$\alpha \geq 1, \beta \geq 1, \gamma \geq 1$$

Where α, β, γ are constants that can be determined by a small grid search. ϕ is a user-specified coefficient regulating the amount of additional resources required for model scaling, while α, β, γ specify how these additional resources can be allocated to network width, depth and resolution, respectively.

EfficientNet-B0 can be scaled up by setting α, β, γ as constants and extending the baseline network with different α to achieve an EfficientNet family (B1, B2, B3 to B7). EfficientNet-B7 achieves state-of-the-art results with a 97.1% top-5 accuracy on ImageNet, while being 8.4× smaller and 6.1× faster on inference than the best existing ConvNets such as SENet [7] and Gpipe [8].

The proposed model is based on EfficientNet-B7. We customized the last convolution layer of the network by adding a Global Average Pooling (GAP) to improve the accuracy and reduce overfitting. After (GAP), we added 2 Dense layers of size 1024 and used a 25% Dropout after Dense layers. Finally, a Softmax layer is added to give the probability prediction scores for detecting CNV, DME, drusen, and normal on OCT images.

3   Experimental Results

Keras Library [5] was used to program the proposed networks and training was carried out in an NVIDIA Quadro P6000 [17] using 200 epochs. Adam [12] was used as optimizer, and L2 regularization was performed on weights with a weight decreasing factor of 0.0005. Batch size is fixed to 32 for the proposed model. The initial learning rate was set to 0.003, and then decreased by 1% after each epoch in order to improve and speed-up convergence while avoiding overshooting. Hyperparameter optimization was conducted on the validation set and the best results were kept for testing. All the experiments are evaluated in terms of sensitivity (SE), specificity (SP), area under curve (AUC) and accuracy (ACC) [4]. The SE and SP show the performance of the proposed approach with respect to 4 classes (CNV, DME, drusen, normal). ACC is used as a statistical measure of how well a multi-class classification test correctly identifies or excludes a condition. AUC computed using ROC curve is a performance measure widely used for medical classification problems. Our performance metrics are computed using one-vs-all.

The proposed fine-tuned EfficientNet-B7 gives interesting results for multi-disease classification. The model achieved an accuracy of 98.46%, a specificity of 98.14%, a sensitivity of 98.97%, and an AUC of 0.998 using 8,406 OCT images for testing. Only 127 from 8,406 were misclassified. Figure 3 shows the ROC curve for the multi-classification problem. We can see the high performance obtained by our model. The confusion matrix is given in Fig. 4.

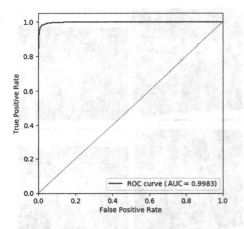

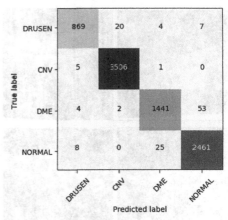

Fig. 3. ROC curve (detection of CNV, DME, drusen and normal).

Fig. 4. Confusion matrix for our model predictions.

To understand how the model learned to detect the pathology signs of CNV, DME and drusen, we developed an explainability technique based on the use of Gradient-weighted Class Activation Mapping (Grad-CAM) [20]. This technique was used to provide a visual description from the results of our proposed CNN model. Grad-CAM uses the gradients of any target, flowing into the final convolutional layer to generate a coarse localization map that highlights important regions in the predictive image. This technique refers to our current CNN model, without any improvements in architecture or retraining. The suggested technique blends Grad-CAM with fine-grained visualizations to construct a high-resolution class-discriminative visualization.

Figure 5 shows positive cases of each disease. We can see that the heatmap is accurately located around the most important symptoms. For the CNV, our model detected the subretinal/outer retinal hyper-reflective materials associated with the pigment epithelial detachment. These regions were identified as the most significant for CNV. For the DME, the model detected the sub/intra-retinal fluid accumulation as the main distinguishing features. The drusen was located as undulations and elevations in the hyper-reflective band of the neurosensory retinal detachment with less reflective material beneath them. In the normal case, the hyper-reflective band was detected as a region of interest.

The performance comparison with recent state-of-the-art methods is presented in Table 1 for the detection of CNV, DME and drusen. As we can see, our model outperforms the model proposed by Lu et al. [15]. In this study the authors used proprietary dataset with non-public images. In Li et al. [13] and Kermany et al. [11], the authors considered the problem of classifying CNV, DME and drusen. Our model outperforms the models proposed in this studies in term of accuracy and sensitivity. They obtained an AUC of 0.998 and 0.999, which is similar to our score of 0.998. However the authors tested their models on a limited number of OCT images (1,000 and 1,536) vs. 8,406 used as a test

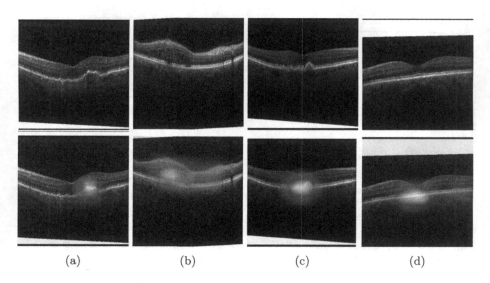

(a)   (b)   (c)   (d)

Fig. 5. Positive cases of each disease. The yellow coloured areas in the retinal images show the signs detected by our deep learning model for predicting multiple diseases on OCT images. (a): Choroidal neovascularization (CNV); (b) Diabetic macular edema (DME); (c) Drusen; (d) Normal. (Color figure online)

set in our work. This shows that our fine-tuned model EfficientNet-B7 is robust and exceeds the performance of the models used in [11,13,15].

Table 1. Performance comparison with state-of-the-art methods using OCT images

Authors	ACC%	SP%	SE%	AUC	# validation sample
Chan *et al.* [2]	93.75	93.75	93.75	–	–
Li *et al.* [13]	97.9	99.4	96.8	0.998	1,536
Awais *et al.* [1]	90.6	81.25	100	–	–
Kermany *et al.* [11]	96.6	97.4	97.8	0.999	1,000
Lu *et al.* [15]	95.9	97.3	94.0	0.984	3,317
Ours	**98.46**	**98.68**	**98.37**	**0.998**	**8,406**

4   Conclusion

In this work, we developed a deep convolutional neural network (CNN) model based on a recently developed EfficientNET-B7 CNN. The model was fine-tuned and trained for a 4-class classification problem to detect choroidal neovascularization (CNV), diabetic macular edema (DME), drusen, and normal cases using Optical Coherence Tomography (OCT) images. The obtained results show that our proposed model surpasses some recent deep learning approaches. We obtained high AUC scores with 0.998 and an accuracy of 98.43% for detecting CNV, DME, drusen and normal on a test set of 8,040 OCT images. Past published work used a small number of images in their experiments. This demonstrates that our model is robust in the classification of CNV, DME, drusen and normal on a large number of images. An explainability algorithm was also developed and showed that our model can be efficient identifying the most important symptoms of the disease. The proposed technique is an interesting contribution in the development of a diagnosis system able to detect multiple eye diseases on OCT images. Future work includes experimenting with more images as well as modifying the model to identify other types of medical images and other health issues.

References

1. Awais, M., Müller, H., Tang, T.B., Meriaudeau, F.: Classification of SD-OCT images using a deep learning approach. In: 2017 IEEE International Conference on Signal and Image Processing Applications (ICSIPA), pp. 489–492. IEEE (2017)
2. Chan, G.C., Kamble, R., Müller, H., Shah, S.A., Tang, T., Mériaudeau, F.: Fusing results of several deep learning architectures for automatic classification of normal and diabetic macular edema in optical coherence tomography. In: 2018 40th Annual International Conference of the IEEE Engineering in Medicine and Biology Society (EMBC), pp. 670–673. IEEE (2018)
3. Chan, G.C., Muhammad, A., Shah, S.A., Tang, T.B., Lu, C.K., Meriaudeau, F.: Transfer learning for diabetic macular edema DME detection on optical coherence tomography OCT images. In: 2017 IEEE International Conference on Signal and Image Processing Applications (ICSIPA), pp. 493–496. IEEE (2017)
4. Chetoui, M., Akhloufi, M.A., Kardouchi, M.: Diabetic retinopathy detection using machine learning and texture features. In: 31st IEEE Canadian Conference on Electrical and Computer Engineering (CCECE 2018) (2018)
5. Chollet, F., et al.: Keras (2015). https://keras.io
6. He, K., Zhang, X., Ren, S., Sun, J.: Deep residual learning for image recognition. In: Proceedings of the IEEE Conference on Computer Vision and Pattern Recognition, pp. 770–778 (2016)
7. Hu, J., Shen, L., Sun, G.: Squeeze-and-excitation networks. In: Proceedings of the IEEE Conference on Computer Vision and Pattern Recognition, pp. 7132–7141 (2018)
8. Huang, Y., et al.: GPipe: efficient training of giant neural networks using pipeline parallelism. In: Advances in Neural Information Processing Systems, pp. 103–112 (2019)
9. Iandola, F.N., Han, S., Moskewicz, M.W., Ashraf, K., Dally, W.J., Keutzer, K.: SqueezeNet: AlexNet-level accuracy with 50x fewer parameters and <0.5 mb model size. arXiv preprint arXiv:1602.07360 1(10) (2016)
10. Kermany, D., Zhang, K., Goldbaum, M.: Labeled optical coherence tomography OCT and chest x-ray images for classification. Mendeley data (2018). https://data.mendeley.com/datasets/rscbjbr9sj/2
11. Kermany, D.S., et al.: Identifying medical diagnoses and treatable diseases by image-based deep learning. Cell 172(5), 1122–1131 (2018)

12. Kingma, D.P., Ba, J.: Adam: a method for stochastic optimization. arXiv preprint arXiv:1412.6980 (2014)
13. Li, F., et al.: Deep learning-based automated detection of retinal diseases using optical coherence tomography images. Biomed. Opt. Express **10**(12), 6204–6226 (2019)
14. Li, F., Chen, H., Liu, Z., Zhang, X., Wu, Z.: Fully automated detection of retinal disorders by image-based deep learning. Graefe's Arch. Clin. Exp. Ophthalmol. **257**(3), 495–505 (2019)
15. Lu, W., Tong, Y., Yu, Y., Xing, Y., Chen, C., Shen, Y.: Deep learning-based automated classification of multi-categorical abnormalities from optical coherence tomography images. Transl. Vis. Sci. Technol. **7**(6), 41–41 (2018)
16. Malik, S., Kanwal, N., Asghar, M.N., Sadiq, M.A.A., Karamat, I., Fleury, M.: Data driven approach for eye disease classification with machine learning. Appl. Sci. **9**(14), 2789 (2019)
17. NVIDIA: QUADRO P6000. https://www.nvidia.com/content/dam/en-zz/Solutio ns/design-visualization/productspage/quadro/quadro-desktop/quadro-pascal-p60 00-data-sheet-us-nv-704590-r1.pdf. Accessed Feb 2020
18. Perdomo, O., Otálora, S., González, F.A., Meriaudeau, F., Müller, H.: OCT-NET: a convolutional network for automatic classification of normal and diabetic macular edema using SD-OCT volumes. In: 2018 IEEE 15th International Symposium on Biomedical Imaging (ISBI 2018), pp. 1423–1426. IEEE (2018)
19. Sandler, M., Howard, A., Zhu, M., Zhmoginov, A., Chen, L.C.: MobileNetv 2: inverted residuals and linear bottlenecks. In: Proceedings of the IEEE Conference on Computer Vision and Pattern Recognition, pp. 4510–4520 (2018)
20. Selvaraju, R.R., Cogswell, M., Das, A., Vedantam, R., Parikh, D., Batra, D.: Grad-CAM: visual explanations from deep networks via gradient-based localization. In: Proceedings of the IEEE International Conference on Computer Vision, pp. 618–626 (2017)
21. Simonyan, K., Zisserman, A.: Very deep convolutional networks for large-scale image recognition. arxiv 2014. arXiv preprint arXiv:1409.1556 1409 (2014)
22. Szegedy, C., et al.: Going deeper with convolutions. In: Proceedings of the IEEE Conference on Computer Vision and Pattern Recognition, pp. 1–9 (2015)
23. Tan, M., et al.: MnasNet: platform-aware neural architecture search for mobile. In: Proceedings of the IEEE Conference on Computer Vision and Pattern Recognition, pp. 2820–2828 (2019)
24. Tan, M., Le, Q.V.: EfficientNet: rethinking model scaling for convolutional neural networks. CoRR abs/1905.11946 (2019). http://arxiv.org/abs/1905.11946

Grand Challenge on Automatic Lung Cancer Patient Management

An Automated Workflow for Lung Nodule Follow-Up Recommendation Using Deep Learning

Krishna Chaitanya Kaluva[✉], Kiran Vaidhya, Abhijith Chunduru, Sambit Tarai,
Sai Prasad Pranav Nadimpalli, and Suthirth Vaidya

Predible Health, Bangalore, India
krishna@prediblehealth.com

Abstract. Early detection of lung cancer increases a patient's survival rate and provides healthcare professionals, valuable time, and information to administer effective treatment. Lung nodules are early signs of lung cancer. Computer-aided diagnostic systems that can identify pulmonary nodules improve early detection as well as provide an independent second opinion. We propose an automated workflow for follow-up recommendation based on low-dose computed tomography (LDCT) images using deep learning, as per 2017 Fleischner Society guidelines. As per guidelines, follow-up is based on size, volume and texture of nodules. In this paper, we present a 5 stage approach for automated follow-up recommendation. The 5 stages are Lung segmentation, Nodule detection and False Positive Reduction (FPR), Texture classification, Nodule segmentation and Follow-up recommendation. Our nodule detection has a sensitivity of **94%** @ 1 false positive per scan. The FPR network improves the specificity of detection to **90%** without changing sensitivity. Nodule segmentation has a Jaccard index of **0.77** on 768 nodules from Lung Nodule Database (LNDb) [1]. Texture classification has a sensitivity of **97%** on solid nodules and a Fleiss-Cohen's Kappa of 0.37 on LNDb data with most errors between sub-solid and solid nodules. Our rule-based follow-up recommendation has a Fleiss-Cohen's Kappa of **0.53** on 236 patients from LNDb. In conclusion, we found that rule-based approach for follow-up alongside deep learning models is the best approach in achieving best results. As we improve the first 4 stages, we foresee that recommendation from AI will become closer to radiologists recommendation.

Keywords: Artificial intelligence · Lung cancer screening · CT · 3D convolutional neural networks (CNN) · Pulmonary nodule detection · Nodule malignancy

1 Introduction

Lung cancer is the most fatal form of cancer in the world, whose triage and treatment presents a grand challenge in modern healthcare. The International Agency for Research on Cancer estimated that globally, lung cancer is the leading cause of cancer deaths

© Springer Nature Switzerland AG 2020
A. Campilho et al. (Eds.): ICIAR 2020, LNCS 12132, pp. 369–377, 2020.
https://doi.org/10.1007/978-3-030-50516-5_32

among both sexes, contributing to 18.6% of all cancer deaths in 2018 [2]. The American Lung Association (ALA) observed a significant decrease in lung cancer mortality in the USA over the decade 2010–2019 [3] which was attributed to widespread adoption of lung cancer screening with low-dose computed tomography (LDCT) scans, which resulted in early detection of lung cancer. The radiologist's performance often suffers due to fatigue, immense workload, and a lack of clear consensus on the diagnosis within the radiology community [4, 5].

Computer Aided diagnosis (CAD) can help reduce radiologist's fatigue and work load as they can easily scan through the entire CT image in less than a minute and present with a lot more information. Automatic lung cancer diagnosis is a hard problem as each of detection, segmentation and classification are 3 dimensional problems which require training complex machine learning models and huge amounts of data. Some of the previous methods use 2D proposals for nodules and combine them to get 3D bounding boxes [6, 7]. Nodules are heterogeneous in size, shape and texture that even experienced radiologists don't agree on some nodules. Cancer diagnosis is highly dependent on the textural features, size and number of the nodules.

We propose a 5 stage automated recommendation system where we first localize the lung region from the entire CT scan followed by predicting nodules in the lung region using a 3D faster R-CNN [8] which are classified as nodule or false positive by a 3D wide residual network [9]. 3D patches are extracted around the nodule centroid and are used to predict the texture and segmentation of the nodule. The segmentation is used to calculate the long and short axis diameters and volume of the nodule. The list of nodules, their texture probabilities and volumes are then passed to the follow-up recommendation system which returns a follow up based on 2017 Fleischner society guidelines for lung cancer screening.

2   Methodology

Figure 1 describes the workflow of the entire algorithm in 5 stages. The following subsections explain each stage in detail.

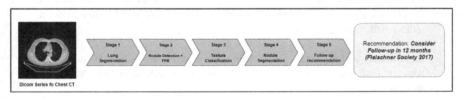

Fig. 1. End-to-end workflow of automated follow-up recommendation system for lung cancer screening

2.1   Data and Image Pre-processing

In all the 5 stages, LNDb (Lung Nodule Database) [1] dataset is used as validation/test set. LNDb dataset consists of 298 CT scans, with 236 scans released for training/validation containing 768 nodules. Other datasets used for training/validation are Lung Image Database Consortium (LIDC-IDRI) [10] with 1010 CT scans and annotations from 4 radiologists, Lung Tissue Research Consortium (LTRC) [11] and National Lung Screening Trial (NLST) [12]. From here on, we shall refer to the training set with 236 scans as SET1 and the test set with 58 scans as SET2.

In every stage, each CT scan is resampled to 1 mm spacing in all axes. A HU (Hounsfield Unit) window of [−1200, 400] is applied followed by normalization to [−1, 1]. Required padding is applied based on the patch size that needs to be extracted.

2.2 Stage1 - Lung Segmentation

A chest CT scan contains anatomical regions other than lung. We use a 3D convolutional neural network with Unet architecture [13] to segment the lung region from the CT scan. Each CT scan is pre-processed as mentioned in Sect. 2.1 and 3D patches are extracted. This network is trained on 600 scans and validated on 172 scans from LTRC dataset.

2.3 Stage2a – Nodule Detection

Our deep learning model for Nodule detection is inspired by the winning solution of DSB2017 [14] where a faster R-CNN [8] architecture is used to predict bounding boxes of nodules in 3D. This network takes a 3D patch and predicts bounding boxes inside it. Each bounding box/detection is modelled as

$$[conf, z, y, x, r]$$

Where

conf - the confidence with which the network calls a detection as nodule
[z, y, x] - centroid of the detection
r - radius of the bounding box

This network is trained on 2000 CT scans from LIDC, NLST and a proprietary dataset and validated on 354 CT scans from LIDC. There are a total of 2608 nodules in the entire data. In the LIDC dataset, we only use nodules with consensus of at least 2/4 radiologists. The model is tested on LNDb SET1 and SET2.

Each CT scan is pre-processed as mentioned in Sect. 2.1 and 3D patches are extracted. Data augmentation and hard negative mining are implemented to make the model robust to artefacts and improve sensitivity. During inference, due to GPU memory constraints, the lung region had to be split into smaller 3D crops and detections are combined using non-maximum suppression. We intend this network to be very sensitive, thereby we set a threshold of 0.2 on confidence of the detections to remove obvious false positives before sending them to stage2b.

2.4 Stage2b – False Positive Reduction (FPR)

Object detection networks tend to predict many false positives (FPs), even at high confidence thresholds. To counter this issue, we use a False Positive Reduction (FPR) model trained to classify and remove the false positives detected by the nodule detection model. A 3D wide residual network [9] with depth of 10 and widen factor of 5 is used for this classification task. A dataset of 10000 nodule proposals predicted from 2500 NLST scans by the Faster RCNN is annotated by a trained personnel for 2 classes - nodules and false positives. After the annotation, we obtained 7600 false positives and 2400 nodules. A validation set with 388 nodules and 380 false positives is used to test the performance of the model.

We extract a three dimensional patch around each detection and feed it to the FPR model. The detections that are classified as false positives are discarded. To account for the class imbalance, focal loss [15] is used as the objective function for optimization.

2.5   Stage 3 – Texture Classification

Texture is an important image characteristic for assessing the malignancy of a nodule. We use a 3D wide residual network for classifying each nodule into one of the 3 categories - GGO, Sub-solid, and Solid. This network is trained on nodules from 1510 CT scans (LIDC - 1010, NLST - 500) and validated on all nodules from LNDb SET1. Since the dataset contained more solid nodules, we sampled only sub-solid and ground glass nodules from NLST dataset.

Each CT image is pre-processed as mentioned in Sect. 2.1 and 3D patches are extracted with nodule at the center. Table 1 gives the summary of the number of nodules used in each category from each dataset.

As seen from Table 1, even after adding sub-solid and GGO nodules from NLST dataset, the class imbalance is high. Hence we employ focal loss [15] which down weighs the well classified examples and gives more weightage to hard examples. Data augmentations is also used to prevent overfitting.

Table 1. Table showing the amount of data from each dataset.

Data	#GGO	# Sub-solid	# Solid	Total
LIDC	155	40	1547	1742
LNDb	38	58	672	768
NLST	287	347	0	634

2.6   Stage 4 – Nodule Segmentation

Early stage lung nodules have small diameters and relatively low volume compared to the lung which makes segmenting nodules tougher. Inspired by the submission made by Özgün et al. [16] for medical image segmentation, we trained a 3D U-Net to learn the dense segmentation of the lung nodules.

The LIDC-IDRI dataset consists of 1010 scans, which are annotated by four radiologists for nodules, texture and segmentation. The intersection of the segmentations of radiologists for each nodule is considered as ground-truth mask. We split the dataset into a training subset consisting of 667 scans with 1982 nodules, and a validation subset consists of 343 scans with 497 nodules. 768 nodules from LNDb SET1 are used as a test set.

Each CT image is resampled to 1 mm spacing. We feed 3D patches of 2 different HU windows - Mediastinum and Lung windows as 2 channels to the network. We do this as nodule boundaries are sometimes more clearly visible in the mediastinum window. Image and label patches are extracted around the nodule centroids. For each scan, we also sampled 10 patches at random from lung regions without nodules, to augment the background class.

2.7   Stage 5 – Follow up Recommendation

Once all the 4 stages are done, we tried 2 different approaches for follow up recommendation.

1. Rule-based recommendation
2. A Support Vector Machine (SVM)

In the first method, a set of rules are written according to the 2017 Fleischner recommendation system, which takes in the list of nodules, their textures, and volumes and gives a Follow-up recommendation.

In the second method, we built a Support Vector Machine (SVM) with the following 7 features as inputs on LNDb SET1

1. Number of nodules
2. Number of solid nodules
3. Number of sub-solid nodules
4. Number of non-solid nodules/GGOs
5. Diameter of largest solid nodule
6. Diameter of largest sub-solid nodule
7. Diameter of largest non-solid nodule/GGO

The SVM's hyperparameters are tuned using a 4 fold cross validation. The 7 inputs are obtained for each case from the predictions from stages 2 to 4.

3   Results

3.1   Lung Segmentation

On the 172 validation scans from LTRC, the 3D Unet has an average dice score of 0.98.

3.2 Nodule Detection

On the validation set of 354 CT scans, the model has a sensitivity of 93% @ 1 false positive per scan (Fp/scan) on nodules > 5 mm and a sensitivity of 91% @ 1 Fp/scan on nodules > 3 mm. Figure 2 shows the Free Receiver Operating Characteristic (FROC) plot on the validation set, the dotted lines indicate 95% confidence interval (CI). The nodule detection sensitivities for solid, sub-solid and GGO categories are tabulated in Table 2.

On the 236 validation CT scans from LNDb SET1, the model has an FROC score of 0.43 on 768 nodules of all sizes. On LNDb SET2, the model has an FROC score of 0.67 on nodules with consensus of atleast 2 radiologists.

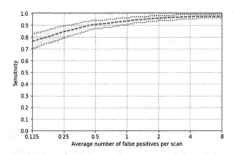

Fig. 2. FROC plot on validation set, dotted lines indicate 95% CI.

Table 2. Sensitivity for each subtype of the nodules

Nodule Type	Detection accuracy
Overall	**93%** at 1 Fp/scan
Solid	95% at 1 Fp/scan
Sub solid	90% at 1 Fp/scan
GGO	84% at 1 Fp/scan

3.3 False Positive Reduction

On the validation set, the 3D wide residual network has a sensitivity of 98% for nodules while removing about 75% of the FPs on the validation set. The 2% of the nodules removed are either small or subjective. Of the remaining false positives majority of them are vessel joints and pleural plaques. The model improves the specificity of detection to 90% without changing sensitivity on 354 validation scans of LIDC from Sect. 3.2.

3.4 Texture Classification

The model achieved an overall accuracy of 87% on the validation set of 236 scans from LNDb dataset. The Fleiss Cohen's Kappa score is 0.37 for the best model and the confusion matrix for the model with best Fleiss Kappa is shown in Fig. 3. The model has 97% sensitivity on solid nodules but performs poorly on sub-solid and ground glass nodules.

	GGO	Sub-solid	Solid
GGO	15	1	22
Sub-solid	5	1	52
Solid	11	12	649

Fig. 3. Confusion matrix for the model with best Fleiss Kappa (Rows – ground truth, Columns – predicted)

3.5 Nodule Segmentation

The model is validated on 497 nodules from LIDC-IDRI. For all these nodules, intersection of segmentations from the radiologists is considered as ground truth. The 3D U-net has an average dice of 0.75 on these nodules. Table 3 gives a summary of metrics on LNDb SET1. Figure 4 shows an example slice of a CT image with nodule marked and corresponding segmentation edge map predicted by the model.

Table 3. Table showing the metrics of 3D Unet on CT scans from LNDb SET1

Hausdorff distance (HD)	9.12
Modified Pearson correlation coefficient (r*)	0.14
Mean average distance (MAD)	5.95
Modified Jaccard index (J*)	**0.78**

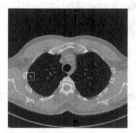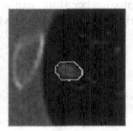

Fig. 4. Figures showing the image and segmentation outlay of a nodule predicted by Unet

3.6 Follow-Up Recommendation

Fleiss Cohen's Kappa is calculated to evaluate both rule-based and SVM methods. The SVM has a kappa of 0.28 whereas the rule-based method has a Kappa of **0.53** on 236 CT scans from LNDb SET1. The rule-based method has a sensitivity of 90% and specificity of 70% for follow-up class 0, with poor sensitivity for other classes. Follow-up class 3 has a specificity of 80%. On LNDb SET2, the rule-based method has a Cohen's Kappa of 0.46 and SVM has as Kappa of 0.35. Overall, the rule-based method is better than the SVM model.

4   Discussion and Conclusion

In this paper, we proposed an automated 5 stage pipeline for follow-up recommendation which involves lung localization, nodule detection, nodule segmentation, texture classification and follow-up recommendation. The proposed pipeline achieves high FROC scores for Nodule detection and high sensitivity for follow up class 0. The False positive reduction network helped in improving the precision of the nodule detection network, also helping us get a better nodule detection performance compared to other participants. Having lesser false positives also helped us get better results for follow up class 0 in the follow-up recommendation task.

In the follow-up recommendation task, the scores for follow up classes 1, 2 and 3 depend a lot on how well the algorithm can segment and classify texture of a nodule. In the nodule segmentation task, considering intersection of segmentations from all the radiologists could have made our model very precise and caused it to under-segment the nodules. We believe, these inaccuracies in the texture classification and nodule segmentation models propagated to reduce the performance for classes 1, 2 and 3 in the follow-up recommendation task.

Future work will involve improving nodule segmentation and texture classification models. Adding more data for under-represented classes, considering majority consensus for nodule contours in the segmentation tasks, using surface based loss measures for segmentation task are a few approaches that could help us improve these models and thereby improve follow-up recommendation as well. On the limited data that we have tested for this competition, the rule based follow-up recommendation seemed to have worked well and also has been more explainable. It would be interesting to see if learnable models outperform the rule based methods as we add more data.

References

1. Pedrosa, J., et al.: LNDb: a lung nodule database on computed tomography. arXiv preprint arXiv:1911.08434 (2019)
2. Bray, F., et al.: Global cancer statistics 2018: GLOBOCAN estimates of incidence and mortality worldwide for 36 cancers in 185 countries. CA Can. J. Clin. **68**, 394–424 (2018). https://doi.org/10.3322/caac.21492
3. https://www.lung.org/our-initiatives/research/monitoring-trends-in-lung-disease/state-of-lung-cancer
4. Parikh, J.R., et al.: Radiologist burnout according to surveyed radiology practice leaders. J. Am. Coll. Radiol. **17**(1), 78–81 (2020)
5. Singh, S., et al.: Evaluation of reader variability in the interpretation of follow-up CT scans at lung cancer screening. Radiology **259**(1), 263–270 (2011)

6. Gulshan, V., et al.: Development and validation of a deep learning algorithm for detection of diabetic retinopathy in retinal fundus photographs. J. Am. Med. Assoc. **316**(22), 2402–2410 (2016)
7. Litjens, G., et al.: A survey on deep learning in medical image analysis. arXiv preprint arXiv:1702.05747 (2017)
8. Ren, S., He, K., et al.: Faster R-CNN: towards real-time object detection with region proposal networks. In: Advances in Neural Information Processing Systems, pp. 91–99 (2015)
9. Zagoruyko, S., et al.: Wide residual networks. arXiv preprint arXiv:1605.07146 (2016)
10. Armato III, S.G., et al.: The lung image database consortium (LIDC) and image database resource initiative (IDRI): a completed reference database of lung nodules on CT scans. Med. Phys. **38**(2), 915–931 (2011)
11. Holmes III, D., Bartholmai, B., Karwoski, R., Zavaletta, V., Robb, R.: The lung tissue research consortium: an extensive open database containing histological clinical and radiological data to study chronic lung disease. In: MICCAI Open Science Workshop (2006)
12. National Lung Screening Trial Research Team: Reduced lung-cancer mortality with low-dose computed tomographic screening. N. Engl. J. Med. **365**(5), 395–409 (2011)
13. Ronneberger, O., Fischer, P., Brox, T.: U-Net: convolutional networks for biomedical image segmentation. In: Navab, N., Hornegger, J., Wells, W.M., Frangi, A.F. (eds.) MICCAI 2015. LNCS, vol. 9351, pp. 234–241. Springer, Cham (2015). https://doi.org/10.1007/978-3-319-24574-4_28
14. Liao, F., et al.: Evaluate the malignancy of pulmonary nodules using the 3-D deep leaky noisy-or network. IEEE Trans. Neural Netw. Learn. Syst. **30**(11), 3484–3495 (2019)
15. Lin, T.Y., et al.: Focal loss for dense object detection. In: Proceedings of the IEEE international Conference on Computer Vision, pp. 2980–2988 (2017)
16. Çiçek, Ö., Abdulkadir, A., Lienkamp, Soeren S., Brox, T., Ronneberger, O.: 3D U-Net: learning dense volumetric segmentation from sparse annotation. In: Ourselin, S., Joskowicz, L., Sabuncu, Mert R., Unal, G., Wells, W. (eds.) MICCAI 2016, Part II. LNCS, vol. 9901, pp. 424–432. Springer, Cham (2016). https://doi.org/10.1007/978-3-319-46723-8_49

Pulmonary-Nodule Detection Using an Ensemble of 3D SE-ResNet18 and DPN68 Models

Or Katz, Dan Presil$^{(\boxtimes)}$, Liz Cohen, Yael Schwartzbard, Sarah Hoch, and Shlomo Kashani

NEC-IRC, Herzliya, Israel
{or.katz,dan.presil,liz.cohen,yael.schwartzbard}@necam.com,
sarahe.hoch@gmail.com, shlomo@deeponcology.ai

Abstract. This short paper describes our contribution to the **LNDb - Grand Challenge on automatic lung cancer patient management** [1]. We only participated in Sub-Challenge A: **Nodule Detection**. The officially stated goal of this challenge is **From chest CT scans, participants must detect pulmonary nodules**. We developed a computer-aided detection (CAD) system for the identification of small pulmonary nodules in screening CT scans. The two main modules of our system consist of a CNN based nodule candidate detection, and a neural classifier for false positive reduction. The preliminary results obtained on the challenge database is discussed.

In this work, we developed an Ensemble learning pipeline using state of the art convolutional neural networks (CNNs) as base detectors. In particular, we utilize the 3D versions of SE-ResNet18 and DPN68. Much like classical bagging, base learners were trained on 10 stratified data-set folds (the LUNA16 patient-level dataset splits) generated by bootstrapping both our training set (LUNA16) and the challenge provided training set. Furthermore, additional variation was introduced by using different CNN architectures. Particularly, we opted for an exhaustive search of the best detectors, consisting mostly of DPN68 [2] and SE-ResNet18 [3] architectures.

We unfortunately joined the competition late, and we did not train our system on the corpus provided by the organizers and therefore we only run inference using our LIDC-IDRI trained model. We do realize this is not the best approach.

1 Introduction

Pulmonary nodules are an important indicator for lethal malignant lung cancer, frequently misdiagnosed as calcification or even left completely undiagnosed. In the United States alone, lung-cancer accounts for an estimated 154,050 deaths per annum [4]. With a 5-year survival rate of 65% [4], early diagnosis and treatment is now more likely and possibly the most suitable means for lung-cancer related death reduction.

© Springer Nature Switzerland AG 2020
A. Campilho et al. (Eds.): ICIAR 2020, LNCS 12132, pp. 378–385, 2020.
https://doi.org/10.1007/978-3-030-50516-5_33

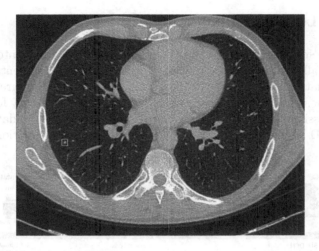

Fig. 1. A nodule from the dataset. An exemplary visualization of the data provided by the organizers, The LNDb dataset contains 294 CT scans collected retrospectively at the Centro Hospitalar e Universitário de São João (CHUSJ) in Porto, Portugal between 2016 and 2018.

Lung cancer most commonly manifests as non-calcified pulmonary nodules. Computer Tomography (CT) has been shown to be the best imaging modality for the detection of small pulmonary nodules, CT [Fig. 1] slices are widely used in the detection and diagnosis of lung-cancer. Radiologists, relying on personal experience, are involved in a laborious task of manually searching CT slices for lesions. Therefore, there is a very real need for automated analysis tools, providing assistance to clinicians screening for lung metastases.

2 Data Collection

For task A, lesion detection, we trained our CAD system on the well-known LIDC-IDRI [5] data set. In total, 1018 CT scans are included. The database also contains annotations which were collected during a two-phase annotation process using 4 experienced radiologists. Each radiologist marked lesions they identified as non-nodule, nodule $< 3\,\mathrm{mm}$, and nodules $\geq 3\,\mathrm{mm}$. Our data consists of all nodules $\geq 3\,\mathrm{mm}$ accepted by at least 3 out of 4 radiologists. For the false positive reduction pipeline, we added the data-set provided by LNDb competition organizers, the training dataset consists of approximately [1351] positive and [1890] negative samples. The respective frequency of each category is reported in Table 1.

Table 1. Nodule frequencies

Category	Frequency LIDC-IDRI	Frequency LNDb
Positive	**1351**	**1033**
Negative	**1890**	**494**

3   Deep Learning Pipeline

Our pipeline starts by normalizing each and every DICOM slice into 1 mm spacing isotropically on all three axes. Subsequently, 3D lobe segmentation is performed so that only 3D patches of the lungs, and not the lobes, are fed into our CNN. For the candidate nodule classification task we used binary focal loss and for the regression task we used L1 regularization. For each candidate nodule we performed 3D segmentation with U-NET [11] for volume estimation (Fig. 2).

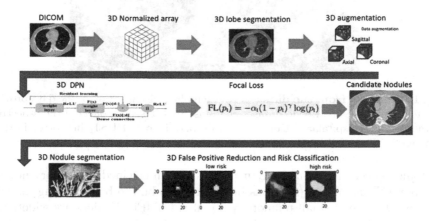

Fig. 2. Our pipeline

3.1   Data Preprocessing

The image pixel values are clipped to $(-1200, 600)$ HU, and subsequently the pixel values are normalized to $[0, 1]$. The scans are resampled on all axes to $1\,mm * 1\,mm * 1\,mm$ and then transposed according to the respective spacing, origin and the transpose matrix that are found in the scan metadata.

3.2   Lobe Segmentation

To find the candidate nodule locations and to accurately crop the lung area we used a 3D U-NET that was trained on patches of $64 \times 128 \times 128$ on the LOLA11 [12] dataset.

3.3   Data Augmentation

Due to the relatively small size of our training set, and the complexity of the visual tasks we are trying to solve, our approach relies heavily on 3D data-augmentation which we implemented in Python on our own. During run-time, we utilized an extensive data augmentation protocol by implementing flipping, scaling, rotation, and random Hounsfield Unit (HU) perturbation transformations. The augmentations are exemplified as follows:

```
1 self.transforms = []
2 if flipping:
3        self.transforms.append(RandomRotate())
4 if flipping:
5        self.transforms.append(flipping())
6 if scaling:
7        self.transforms.append(scaling())
8 if rotation:
9        self.transforms.append(rotation())
10 if Hounsfield:
11        self.transforms.append(Hounsfield())
```

3.4   CNN Model Generation

The nodule detection framework contains a 3D U-net with two backbones for the encoder - SeNet18 and DPN68. The model utilizes 5 anchors [5, 10, 15, 20, 25] mm with Binary focal loss for the classification task and L1 smooth loss for the regression task.

Model selection was implemented by saving the network achieving the highest validation accuracy (during and) at the end of training on the 10 pre-defined LUNA16 splits. During training, whenever the validation accuracy surpassed a pre-determined threshold, a checkpoint for the currently running CNN was saved noting the respective metrics epoch. Lastly, generalization performance was evaluated by testing the saved CNN on an independent test set that we kept aside.

3.5   False Positive Reduction

In our processing pipeline, 3D volumes (16 mm, 24 mm, 48 mm) are fed into CNN model resulting in a 6 Fully-Connected (FC) output, indicating a categorical output (Nodule/Not Nodule). Usually, only Nodule/Not Nodule is used in the literature which we demonstrate is inferior to our solution.

Our network learns local features directly from the nodule itself via a small Cube, while simultaneously amalgamating global contextual information from the surrounding tissues via two larger cubes, similar to Pyramidal CNN. Subsequently, each FC skip connection is fused to form the final classification layer via a majority vote Fig. 3.

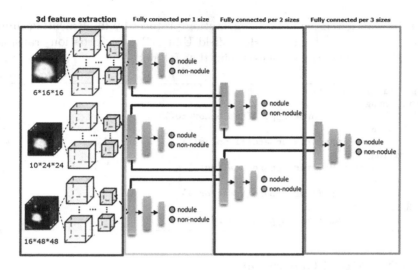

Fig. 3. False positive reduction.

4   Results

The results of our best **single** performance assessment, for each architecture are depicted in Table 2 (detection report).

4.1   Single Model Scores

The FROC, on test data is presented in Fig. 4.

Table 2. Best single model performance metrics across CNNs. Validation and test sensitivities are reported for each of the base CNN learners. **Note: the scores represent sensitivity on the validation set provided by the competition organizers;**

Model	Validation SN/Loss	Validation SN (SUBMISSION)
DPN68	85.4%/0.54	82.4%
SE-ResNet18	88.7%/0.35	83.38%

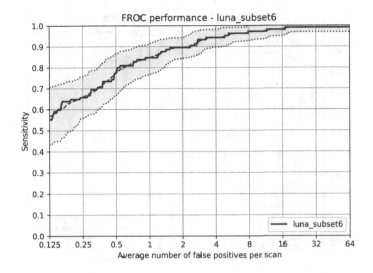

Fig. 4. FROC performance

The competition organizers allowed a single validation set prediction submissions per day with an immediate feedback on the respective resulting metrics. For instance, for our CNN checkpoint model entitled, *0.853_dpn107_finetune_87. 8648_0.3777.csv*, the submission resulted in a score of 0.853, which is one of our highest ranking single CNN models. The leader-board score for the DPN107 CNN is presented in Fig. 5.

4.2 Ensemble Scores

We developed an ensemble system using the CNNs in Table 2 as base detectors. All ensemble members were trained as described in Sect. 3. We did not have time to investigate more sophisticated ensembling methods such as meta-learners. The leader-board score for the Ensemble CNN is presented in Fig. 5.

Compared to the base learner's results our ensemble system was capable of boosting the FROC and we were able to improve performance of each base classifier, achieving an overall FROC of 0.6018 on the validation leader-board (Table 3).

Table 3. SE-ResNet18 model ensemble with DPN68 - Performance on the LUNA dataset

Precision	Recall	F Measure
0.9537	0.9769	0.9643

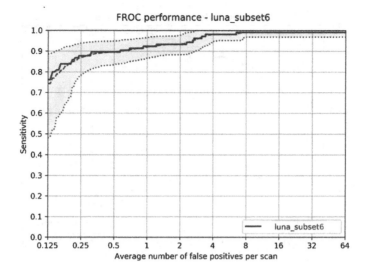

Fig. 5. FROC performance after ensemble

5   Conclusion

As suggested in [13], we utilized transfer-learning from the source dataset (LIDC) with no additional training on the target dataset. Though our system was trained only on a subset of the LIDC dataset, it was able to generalize very well-beyond the original corpus, resulting in a FROC, evaluated on the validation set of the LNDB dataset, which landed us at the 3rd place on Sub-Challenge A: Nodule detection. This demonstrated the significance of a system that can generalize well between datasets. We believe that other contestants who trained directly only on the LNDB dataset and surpassed us on the leaderboard, may have a lower FROC on the real test set due to overfitting on the training set.

References

1. Pedrosa, J., et al.: LNDb: a lung nodule database on computed tomography. arXiv:1911.08434 [eess.IV] (2019)
2. Chen, Y., et al.: Dual path networks. arXiv:1707.01629 [cs.CV] (2017)
3. He, K., et al.: Deep residual learning for image recognition. arXiv: 1512.03385 [cs.CV] (2015)
4. The American Lung Association - Lung Cancer Fact Sheet. https://www.lung.org/lung-health-and-diseases/lung-disease-lookup/lung-cancer/resource-library/lung-cancer-fact-sheet.html
5. Bidaut, L., et al.: Armato SG III McLennan G. The Lung Image Database Consortium (LIDC) and Image Database Resource Initiative (IDRI): A completed reference database of lung nodules on CT scans
6. Zhu, W., et al.: DeepLung: deep 3D dual path nets for automated pulmonary nodule detection and classification. arXiv:1801.09555 [cs.CV] (2018)

7. Huang, G., et al.: Densely connected convolutional networks. arXiv:1608.06993 [cs.CV] (2016)
8. Xie, S., et al.: Aggregated residual transformations for deep neural networks. arXiv:1611.05431 [cs.CV] (2016)
9. Arindra, A., Setio, A., et al.: Validation, comparison, and combination of algorithms for automatic detection of pulmonary nodules in computed tomography images: the LUNA16challenge. Med. Image Anal. **42**, 1–13 (2017). ISSN: 1361-8415. https://doi.org/10.1016/j.media.2017.06.015. http://dx.doi.org/10.1016/j.media.2017.06.015
10. Chu, B., Madhavan, V., Beijbom, O., Hoffman, J., Darrell, T.: Best practices for fine-tuning visual classifiers to new domains. In: Hua, G., Jégou, H. (eds.) ECCV 2016. LNCS, vol. 9915, pp. 435–442. Springer, Cham (2016). https://doi.org/10.1007/978-3-319-49409-8_34
11. Ronneberger, O., Fischer, P., Brox, T.: U-Net: convolutional networks for biomedical image segmentation. arXiv:1505.04597 [cs.CV] (2015)
12. LObe and Lung Analysis 2011 (LOLA11). https://lola11.grand-challenge.org/Home/ (2011)
13. Kornblith, S., Shlens, J., Le, Q.V.: Do Better ImageNet Models Transfer Better? arXiv:1805.08974 [cs.CV] (2018)

3DCNN for Pulmonary Nodule Segmentation and Classification

Zhenhuan Tian[1], Yizhuan Jia[2], Xuejun Men[2], and Zhongwei Sun[2(✉)]

[1] Department of Thoracic Surgery, Peking Union Medical College Hospital, Peking Union Medical College, Beijing, China
[2] Mediclouds Medical Technology, Beijing, China
szw@mediclouds.cn

Abstract. Lung cancer is the leading cause of cancer-related death. Early stage lung cancer detection using computed tomography (CT) could prevent patient death effectively. However, millions of CT scans will have to be analyzed thoroughly worldwide, which represents an enormous burden for radiologists. Therefore, there is significant interest in the development of computer algorithms to optimize this clinical process.

In the paper, we developed an algorithm for segmentation and classification of Pulmonary Nodule. Firstly, we established a CT pulmonary nodule annotation database using three major public databases. Secondly, we adopt state-of-the-art algorithms of deep learning method in medical imaging processing. The proposed algorithm shows the superior performance on the LNDb dataset. We got an outstanding accuracy in the segmentation and classification tasks which reached 77.8% (Dice Coefficient), 78.7% respectively.

Keywords: Pulmonary nodule segmentation · Pulmonary nodule classification · Deep learning

1 Introduction

Lung cancer, causing millions of deaths every year, is a leading cause of cancer death worldwide [1]. Diagnosis and treatment at an early stage are imperative for patients suffering from this disease.

A vital first step in the analysis of lung cancer screening CT scans is the detection of pulmonary nodules, which may or may not represent early stage lung cancer [2]. However, manually identifying nodules in CT scans is obviously time-consuming and boring work, because a radiologist needs to read the CT scans slice by slice, and a chest CT may contain hundreds of slices. For this reason, during recent years, many Computer-Aided Detection (CAD) systems have been proposed [3–6] to reduce the burden on the radiologists and provide an independent second-opinion.

Until now, most of the developed computer-aided algorithms are focus on the single task of nodule detection. However, in the clinical practice of radiology department, there is a real need for comprehensive primary diagnosis including the sizes and features of

© Springer Nature Switzerland AG 2020
A. Campilho et al. (Eds.): ICIAR 2020, LNCS 12132, pp. 386–395, 2020.
https://doi.org/10.1007/978-3-030-50516-5_34

the nodules. Furthermore, for a respiratory physician, it would be extremely useful if a reference solution for further diagnostic examination and treatment would be provided by computer-aided algorithms.

One of such possible situations would be the application in 2017 Fleischner society pulmonary nodule guidelines (Fleischner guidelines) [7]. The Fleischner guidelines are widely used for patient management in the case of nodule findings, and are composed of 4 classes, taking into account the number of nodules (single or multiple), their volume ($<$ 100 mm^3, 100–250 mm^3 and $\geqslant$250 mm^3) and texture (solid, part solid and ground glass opacities (GGO)). The purpose of these guidelines is to reduce the number of unnecessary follow-up examinations while providing greater discretion to the radiologist, clinician, and patient to make management decisions. The main goal of the LNDb challenge is to automatically obtain the recommendations according to the Fleichner guidelines, which means that the number of nodules, their volume and texture, should all be available from the algorithms.

2 Data Preprocessing

The LNDb dataset contains CT scans collected retrospectively at the Centro Hospitalar e Universitário de São João (CHUSJ) in Porto, Portugal between 2016 and 2018 [8]. Each CT scan was read by at least one radiologist at CHUSJ to identify pulmonary nodules and other suspicious lesions. CT data is available on metaImage (.mhd/.raw) format. We used SimpleITK as the tool for reading the raw data. Individual nodule annotations are available on a csv file that contains one finding marked by a radiologist per line. Nodule annotations were obtained by reading the csv files. After that the world coordinates were transformed to image coordinates by using the information of voxel size, coordinate origin and rotation matrix. Each annotated nodule was then resized and a cube with a uniform side length of 5 mm (80 pixel) was extracted. The image pixel values in the cube was adjusted using the lung window(window center: −400HU, window width 1500HU, HU: hounsfield unit), then normalized to the range of 0–1. Accordingly, the annotation cube was extracted at the same anatomical position with the same size and the segmentation was represented by the Boolean value of True of False. All the CT image nodule and annotation cubes were saved as .npy file for further use.

3 Pulmonary Nodule Annotation Database

3.1 Importance of Annotations

The performance of medical image analysis algorithms relies heavily on high-quality annotations of large volume, especially for deep learning and convolutional neural network (CNN). Such systems require large volume of data to be annotated by experts with years of training, especially when diagnostic decisions are involved such as in the LNDb challenge. This kind of datasets are thus hard to since they rely on such time-consuming, high strength and experience based work. As far as we know, there was no other specially designed public database other than LNDb, which would have decisive influence on the performance of deep learning algorithms. For this reason we decided to establish our own pulmonary nodule segmentation and classification database. An image analyst team was set up based on our imaging core laboratory, consisting of two radiologists, one thoracic surgeon from first-class hospital in China and four experienced medical image analysts. A series of training courses and examinations were held to make an adequate preparation. Two databases were chosen as our image sources: the LUNA16 database [9] and TIANCHI database [10].

3.2 LUNA2016 Database

The LUNA2016 challenge is perhaps the most famous challenge focused on a large-scale evaluation of automatic nodule detection algorithms on the publicly available LIDC/IDRI dataset [11]. LUNA2016 consisted of two sub challenges: Nodule detection, to identify locations of possible nodules, and to assign a probability for being a nodule to each location; False positive reduction: given a set of candidate locations, the goal was to assign a probability for being a nodule to each candidate location. Therefore, no segmentation and classification annotations were included. In total, 888 CT images of LUNA2016 database, 1176 pulmonary nodules were extracted by our team for further use.

3.3 TIANCHI Database

Similar to LUNA16 challenge, TIANCHI contained 1000 CT cases and was also organized for the development and validation of pulmonary nodule detection algorithms. Therefore, no segmentation and classification annotations were included. 1237 pulmonary nodules were extracted by our team for further use.

3.4 Standard Operation Procedure of Annotation Database Establishment

All the extracted nodules and reference CT images(including the LNDb database) in both lung and chest windows were provided to the image analyst team for segmentation and classification. At the very beginning, 10% of the data was used for training, team members were regard as qualified until they pass the exam according to our core lab consensus. Consensus of analysis method achieved until: 1) For segmentation, intra-observer and inter-observer difference of nodule volume within 20% (averaged volume difference/averaged volume); 2) For classification, intra-observer and inter-observer difference of classification within 10%(disagreements/total nodule number). Then first 30% of annotations was 100% reviewed by radiologists and thoracic surgeon in our team to achieve relatively low intra-observer and inter- observer variabilities. After discussion and revision of ambiguous or mistaken annotated nodules, the final results were handed over to the algorithm team (See Fig. 1). An example of pixel-level annotation is shown in Fig. 2.

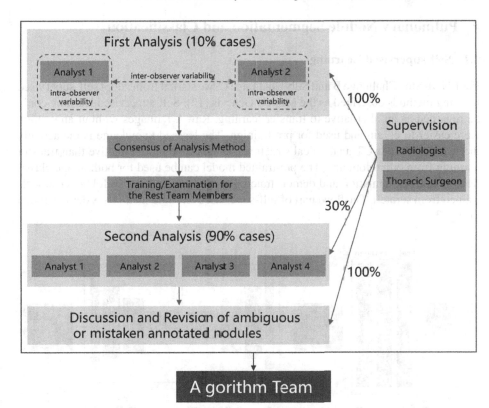

Fig. 1. Flowchart of annotation procedures.

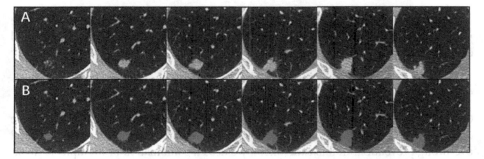

Fig. 2. Pixel-level manual annotation of pulmonary nodule. A. original CT images of a solid pulmonary nodule; B. annotations, red color - RGB [255, 0, 0] was used to stand for solid nodule. (Color figure online)

4 Pulmonary Nodule Segmentation and Classification

4.1 Self-supervised Learning

For LNDb Sub-Challenge B and Sub-Challenge C, we started from adopt self-supervised learning methods described as the Models Genesis [12]. Self-supervised learning could be regarded as an alternative to transfer learning. Raw CT images without annotation were easier to obtain and used for pre-training. The learned knowledge is the actually the feature of chest CT anatomical structures, so could be more effective than transfer learning from other domains. The pre-trained model can be used for both image classification and segmentation and demonstrated 3 to 5 points increase in IoU over models trained from scratch. The flowchart of self-supervised learning methods is demonstrated in Fig. 3.

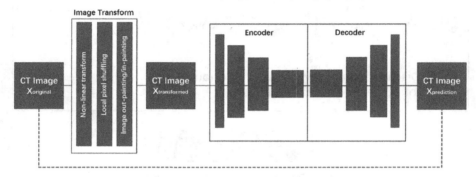

Fig. 3. Self-supervised learning framework.

We used all the CT images in LUNA16, TIANCHI and LNDb database to do the pre-training then got the pretrained weights for pulmonary nodule segmentation and classification task.

4.2 Data Augmentation

Considering the variety of nodule structure, despite using self-annotated data and public datasets, data augmentation is still necessary for the enhancement of robustness and overfitting prevention. In this experiment, we used a combination of multiple augmentation processes, which means increased the number of nodules would be generated by randomly applying one or more of following augmentation steps: 1) Flip the cube along a random axis. 2) Zoom the cube randomly along one or more axis. 3) Shift the cube along a random axis. 4) Rotate (0–360°) the cube along one or two axes randomly. Figure 4 gives an illustration of 3D data augmentation.

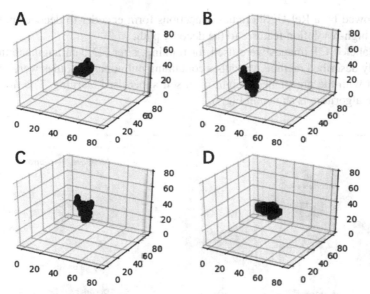

Fig. 4. 3D data augmentation. A. original nodule; B. nodule after flipping and shift; C. nodule after flipping, shift and zoom; D. nodule after flipping, shift, zoom and rotation.

4.3 Pulmonary Nodule Segmentation

A series of studies [13–15] have shown that 3D networks perform better than a 2D network as they also captures the vertical information of nodules. As such, we adopted 3D U-net [16] (Fig. 5) as the network architecture for nodule segmentation. The 3D U-Net architecture is similar to the 2D U-Net, It comprises of an encoder (left) and a decoder path (right). For encoder, each layer contains two $3 \times 3 \times 3$ convolutions followed by a rectified linear unit(ReLU), and then a $2 \times 2 \times 2$ max pooling with strides of two in each dimension. For the decoder, each layer consists of an up-convolution of $2 \times 2 \times 2$ by strides of two in each dimension, followed by two $3 \times 3 \times 3$ convolutions

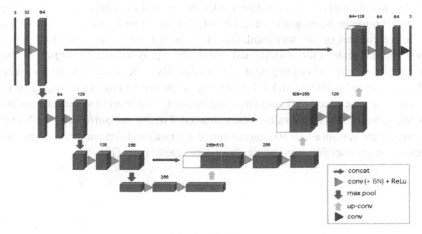

Fig. 5. 3D U-net.

each followed by a ReLU. Shortcut connections form encoder to decoder provide the essential high-resolution features to the decoder path.

The labeled CT images were used for training, after 3D data augmentation using previously described method, the cubes containing pulmonary nodules were sent to the 3D U-net with the pre-trained weight from self-supervised learning. The flowchart of the whole algorithm is shown in Fig. 6.

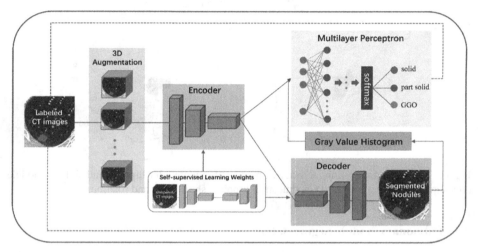

Fig. 6. Pulmonary nodule segmentation and classification flowchart.

4.4 Pulmonary Nodule Classification

In consideration of the extremely imbalanced dataset for lung nodule classification, with merely 5% GGO nodules, 10% part-solid nodules, we added traditional image feature extraction methods to improve the performance and prevent the overfitting.

To be specific, our classification network is based on the pretrained encoder and gray value distribution in the lung nodule region which is the main feature difference between 3 classes of nodules (solid, part-solid, GGO). In the first step, we counted all the pixels of the nodule region as the volume and extract their gray values, the region of nodule is the predicted mask of our segmentation model. Then the window state and window width were set to −600 HU and 1500 HU, and a uniformed gray value histogram with 50 ranges was calculated to demonstrate the intensity distribution within the nodule, in addition, variance and mean value were extracted. Finally, we uniformed the 53 values (volume, mean, variance and 50 values from histogram) and concatenated them into the output of the encoder followed by the fully connected layers (Fig. 5).

5   Results

In the training process based on the methods proposed above, the whole dataset was divided into 6 groups randomly. Each 5 of them were used as training set alternatively, and the last group was used as validation.

Finally, we got an outstanding accuracy in the segmentation and classification tasks which have reached 77.8% (Dice Coefficient), 78.7% respectively. A example of comparison between ground truth and prediction segmentation results is shown in Fig. 7.

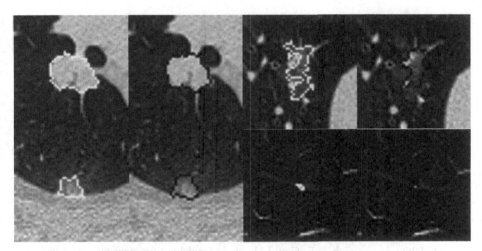

Fig. 7. Segmentation results. Red. ground truth; blue. prediction. (Color figure online)

Relying on our own annotations with a number of close to 3000 nodules and the pretrained weights using large volume of lung CT images which we achieved in the previous steps, we got relatively excellent performance in the LNDb Sub-Challenge B and C. In the Challenge B, we took the second place, with a Challenge B Score(average value of six normalized indices including modified Jaccard index, mean average distance, hausdorff distance, modified pearson correlation coefficient, bias and Standard deviation) of 0.748, which is slightly lower than the first place result(0.758). In the Challenge C, we took the third place, with Challenge C Score (Fleiss-Cohen weighted Cohen's kappa [17]) of 0.685.

6   Discussion and Conclusion

In this paper, we proposed an effective 3D framework for pulmonary nodule segmentation and classification. Firstly, based on our knowledge, in the field of medical image processing using deep learning method, the most important determinant of the final performance is always the quantity and quality used for training. From this point, one of the most important works we have done was to establish a segmentation and classification annotation database of our own. The quality of annotations was guaranteed by adoption of standard operation procedure which was wild used in image core lab. Secondly, integrated application of self-supervised learning, 3D data augmentation and 3D U-net backbone had provided extra contribution to the performance of the final results. The first limitation of our work is the accuracy of classification is not high enough and thus is insufficient for clinical use. We believe that the 3D information of pulmonary nodule and adjacent tissue has not been well represented and fully used. Secondly, there is still space for improvement of the segmentation algorithm if more well annotated data (especially GGO and part-solid nodules) could be obtained.

References

1. Torre, L.A., et al.: Lung Cancer Statistics. In: Ahmad, A., Gadgeel, S. (eds.) Lung Cancer and Personalized Medicine, pp. 1–19. Springer, Cham (2016). https://doi.org/10.1007/978-3-319-24223-1
2. ur Rehman, M.Z., et al.: An appraisal of nodules detection techniques for lung cancer in CT images. Biomed. Signal Process. Control, **41**, 140–151 (2018)
3. Litjens, G., Kooi, T., Bejnordi, B.E., et al.: A survey on deep learning in medical image analysis. Medical Image Analysis **42**, 60–88 (2017)
4. Taghanaki, S.A., Abhishek, K., Cohen, J.P., et al.: Deep semantic segmentation of natural and medical images: a review (2019). arXiv preprint arXiv:1910.07655
5. Ronneberger, O., Fischer, P., Brox, T.: U-Net: convolutional networks for biomedical image segmentation. arXiv (2015): 1505.04597. http://arxiv.org/abs/1505.04597. arXiv: 1505.04597
6. Cheplygina, V., de Bruijne, M., Pluim, J.P.: Not-so-supervised: a survey of semi-supervised, multi-instance, and transfer learning in medical image analysis. Med. Image Anal. **54**, 280–296 (2019)
7. MacMahon, H., Naidich, D.P., Goo, J.M., et al.: Guidelines for management of incidental pulmonary nodules detected on CT images: from the Fleischner Society 2017. Radiology **284**(1), 228–243 (2017)
8. Pedrosa, J., Aresta, G., Ferreira, C., et al.: Lndb: a lung nodule database on computed tomography. arXiv preprint arXiv:1911.08434 (2019)
9. Setio, A.A.A., et al.: Validation, comparison, and combination of algorithms for automatic detection of pulmonary nodules in computed tomography images: the LUNA16 challenge. Med. Image Anal. **42**, 1–13 (2017)
10. TIANCHI. https://tianchi.aliyun.com/competition/entrance/231601/introduction
11. Armato III, S.G., McLennan, G., Bidaut, L., et al.: The lung image database consortium (lidc) and image database resource initiative (idri): a completed reference database of lung nodules on ct scans. Med. Phys. **38**(2), 915–931 (2011)
12. Zhou, Z., et al.: Models Genesis: Generic Autodidactic Models for 3D Medical Image Analysis. In: Shen, D., et al. (eds.) MICCAI 2019. LNCS, vol. 11767, pp. 384–393. Springer, Cham (2019). https://doi.org/10.1007/978-3-030-32251-9_42
13. Chen, H., Dou, Q., Wang, X., Qin, J., Cheng, J.C.Y., Heng, P.A.: 3D fully convolutional networks for intervertebral disc localization and segmentation. In: Zheng, G., Liao, H., Jannin, P., Cattin, P., Lee, S. (eds.) MIAR 2016. LNCS, vol. 9805, pp. 375–382. Springer, Cham (2016). https://doi.org/10.1007/978-3-319-43775-0_34
14. Baumgartner, C.F., Koch, L.M., Pollefeys, M., et al.: An exploration of 2D and 3D deep learning techniques for cardiac MR image segmentation. In: Pop, M., et al. (eds.) STACOM 2017. LNCS, vol. 10663, pp. 111–119. Springer, Cham (2018). https://doi.org/10.1007/978-3-319-75541-0_12

15. Zhou, X., Takayama, R., Wang, S., et al.: Deep learning of the sectional appearances of 3D CT images for anatomical structure segmentation based on an FCN voting method. Med. Phys. **44**(10), 5221–5233 (2017)
16. Çiçek, Ö., Abdulkadir, A., Lienkamp, S.S., Brox, T., Ronneberger, Olaf: 3D U-Net: learning dense volumetric segmentation from sparse annotation. In: Ourselin, S., Joskowicz, L., Sabuncu, M.R., Unal, G., Wells, W. (eds.) MICCAI 2016. LNCS, vol. 9901, pp. 424–432. Springer, Cham (2016). https://doi.org/10.1007/978-3-319-46723-8_49
17. Spitzer, R.L., Cohen, J., Fleiss, J.L., et al.: Quantification of agreement in psychiatric diagnosis. Arch. Gen. Psychiatry **17**, 83–87 (1967)

Residual Networks for Pulmonary Nodule Segmentation and Texture Characterization

Adrian Galdran$^{(\boxtimes)}$ and Hamid Bouchachia$^{(\boxtimes)}$

Department of Computing and Informatics, Bournemouth University, Poole, UK
{agaldran,abouchachia}@bournemouth.ac.uk

Abstract. The automated analysis of Computed Tomography scans of the lung holds great potential to enhance current clinical workflows for the screening of lung cancer. Among the tasks of interest in such analysis this paper is concerned with the segmentation of lung nodules and their characterization in terms of texture. This paper describes our solution for these two problems in the context of the LNdB challenge, held jointly with ICIAR 2020. We propose a) the optimization of a standard 2D Residual Network, but with a regularization technique adapted for the particular problem of texture classification, and b) a 3D U-Net architecture endowed with residual connections within each block and also connecting the downsampling and the upsampling paths. Cross-validation results indicate that our approach is specially effective for the task of texture classification. In the test set withheld by the organization, the presented method ranked 4th in texture classification and 3rd in the nodule segmentation tasks. Code to reproduce our results is made available at http://www.github.com/agaldran/lndb.

Keywords: Lung nodule segmentation · Texture classification · Imbalanced classification · Label smoothing

1 Introduction

Pulmonary cancer is known to be among the most lethal types of cancer worldwide [14]. Early detection of lung cancer may have a great impact in mortality rates, and Computed Tomography (CT) is recognized as a promising screening test for this purpose [1]. Large-scale screening programs are susceptible of becoming more effective and efficient by the deployment of Computer-Aided Diagnosis (CAD) tools. Such tools might bring standardization to a problem that suffers from great interobserver variability, and they also have the potential of reducing the workload of specialists by assisting them with complementary decisions.

The main task related with CAD in the processing of pulmonary CT scans is the automated analysis of lung nodules. This comprises several sub-tasks, namely lung nodule detection (localization of lesions within the scan), nodule segmentation (delineation of lesion borders), and lung nodule characterization (classification of each nodule into different categories, *e.g.* malignancy or texture). This

© Springer Nature Switzerland AG 2020
A. Campilho et al. (Eds.): ICIAR 2020, LNCS 12132, pp. 396–405, 2020.
https://doi.org/10.1007/978-3-030-50516-5_35

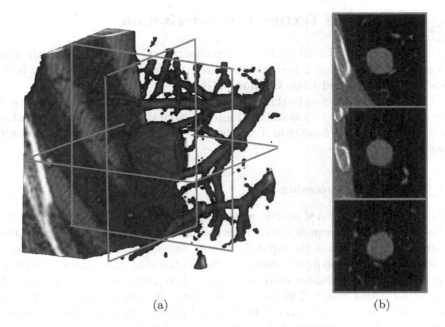

(a)   (b)

Fig. 1. (a) 3-D visualization of a lung nodule from the LNdB dataset (b) Three 2-D orthogonal views of (a), used here to train a model for texture classification.

array of problems has attracted considerable attention from the medical image analysis community in the last years [16]. In particular since the advent of Deep Learning techniques, a wide range of approaches based on Convolutional Neural Networks has been proposed for lung nodule detection [3], segmentation [4], characterization [5,7], or direct end-to-end screening [2] with remarkable success.

This paper describes a solution to the LNdB challenge, held jointly with ICIAR 2020. This challenge is built around the release of a new database of lung CT scans (termed itself LNdB), accompanied with manual ground-truth related to lung nodule localization, segmentation, texture categorization, and follow-up recommendation based on 2017 Fleischner society guidelines [12]. A sample of one of the nodules from the LNdB database is shown in Fig. 1.

Our solution is solely concerned with the tasks of lung nodule segmentation and texture characterization. In the remaining of this paper, we describe our approach to each of these tasks, which is based on the effective training of two residual networks. For the texture categorization scenario, we adopt a regularization scheme based on a custom manipulation of manual labels that better optimizes the κ score in this kind of problems. For the nodule segmentation problem, we construct a modified UNet architecture by adding residual connections inside each of its blocks, and also from the downsampling path to the upsampling path. The reported cross-validation results are promising, specially in the texture classification task, where our approach seems to be able to successfully overcome the difficulties associated to a highly imbalanced dataset.

2   Lung Nodule Texture Characterization

In the context of the LNdB challenge, sub-challenge C corresponded to the classification of lung nodule's texture into three distinct categories, namely Solid, Sub-Solid, and Ground-Glass Opaque (GGO).

The solution proposed in this paper was based on four main components: 1) input pre-processing, 2) a standard Residual Neural Network, 2) a specialized label smoothing technique, and 3) application of oversampling on the minority classes.

2.1   Input Pre-processing

Initially, for each provided nodule center a cubic volume of size $64 \times 64 \times 64$ was extracted and stored separately to facilitate model training. In addition, instead of attempting to process the input data by means of three-dimensional convolutions, we simplified the input volumes by first extracting three orthogonal planes of dimension 64×64 centered around each nodule, and then stacking them into a single 3-channel image. This turned the inputs into tensors amenable to standard 2D-Convolutional Neural Networks, and resulted in a 95% dimensionality reduction. A representation of this process is displayed in Fig. 1.

2.2   Convolutional Neural Network

We experimented with Residual Networks of different depths (18-layers, 50-layers, and 101-layers depth networks). Several modifications were also tested, namely the size of the filters in the very first layer was reduced from 7×7 to 3×3, and an initial Batch-Normalization/Instance Normalization layer was inserted in each architecture. In addition, initialization with weights pre-trained on the ImageNet database was also tested. Eventually, our best configuration based on cross-validation analysis was a 50-layer Residual Network, trained from random weights and with no initial normalization layer.

2.3   Gaussian Label Smoothing

A simple but powerful approach to regularization in CNNs consists of performing Label Smoothing (LS) [15]. LS is often applied for multi-class classification tasks, where the Cross-Entropy loss function is employed, and annotations are available in the form of one-hot encoding. The idea of LS consists of replacing these original one-hot encoded labels by a smoothed version of them, where part of their value is redistributed uniformly among the rest of the categories. LS has been proven useful to prevent neural networks from becoming over-confident and avoid overfitting in a wide range of problems [11].

Recently, a modified version of LS has been introduced in the context of Diabetic Retinopathy Grading from retinal fundus images [8], termed Gaussian Label Smoothing (GLS). The main assumption of GLS is that in a scenario

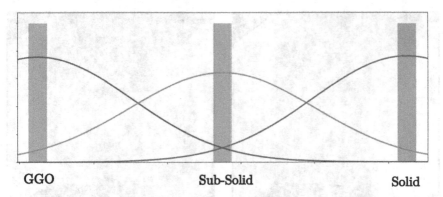

GGO   **Sub-Solid**   **Solid**

Fig. 2. Gaussian Label Smoothing technique (GLS) applied to the texture characterization problem. Light-blue bars represent the original one-hot encoded labels for each
category, whereas the mean of the Gaussian curves represented in blue, orange, and
green represent the corresponding smoothed labels. (Color figure online)

where labels are not independent, but reflect some underlying "ordering", it
can be better to replace the uniform smoothing process in LS by a weighted
smoothing, where neighboring categories receive more weight than further away
ones. It was shown in [8] that, for ordinal classification of Diabetic Retinopathy
grades, GLS outperformed standard LS.

For the problem of texture classification, we use the standard cross-entropy
loss and we adapt the GLS technique to three classes:

$$\mathcal{L}(y, \mathrm{gls}(\hat{y})) = \sum_{k=1}^{3} y^{(k)} \log(\mathrm{gls}(\hat{y}^{(k)})), \tag{1}$$

where y is the output of the CNN followed by a soft-max mapping, $\hat{y}$ is the
original one-hot encoded label, and $\mathrm{gls}(\hat{y}) = G \circ \hat{y}$ is the transformation of $\hat{y}$
by a GLS mapping. As an example, a nodule belonging to the solid category
is no longer represented by a one-hot encoded vector of the form $(1, 0, 0)$ but
rather by a vector close to $(0.80, 0.18, 0.02)$, as shown in Fig. 2. As opposed to
this strategy, the LS technique would encode the label as a vector of the form
$(0.80, 0.10, 0.10)$. The effect of GLS is to induce a larger penalization when the
prediction of the network is far away from the true class, promoting decisions
closer to the actual label. This is particularly useful for the LNdB challenge,
where the evaluation metric is the Quadratic-Weighted Kappa score.

2.4   Minority Class Oversampling

Given the relatively low amount of examples and high ratios of class imbalance
(a proportion of approximately 5%/7%/88% for each class respectively), special
care was taken when considering the sampling of the training set during model
training. Our experiments revealed that the optimal strategy in this setting was

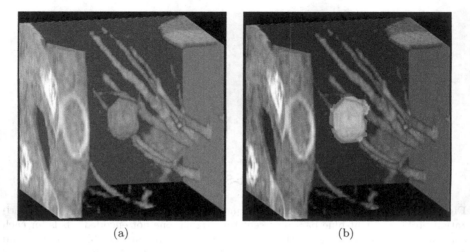

(a)   (b)

Fig. 3. (a) Three-dimensional visualization of a lung nodule from the training set, and (b) same nodule as in (a) with an overlayed manual segmentation.

to perform oversampling on the two minority classes, which is consistent with previous works [6]. Solid nodules were oversampled by a factor of 9 and sub-solid by a factor of 6, which resulted in a class ratio of 25%/25%/50% during training. It must be noticed that this approach lends itself to easily overfitting minority examples, which are shown to the model much more frequently. However, we observed that the application of Test-Augmentation Techniques mitigated this effect considerably, as explained in Sect. 4.

3   Lung Nodule Segmentation

Sub-challenge C corresponded to the task of segmenting lung nodules from a CT scan, given the location of their centroids. An example of a nodule and the associated manual segmentation is shown in Fig. 3. For this task, we implemented a standard U-Net taking as input three-dimensional volumes of $80 \times 80 \times 80$ resolution. We introduced several modifications to the architecture presented in [13]: 1) All 2-dimensional learnable filters were replaced by 3 dimensional filters 2) Batch-normalization layers were inserted in between every convolutional block, and also prior to the first layer in the architecture, 3) Skip connections were added to every convolutional blocks, and 4) Convolutional layers connecting the downsampling path in the architecture with the upsampling path were also added. Note that some of these modifications have been explored in previous works developing enhancements of the standard U-Net architecture [18]. A representation of the resulting architecture is provided in Fig. 4.

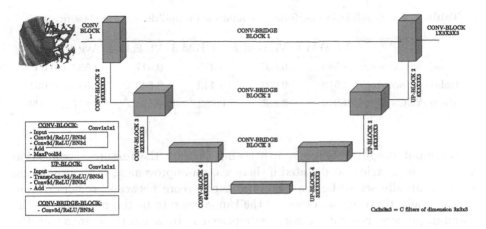

Fig. 4. A description of the 3d-unet used for this project.

The loss function we minimized in this case was the channel-wise Dice soft function, as suggested in [10]:

$$\mathcal{L}(y, \hat{y}) = \frac{2 \sum_{i=1}^{n} y_i \cdot \hat{y}_i}{\sum_{i=1}^{n} y_i^2 + \hat{y}_i^2 + \epsilon}, \quad (2)$$

where y_i is the output of the CNN at each location i, $\hat{y}_i$ is the binary label associated to each voxel, and ϵ is a small constant to prevent division by zero.

4 Training Details

The training of the CNN, both for texture classification and nodule segmentation, followed similar stages. In both cases, an initial learning rate of 0.01 was set, and the weights were updated by means of the Adam optimizer so that the corresponding loss was minimized. In the nodule segmentation problem, the optimizer was wrapped by the look-ahead algorithm [17], since severe instabilities were observed when using the original Adam optimizer. Regarding batch sizes, for the texture classification task a batch size of 8 samples was applied, whereas for the segmentation problem a reduced batch size of 4 had to be employed due to computational constraints Both models were trained for 500 epochs, but training was stopped after no performance improvement was observed in the validation set during 25 epochs. In addition, after no improvement in 15 epochs, the learning rate was decreased by a factor of 10.

As for data augmentation techniques, we performed random reflection along each of the three axis of a given volume, as well as random small offsets, scalings and rotation. It is important to note that, even if the texture classification model was trained on two-dimensional images with three intensity channels (the three orthogonal views depicted in Fig. 1), in this case we also performed data augmentation on three dimensional volumes before sampling the three planes

Table 1. Cross-validation performance analysis on nodule texture classification.

	Vl. Fold 1	Vl. Fold 2	Vl. Fold 3	Vl. Fold 4	Avg $\pm$ Std
Quad. kappa score	0.542	0.607	0.475	0.617	0.560 $\pm$ 0.057
Balanced accuracy	0.518	0.621	0.449	0.543	0.533 $\pm$ 0.061
Mean AUC	0.849	0.885	0.852	0.803	0.847 $\pm$ 0.029

of the input image. It is also worth noting that for the texture classification scenario, the metric that dictated if there was an improvement in the validation set was initially set to be the quadratic kappa score between predictions and actual labels. However, we observed the kappa score to be too noisy during the training process. For this reason, we replaced it by a metric aggregating the average Area Under the Curve for three classes and the kappa score itself. For the segmentation case, the validation metric was the dice score computed over each scan after thresholding the network output by means of the Otsu algorithm (this was done due to the high variability of results depending on the threshold selection), computed per volume and averaged afterwards.

To reduce overfitting and improve the performance of both networks at inference time, we also implemented a straightforward Test-Time Augmentation strategy. Besides considering the prediction on a given volume, such volume also went through a reflection over each axis from the set $\{x, y, z, xy, xz, yz, xyz\}$, predictions were computed on the modified volume and the same reflection was applied again on the predictions in the segmentation case. We observed a considerable benefit when applying this strategy, specially in mitigating the overfitting that arised from the heavy oversampling of minority classes in the texture classification scenario, as described in Sect. 2.4.

5   Results

In this section we report numerical results for cross-validation performance as well as performance in the final test set.

5.1   Lung Nodule Texture Classification

The organization of this challenge provided an official split of all the training scans. This represented a set of 768 nodules, that were split in four subsets of 200, 194, 186 and 192 respectively. Each of this subset was used for validation purposes once, while a model would be trained in the union of the remaining three subsets, which resulted in four different models being trained. Table 1 reports the quadratic weighted kappa score, and other metrics of interest, for each of this folds. We also display confusion matrices for each fold in Fig. 5.

For testing purposes, each nodule in the test set was run through each of the above four models, and the resulting probabilities were averaged to build our final submission. This produced a quadratic-weighted kappa score of $\kappa = 0.6134$,

T/P	GGO	S-S	Solid
GGO	4	3	4
S-S	0	3	10
Solid	0	7	168

(a)

T/P	GGO	S-S	Solid
GGO	5	1	2
S-S	2	4	8
Solid	2	6	163

(b)

T/P	Solid	S-S	GGO
GGO	2	3	5
S-S	1	2	10
Solid	0	1	161

(c)

T/P	GGO	S-S	Solid
GGO	7	1	4
S-S	1	1	8
Solid	0	9	160

(d)

Fig. 5. (a)–(d): Confusion matrices corresponding to each of the validation folds.

Table 2. Validation (top) and test (bottom) results as provided by the organization

	J^*	MAD	HD	Inv. Pearson CC	Bias	Std. Dev.
Validation	0.4321	0.4576	2.2364	0.1216	125.4638	706.6704
Test	0.4447	0.4115	2.0618	0.1452	41.4341	129.47

which ranked fourth in this competition. The greater performance in the test set can likely be attributed to differences in class proportions between this and the several validation sets used in Table 1.

5.2   Lung Nodule Segmentation

In sub-challenge B, the nodule segmentation task was evaluated under a number of different metrics, including Modified Jaccard index (J^*), Mean average distance (MAD), Mean Hausdorff distance (HD), Inverted Pearson correlation coefficient, Bias, and Standard deviation[1]. In addition, the organization considered the largest interconnected object as the final segmentation in each case.

Since the functionalities to compute the above metrics were not provided by the organizers, we were unable to perform an analysis similar to the one in the previous section for our cross-validation analysis. For this reason, we report in the first row of Table 2 results obtained by predicting each nodule with the corresponding model trained on the appropriate split of the training set, aggregating those predictions, and computing an overall score over the entire training set this way.

Our final submission was again built by averaging the predictions of each of our four models trained in different folds of the training set. The numerical analysis corresponding to our segmentations in the test set is shown in the bottom row of Table 2. Our approach ranked third in the official challenge leaderboard.

6   Discussion and Conclusion

From the results presented above, it can be concluded that both the texture classification and the nodule segmentation tasks were solved to a reasonable level.

---

[1] The challenge website at https://lndb.grand-challenge.org/Evaluation/ contains rigorous definitions of each of these quantities.

In particular, results in Table 1 are well-aligned with inter-observer variability in this dataset among radiologists, as reported in [12].

Another interesting conclusion to be extracted from Table 1 is the observation that the quadratic-weighted κ score captures different properties of a solution when compared with the averaged AUC metric (this was computed adjusting for the support of each class). For instance, the worst results in terms of κ score were obtained in fold 3, but this fold also had the second best average AUC. In our opinion, the inclusion of the average AUC together with the κ score as the monitoring metric based on which we early-stopped the training of the network was greatly beneficial to avoid falling in a sharp local minima during the optimization process.

Despite an overall better ranking, results for nodule segmentation were slightly poorer when compared with texture classification. A reason for this may have been our approach based on directly segmenting the 3D volume, instead of sampling 2D planes and learning from these. While a 3D model had far less learnable parameters in this case, it was much more computationally intensive in terms of number of operations performed by the network, which led to a slow hyperparameter tuning process.

In addition, we observed that the selection of the binarizing approach once the network had been trained had a great impact in the resulting segmentation, as confirmed by the large standard deviation in Table 2. In our experiments, we observed that if an optimal threshold was selected for each prediction (as opposed to a single threshold for all predictions, or even the adaptive threshold selection algorithm based on Otsu's technique we ended up using), results were much better. In other words, a reasonable binarizing threshold for a particular probabilistic prediction turned out to be very poor when applied to another prediction. We believe future work may focus on a better strategy to select a thresholding value in a per-volume basis, as has been suggested in other medical image segmentation problems [9].

References

1. National Lung Screening Trial Research Team: Reduced lung-cancer mortality with low-dose computed tomographic screening. N. Engl. J. Med. **365**(5), 395–409 (2011)
2. Aresta, G., et al.: Towards an automatic lung cancer screening system in low dose computed tomography. In: Stoyanov, D., et al. (eds.) RAMBO/BIA/TIA 2018. LNCS, vol. 11040, pp. 310–318. Springer, Cham (2018). https://doi.org/10.1007/978-3-030-00946-5_31
3. Aresta, G., Cunha, A., Campilho, A.: Detection of juxta-pleural lung nodules in computed tomography images. In: Medical Imaging 2017: Computer-Aided Diagnosis, vol. 10134, p. 101343N. International Society for Optics and Photonics, March 2017
4. Aresta, G., et al.: iW-Net: an automatic and minimalistic interactive lung nodule segmentation deep network. Sci. Rep. **9**(1), 1–9 (2019)

5. Bonavita, I., Rafael-Palou, X., Ceresa, M., Piella, G., Ribas, V., González Ballester, M.A.: Integration of convolutional neural networks for pulmonary nodule malignancy assessment in a lung cancer classification pipeline. Comput. Methods Programs Biomed. **185**, 105172 (2020)
6. Buda, M., Maki, A., Mazurowski, M.A.: A systematic study of the class imbalance problem in convolutional neural networks. Neural Netw. **106**, 249–259 (2018)
7. Ferreira, C.A., Cunha, A., Mendonça, A.M., Campilho, A.: Convolutional neural network architectures for texture classification of pulmonary nodules. In: Vera-Rodriguez, R., Fierrez, J., Morales, A. (eds.) CIARP 2018. LNCS, vol. 11401, pp. 783–791. Springer, Cham (2019). https://doi.org/10.1007/978-3-030-13469-3_91
8. Galdran, A., et al.: Non-uniform label smoothing for diabetic retinopathy grading from retinal fundus images with deep neural networks. Translational Vision Science and Technology, June 2020
9. Galdran, A., Costa, P., Bria, A., Araújo, T., Mendonça, A.M., Campilho, A.: A no-reference quality metric for retinal vessel tree segmentation. In: Frangi, A.F., Schnabel, J.A., Davatzikos, C., Alberola-López, C., Fichtinger, G. (eds.) MICCAI 2018. LNCS, vol. 11070, pp. 82–90. Springer, Cham (2018). https://doi.org/10.1007/978-3-030-00928-1_10
10. Milletari, F., Navab, N., Ahmadi, S.A.: V-Net: fully convolutional neural networks for volumetric medical image segmentation. In: 2016 4th International Conference on 3D Vision (3DV), pp. 565–571, October 2016
11. Müller, R., Kornblith, S., Hinton, G.E.: When does label smoothing help? In: Wallach, H., Larochelle, H., Beygelzimer, A., d'Alché-Buc, F., Fox, E., Garnett, R. (eds.) Advances in Neural Information Processing Systems 32, pp. 4696–4705. Curran Associates, Inc. (2019)
12. Pedrosa, J., et al.: LNDb: a lung nodule database on computed tomography. arXiv:1911.08434 [cs, eess], December 2019. http://arxiv.org/abs/1911.08434
13. Ronneberger, O., Fischer, P., Brox, T.: U-Net: convolutional networks for biomedical image segmentation. In: Navab, N., Hornegger, J., Wells, W.M., Frangi, A.F. (eds.) MICCAI 2015. LNCS, vol. 9351, pp. 234–241. Springer, Cham (2015). https://doi.org/10.1007/978-3-319-24574-4_28
14. Siegel, R.L., Miller, K.D., Jemal, A.: Cancer statistics, 2019. CA Cancer J. Cin. **69**(1), 7–34 (2019)
15. Szegedy, C., Vanhoucke, V., Ioffe, S., Shlens, J., Wojna, Z.: Rethinking the inception architecture for computer vision. In: Proceedings of the IEEE Conference on Computer Vision and Pattern Recognition (CVPR), pp. 2818–2826, June 2016
16. Wu, J., Qian, T.: A survey of pulmonary nodule detection, segmentation and classification in computed tomography with deep learning techniques. J. Med. Artif. Intell. **2** (2019)
17. Zhang, M., Lucas, J., Ba, J., Hinton, G.E.: Lookahead optimizer: k steps forward, 1 step back. In: Wallach, H., Larochelle, H., Beygelzimer, A., d'Alché-Buc, F., Fox, E., Garnett, R. (eds.) Advances in Neural Information Processing Systems 32, pp. 9593–9604. Curran Associates, Inc. (2019)
18. Zhou, Z., Siddiquee, M.M.R., Tajbakhsh, N., Liang, J.: UNet++: redesigning skip connections to exploit multiscale features in image segmentation. IEEE Trans. Med. Imaging **39**, 1856–1867 (2020)

Automatic Lung Cancer Follow-Up Recommendation with 3D Deep Learning

Gurraj Atwal[✉] and Hady Ahmady Phoulady

California State University, Sacramento, CA 95819, USA
{ga584,phoulady}@csus.edu

Abstract. Lung cancer is the most common form of cancer in the world affecting millions yearly. Early detection and treatment is critical in driving down mortality rates for this disease. A traditional form of early detection involves radiologists manually screening low dose computed tomography scans which can be tedious and time consuming. We propose an automatic system of deep learning methods for the detection, segmentation, and classification of pulmonary nodules. The system is composed of 3D convolutional neural networks based on VGG and U-Net architectures. Chest scans are received as input and, through a series of patch-wise predictions, patient follow-up recommendations are predicted based on the 2017 Fleischner society pulmonary nodule guidelines. The system was developed as part of the LNDb challenge and participated in the main challenge as well as all sub-challenges. While the proposed method struggled with false positives for the detection task and a class imbalance for the texture characterization task, it presents a baseline for future work.

Keywords: Lung cancer · Low dose computed tomography · Pulmonary nodules · Deep learning · Detection · Segmentation · Classification

1 Introduction

According to the World Health Organization, "cancer is a leading cause of death worldwide, accounting for an estimated 9.6 million deaths in 2018" [1]. Lung cancer is by far the most common form of cancer which makes early detection and treatment even more impactful. Early detection can lead to greater probability of survival, less morbidity, and less expensive treatment. A traditional form of early detection involves radiologists manually screening low dose computed tomography (CT) scans in search of potentially cancerous lesions or nodules in lungs. The process can be tedious, time consuming, and even error prone. We propose an automatic system of deep learning methods for the detection, segmentation, and classification of pulmonary nodules to potentially take some of the burden off of health professionals. The system was developed as part of the LNDb challenge which is composed of a main challenge and three sub-challenges [2].

The main challenge involves predicting a patient's follow-up recommendation according to the 2017 Fleischner society pulmonary nodule guidelines [3]. Given a chest CT, the system must predict one of four follow-up recommendation classes: 0) No

© Springer Nature Switzerland AG 2020
A. Campilho et al. (Eds.): ICIAR 2020, LNCS 12132, pp. 406–418, 2020.
https://doi.org/10.1007/978-3-030-50516-5_36

routing follow-up required or optional CT at 12 months according to patient risk; 1) CT at 6–12 months required; 2) CT at 3–6 months required; 3) CT, PET/CT or tissue sampling at 3 months required [2]. The four recommendations take into account the number of nodules in a chest CT, their volume, and their texture. The three sub-challenges focus on predicting these three attributes. The first sub-challenge, nodule detection, detects pulmonary nodules from chest CTs. The second sub-challenge, nodule segmentation, segments pulmonary nodules for the purpose of calculating volumes. Lastly, the third sub-challenge, nodule texture characterization, classifies nodules into one of three texture classes [2]. The method we propose participated in all three sub-challenges as well as the main challenge to produce patient follow-up recommendations.

2   Dataset

The LNDb dataset consists of 294 chest CTs along with radiologist annotations. Each CT contains at most six nodules and all nodules have in-slice diameters of at most 30 mm. The annotations provide centroid coordinates and texture ratings for all nodules and segmentations for nodules greater than 3 mm. Of the 294 CTs, 58 CTs and their associated annotations have been withheld for the test dataset by the LNDb challenge [2] (Fig. 1).

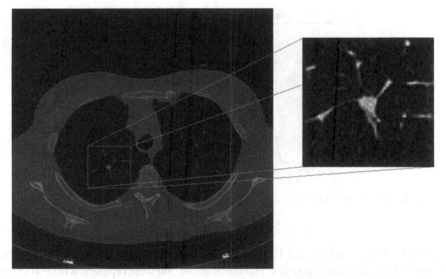

Fig. 1. A sample CT from the dataset with a nodule identified using radiologist annotations.

Automatic analysis of CTs pose a unique challenge considering CTs are essentially 3D arrays that vary in size. The number of slices as well as the size of slices vary across different CTs. Each value in the 3D array is a single-channel Hounsfield unit (HU) (instead of a typical RGB model). These characteristics along with the fact that nodules are at most 30 mm in diameter drove the pre-processing steps we took.

3 Method

3.1 Data Exploration and Pre-processing

A graphics card with 8 GB of random-access memory was used to train three neural networks. Given the memory constraints, the networks had to be trained on smaller 3D patches of CTs. As such, a second dataset of 3D CT patches was derived from the original dataset. A number of pre-processing steps were taken to create the new dataset.

Resampling. As part of data exploration, we found that the 3D pixel size, or voxel size, varied across CTs: 0.4328 mm–0.8945 mm for the x-axis, 0.4328 mm–0.8945 mm for the y-axis, and 0.5000 mm–1.4000 mm for the z-axis. To avoid issues when training our convolutional neural networks (CNN), all CTs were resampled so that each voxel had an isotropic size of 0.6375 mm × 0.6375 mm × 0.6375 mm. This size was chosen because it aligns with what the LNDb challenge expects for the segmentation task submission and because it is a middle ground for each axis.

Normalization. As mentioned earlier, CTs are 3D arrays of HU. We plotted the relative frequency distribution of HU for all CT overlaid with the relative frequency distribution of HU for all radiologist-segmented nodules. The latter was calculated by applying masks found in the original dataset to each CT and then counting HU for only the masked voxels.

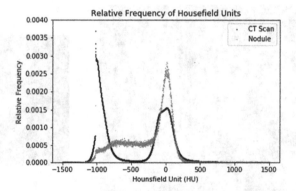

Fig. 2. The distribution of HU of CTs overlaid with the distribution of HU of nodules.

Nodules had an affinity for HU between −1000 and 500 (Fig. 2). As a result, all resampled CTs were clipped so that HU less than −1000 were replaced with −1000 and HU greater than 500 were replaced with 500. Next, the mean HU value was calculated across CTs. All CTs were zero-centered by subtracting the mean value, −477.88. Finally, to save disk space, all values were min-max normalized between 0 and 255 so that they could be stored as unsigned 8-bit integers.

Patch Extraction. The final step in creating the derived dataset was to extract equal-sized patches from CTs. By extracting patches and storing them on disk, our neural networks would be able to load them in real-time when training. We chose patches of size 51 mm × 51 mm × 51 mm because they encapsulated the largest nodules (30 mm), fit in GPU memory, and aligned with what the LNDb challenge expects for the segmentation task submission.

We first extracted all patches that had radiologist annotations: 768 patches containing nodules and 451 patches containing non-nodule pulmonary lesions. For each annotated patch, we also extracted the corresponding segmentation patches to be used for the segmentation task. For the detection task, we needed more negative patches containing no nodules so that the dataset better represented a full CT. Patches inside lungs are harder for a detection neural network to classify when compared to patches outside of lungs. We leveraged the provided centroid coordinates of nodules to extract negative patches at varying distance thresholds from positive patches with varying probabilities of acceptance.

In other words, centroid coordinate candidates were generated for each CT with a stride of 12.75 mm. This yielded over 15,000 negative patch candidates per CT. For each candidate, we calculated the distance to the closest positive nodule centroid. Candidates less than 70 mm from a nodule were randomly accepted 10% of the time, candidates greater than 70 mm but less than 100 mm from a nodule were randomly accepted 5% of the time, and candidates greater than 100 mm from a nodule were accepted 1% of the time. The thresholds were determined by visually analyzing where negative patches were being sourced from. The goal was to have most negative patches sourced from within the lung but still maintain representation outside of the lung (Fig. 3).

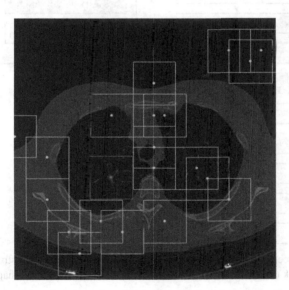

Fig. 3. Positive patches (green) and a subset of randomly selected negative patches (red).

3.2 Nodule Detection

The first step in calculating a follow-up recommendation is nodule detection. We need to be able to determine how many nodules exist in a given CT. As mentioned earlier, a full CT cannot be used as input for a neural network due to hardware constraints. Instead, we trained a binary classifier that classifies 51 mm 3D patches of CT as either containing a nodule or not containing a nodule.

Training. Of the 236 available CTs, 80% or 188 CTs were used for training and the remaining 48 CTs were used for validation. In terms of patches, 768 patches containing a nodule and 7,785 patches containing no nodules were extracted from the CTs and used for training a CNN based on the VGG network architecture [4]. We chose this 1:9 ratio of classes as opposed to a more balanced ratio because it better represents a full CT. To compensate for the class imbalance, we used balanced class weights for the binary cross-entropy loss function.

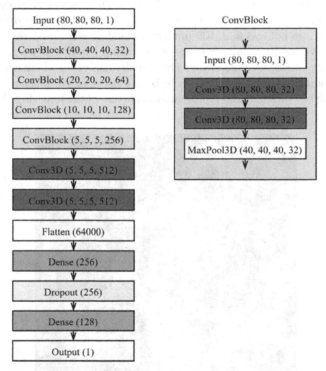

Fig. 4. The architecture for the detection neural network (left) and the structure of the first convolutional block (right). The numbers in parenthesis refer to the output shape of the layer.

To avoid any overfitting, a dropout layer with a rate of 0.4 was placed between the two fully-connected layers. We also applied small amounts of training-time augmentation to all axis. Each batch, consisting of 8 patches (due to hardware constraints), was randomly shifted ±6.4 mm, rotated ±10°, and flipped.

The learning rate was slowly decreased as learning plateaued (Fig. 5). Epochs 1–120 used a learning rate of 1×10^{-5}, epochs 121–140 used a learning rate of 1×10^{-6}, and epochs 141–150 used a learning rate of 1×10^{-7}. In total, training took over 50 h due to the 3D nature of the data, the large number of negative patches, and the hardware.

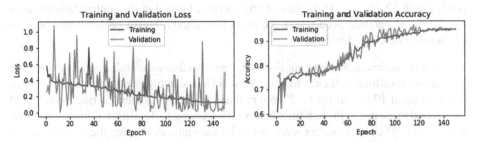

Fig. 5. The loss (left) and accuracy (right) for the detection neural network.

Training Results. Epoch 146 produced the best weights with respect to the validation dataset results. Results between the train dataset and the validation dataset were similar. While sensitivity was high, precision was much lower (Table 1).

Table 1. The results for the detection neural network for the positive nodule class.

Dataset	Precision	Sensitivity	F1 Score
Train	0.6132	0.9808	0.7546
Validation	0.6227	0.9514	0.7527

A confusion matrix (Fig. 6) helps to illustrate the high number of false positives.

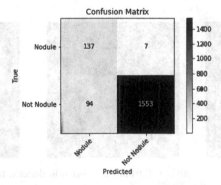

Fig. 6. The confusion matrix for the detection neural network on the validation dataset.

Applying the Classifier. To detect nodules within a given CT, we applied the binary classifier on thousands of overlapping 51 mm 3D patches extracted across the entire CT. A stride of 12.75 mm was used for selecting patches because it provided a good balance of high resolution and low calculation time. A mask was generated as predictions were made regarding the probabilities of patches containing a nodule. The mask was updated with 12.75 mm 3D patches of probabilities located at the centroid coordinates of the original patches. As a result, the mask contained a higher probability wherever the classifier detected a nodule.

A few transformations were applied to the mask so that we could ultimately calculate the centroid coordinates of each module. First, a Gaussian filter was applied to the mask with a sigma of 10 to blur the 12.75 mm probability patches. After blurring, a threshold was applied to the mask. Probabilities below 50% were set to 0 and probabilities over 50% were set to 1. These parameters were picked by manually evaluating their effectiveness on the train dataset. The result is a mask that segments nodules for a full CT (Fig. 7).

Fig. 7. From left to right: the mask, the mask after applying a Gaussian filter, and the mask after applying a threshold.

After producing the mask, the centroid coordinates of each distinct segmentation are determined. Then, 51 mm 3D patches are extracted for each predicted coordinate and classified once more using the binary classifier to get a final probability (Fig. 8).

Fig. 8. From left to right: A sample CT from the validation dataset, the associated radiologist mask, and our predicted mask.

3.3 Nodule Segmentation

Next, we trained a CNN based on the U-Net architecture [5] for predicting nodule segmentations and in turn nodule volumes. The only difference between our architecture and the architecture described in [5] is that we used 3D layers instead of 2D and we halved the number of channels for each convolutional layer. The network was trained such that it takes a 51 mm 3D patch as input and predicts a segmentation in the same form, a 51 mm 3D patch. The output is populated with ones and zeroes based on whether it is part of a nodule or not.

Training. Of the 236 available CTs, 177 CTs were used for training and 59 CTs were used for validation. From the 236 available CTs, 768 patches containing nodules and 451 patches containing pulmonary lesions were used to train and validate the network. A combination of binary cross-entropy and a negative weighted dice coefficient was used as the loss function.

$$loss = binary\_crossentropy + 2 * (1 - dice\_coefficient) \tag{1}$$

The network was trained for 200 epochs at a learning rate of 1×10^{-5} (Fig. 9) with a batch size of 4 (due to hardware constraints). Similar to the detection task, batches were generated and augmented on the fly as the network was trained. The augmentations were applied to each axis and consisted of small random 3D shifts of ± 3.2 mm, rotations of $\pm 5°$, and flips. The dice coefficient by itself was used to evaluate performance on the validation set as the network was trained.

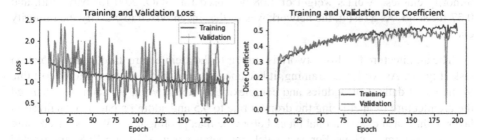

Fig. 9. The loss (left) and dice coefficient (right) for the segmentation neural network.

Training Results. The weights at epoch 200 were used for evaluation. The dice coefficients for the train dataset and validation dataset were similar: 0.4972 and 0.4851 respectively (Fig. 10).

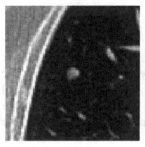

Fig. 10. From left to right: A sample patch from the validation dataset, the associated radiologist mask, and our predicted mask.

3.4 Texture Characterization

The last CNN classifies 51 mm 3D patches of nodules into one of the three texture classes: ground glass opacities (GGO), part solid, and solid.

Training. Similar to the segmentation task, 177 CTs were used for training and 59 CTs were used for validation. From the 236 available CTs, 768 patches containing nodules were used to train and validate the network. However, there was a sizeable class imbalance that needed to be addressed. The GGO class was represented 38 times, part solid was represented 58 times, and solid was represented 672 times. To workaround the class imbalance, we oversampled the two minority classes by a factor of 5. We also skewed the class weights for the categorical cross-entropy loss function to favor the two minority classes. A class weight of 4.08 was used for GGO, 2.60 for part solid, and 0.57 for solid. As a result, the network was further incentivized to classify the minority classes correctly.

The architecture for this network is similar to the architecture used for the detection task (Fig. 4). As we began training, it became apparent that it was easy to overfit due to the small dataset of nodules and complexity of the network. As such, we changed the architecture by increasing the dropout rate to 0.5 and adding a batch normalization layer within each convolutional block before the max pooling layer. We also used more aggressive augmentations. For each batch, the patches were shifted ± 12.8 mm, rotated $\pm 20°$, and flipped.

We found that the network struggled to learn with a learning rate of 1×10^{-5} (Fig. 11). Instead, an initial learning rate of 1×10^{-4} was used to train the first 200 epochs. Performance was evaluated every 100 epochs. Once learning began to plateau, the learning rate was decreased to 1×10^{-5} for epochs 201–300 and again to 1×10^{-6} for epochs 301–350.

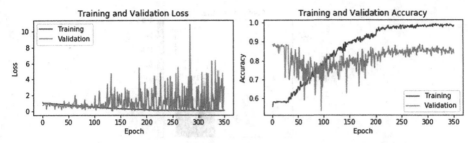

Fig. 11. The loss (left) and accuracy (right) for the classification neural network.

Training Results. The weights at epoch 318 were used for evaluation. Despite our attempts to mitigate overfitting, the results for the validation dataset were much worse than the results for the train dataset especially for the two minority classes (Tables 2, 3 and 4).

Table 2. The results for the classification neural network for the GGO class.

Dataset	Precision	Sensitivity	F1 Score
Train	0.9929	1.0000	0.9964
Validation	0.4000	0.2000	0.2667

Table 3. The results for the classification neural network for the Part Solid class.

Dataset	Precision	Sensitivity	F1 Score
Train	0.9804	0.9091	0.9434
Validation	0.1176	0.1429	0.1290

Table 4. The results for the classification neural network for the Solid class.

Dataset	Precision	Sensitivity	F1 Score
Train	0.9609	0.9899	0.9752
Validation	0.8283	0.8291	0.8269

Again, a confusion matrix (Fig. 12) illustrates the fact that the network had a hard time learning the GGO and part solid classes.

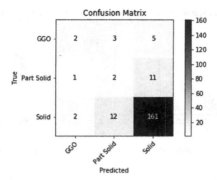

Fig. 12. The confusion matrix for the classification neural network on the validation dataset.

3.5   Fleischner Classification

The final step combines the detection, segmentation, and texture characterization tasks to produce a Fleishner score which maps to a follow-up recommendation. The same process as described earlier was used to detect nodules and their centroid coordinates. After detecting nodules, 51 mm 3D patches were extracted for each coordinate. Segmentations were produced for the extracted patches. To calculate volumes, the number of 1 s in a given mask were summed and then multiplied by the volume of a single voxel: 0.6375 mm^3. Next, texture classes were predicted for each nodule using the classification network. These features were combined using a script provided by the LNDb challenge to produce the predicted probability of a CT belonging to each of the four Fleischner classes.

4   Results

The LNDb challenge provides a means to separately evaluate the performance of our method for each of the four tasks: Fleischner classification, nodule detection, nodule segmentation, and nodule texture characterization. We produced a submission containing predictions for each of these tasks for the 58 test CTs. For the nodule segmentation and texture characterization tasks, the LNDb challenge provides annotations for centroid coordinates. No such annotations were provided nor used for the Fleischner classification or nodule detection tasks.

Nodule Detection. Nodule detection predictions were evaluated on the free receiver operating characteristic (FROC) curve. Average sensitivity was computed at two different agreement levels: all nodules and nodules marked by at least two radiologists. The sensitivities are averaged to produce a final score [2] (Table 5).

Nodule Segmentation. Nodule segmentation predictions were scored based on six different metrics: a modified Jaccard index, mean average distance (MAD), Hausdorff distance (HD), modified Pearson correlation coefficient, bias, and standard deviation [2] (Table 6).

Table 5. The LNDb challenge results for the detection task.

Score	FROC AgrLvl1	FROC AgrLvl2
0.1743	0.1453	0.2034

Table 6. The LNDb challenge results for the segmentation task.

Modified Jaccard	MAD	HD	Modified Pearson	Bias	Standard deviation
0.5438	0.7741	2.5804	0.1957	62.5117	149.6920

Texture Characterization. Texture predictions were compared to the ground truth and agreement was computed according to Fleiss-Cohen weighted Cohen's kappa [2]. Our texture characterization submission received a score of 0.3342.

Fleischner Classification. Fleischner score predictions were compared to the ground truth and agreement was computed according to Fleiss-Cohen weighted Cohen's kappa [2]. Our Fleischner classification submission received a score of 0.5092.

5 Conclusion

Our system struggled with high numbers of false positives for the nodule detection task. One solution would be to train another network specifically for mitigating false positives. The ensemble would work together to produce more precise predictions. The other problem our system struggled with was an inability to learn the minority classes for the texture characterization task. A larger dataset, a different loss function, more sophisticated oversampling or augmentation techniques, or a different network architecture could improve performance for the minority classes. Lastly, for the nodule segmentation task, the more recently introduced UNet++ architecture [6] may improve performance even further.

The LNDb challenge poses unique problems with the potential to improve the early screening process of patients for lung cancer. While the method described in this paper did not have the best performance on the test dataset, it presents a baseline for further work.

References

1. Cancer. World Health Organization (2018). https://www.who.int/news-room/factsheets/detail/cancer
2. Pedrosa, J., et al.: LNDb: A Lung Nodule Database on Computed Tomography. arXiv:1911.08434 [eess.IV] (2019)
3. MacMahon, H., et al.: Guidelines for management of incidental pulmonary nodules detected on CT images: from the Fleischner Society 2017. Radiology **284**(1), 228–243 (2017)

4. Simonyan, K., Zisserman, A.: Very deep convolutional networks for large-scale image recognition. arXiv:1409.1556 [cs.CV] (2015)
5. Ronneberger, O., Fischer, P., Brox, T.: U-Net: convolutional networks for biomedical image segmentation. arXiv:1505.04597 [cs.CV] (2015)
6. Zhou, Z., Siddiquee, M.M.R., Tajbakhsh, N., Liang, J.: UNet ++: redesigning skip connections to exploit multiscale feature in image segmentation. arXiv:1912.05074 [eess.IV] (2020)

Deep Residual 3D U-Net for Joint Segmentation and Texture Classification of Nodules in Lung

Alexandr Rassadin[✉] [iD]

Xperience.ai, Maxim Gorky street 262, 603155 Nizhny Novgorod, Russia
`alexander.rassadin@xperience.ai`

Abstract. In this work we present a method for lung nodules segmentation, their texture classification and subsequent follow-up recommendation from the CT image of lung. Our method consists of neural network model based on popular U-Net architecture family but modified for the joint nodule segmentation and its texture classification tasks and an ensemble-based model for the follow-up recommendation. This solution was evaluated within the LNDb 2020 medical imaging challenge and produced the best nodule segmentation result on the final leaderboard.

Keywords: Deep learning · Medical imaging · Semantic segmentation · U-Net

1 Introduction

Lung cancer is an important disease which can lead to death. Fortunately, nowadays we have early screening procedures for a timely diagnosis of this disease. Early screening with low-dose CT (computed tomography) can reduce mortality by 20% [1]. Unfortunately, global screening of the population would lead to a medical personnel overload. Because of the high patients' flow, doctors lose the ability of steadfast investigation of CT results which can lead to errors in the diagnosis. Nowadays, when artificial intelligence has proven its applicability in many areas of life, shifting routine medical tasks from humans to AI looks like a very desirable option. One of such tasks can be the detection of nodules in the lungs from CT images for follow-up procedures recommendation.

The LNDb 2020 challenge [1] consisted of 4 tracks:

- The detection of nodules in lungs from CT images. All nodules from an entire human lung image should be localized.
- The segmentation of the nodules from CT images. Provided with the potential center of the nodule, one should provide accurate voxel-wise binary segmentation of the nodule (if it exists).
- The classification of the texture of found nodules. Provided with the center of the potential nodule, one should classify one of the three types of texture: ground glass opacities, part-solid or solid.
- The main challenge track consisted of making a follow-up recommendation based on a CT image according to the 2017 Fleischner society pulmonary nodule guidelines [2].

© Springer Nature Switzerland AG 2020
A. Campilho et al. (Eds.): ICIAR 2020, LNCS 12132, pp. 419–427, 2020.
https://doi.org/10.1007/978-3-030-50516-5_37

We participated in all the tracks, except the nodule detection track. The next section describes our approach.

2   Method Overview

2.1   Nodule Segmentation

We started our experiments with the SSCN [3] U-Net [4] which established itself in a number of 3D segmentation tasks. The advantage of this family of architectures is that it allows the usage of larger batches because of exploiting the sparse nature of the data. Unfortunately, 3D sparse U-Net had a low prediction performance in our setup and our choice was to fall back to the plain 3D convolutions. An analogous setup with 4-stage (i.e. 4 poolings in the encoder part followed by 4 upsamplings in the decoder part) U-Net showed its supremacy over the SSCN counterpart, which determined the direction of further experiments.

The next thing we did was the implementation of residual connections [5], which established itself in many computer vision tasks. Following [6], we replaced the standard ReLU activation with ELU [7] and then the batch normalization [8] with the group normalization [9], which, in combination, gave us a sufficient increase in the segmentation quality. Our final encoder/ decoder block is depicted in Fig. 1.

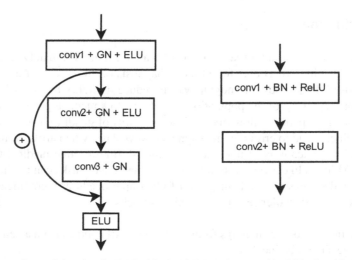

Fig. 1. Comparison of the simple U-Net block (right) and our residual block with GroupNorm and ELU (left).

We followed the standard procedure of encoder construction: twice reducing the spatial dimension after every next block while twice increasing the feature dimension (and vice-versa for the decoder). Later we added an additional 5th block to the encoder and removed the pooling layer after the first block to match the encoder and decoder dimensions, as in [10].

We used the popular attention mechanism CBAM [11] in both the encoder and decoder of our U-Net but adapted it for the 3-dimensional nature of LNDb data. Our experiments show that CBAM leads to a slightly better segmentation quality, which is to be expected, while sacrificing an extra amount of the training time. Here again we see that ELU [7] activation (inside the attention module) provides better results than its ReLU counterpart.

2.2 Nodule Texture Classification

As well as for the segmentation track, we started from the SSCN-based recognizer following VGG [12] architecture, but in this case, sparse convolutions failed again. We continued our experiments with the same U-Net encoder which we used for the segmentation but attached a classification head instead of a decoder. The benefit of this approach is that we can start from the already pre-trained for the nodule segmentation weights, which, by our observations, is crucial for training such a classifier. Our classification head starts from the global average pooling followed by two fully connected layers with ELU activation and ends up with the final classification layer with the softmax activation.

Unfortunately, this approach still gave us a low classification accuracy. So, we came up with the idea of training a *joint segmentation and texture classification* network, which is the main contribution of this work. This approach gave us a sufficient boost of the classification accuracy and also increased our nodule segmentation quality because of exploiting so-called multi-task learning.

It is also worth noting that apart from the joint nodule segmentation and texture classification network, we have tried a simple ensemble-based model (namely, Random Forest) upon encoder features from the pre-trained nodule segmentation model. This is a working option but still less accurate than the joint end-to-end model. Also, using a second-stage predictor negatively impacts the performance of the final solution and overcomplicates it.

2.3 Joint Nodule Segmentation and Texture Classification

We did some experiments with the texture classification head configuration. Upon our observations, usage of convolutions instead of fully connected layers gives no advantage, neither in terms of classification accuracy nor nodule segmentation quality while the model becomes slightly less robust to the overfitting. For the train/val submission we used a dropout rate 0.6 while for the test set submission a dropout of 0.4 was preferable.

Besides group normalization, we experimented with the switchable normalization [13] and the result was dubious. With the same training procedure, the model with switchable normalization behaves noisier in terms of segmentation and texture classification metrics, slightly worse on average but with few high peaks (see Fig. 2) and it is also more prone to overfit. Such behavior, along with the fact that training with SwitchNorm increases the optimization time, made us fall back to the GroupNorm. We found that the optimal number of groups is 8.

Usage of attention in the encoder and decoder of the joint nodule segmentation and texture classification model has led to the earlier overfeat of the classification head: better segmentation results can be obtained only by sacrificing the texture classification

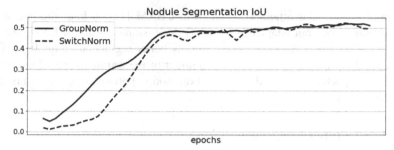

Fig. 2. Training curves of two joint nodule segmentation and texture classification networks: one with group normalization and one with switchable normalization.

quality. We attempted to overcome this by incorporating some attention mechanism also within the classification head. We tried an approach from [14], again, adapted to the 3D nature of the data, but unfortunately, this approach only decreased the overall quality of the model. So, for the joint model we didn't use any attention mechanism.

The maximum feature size in the first encoder block (see Sect. 2.1) which fits our hardware ($2\times$ Nvidia GTX 1080 Ti) was 40. Unfortunately, the training time of such a model was too high for the limited time budget of the challenge and we decided to use the feature size 32 which still gives us a high enough segmentation quality (see Fig. 3) to provide us with the necessary performance.

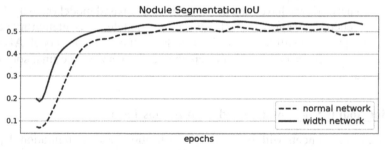

Fig. 3. Training curves of normal and width joint nodule segmentation and texture classification networks.

By the nature of the challenge data, participants were able to choose whether to train their models on 5 classes (6 with non-nodule class) or 3 classes (4 with non-nodule class). In our experiments, we clearly observed that the fewer classes to train on – the better final accuracy, so we used a 3-class model in our approach.

2.4   Fleischner Classification

As for the main target [2] prediction, we used a relatively simple idea. From the challenge guidelines, we know that the follow-up recommendation can be estimated directly from the nodule annotation considering:

- the number of nodules for the patient (single or multiple),
- their volumes,
- their textures.

This means that information about the number of found nodules, their volumes and textures is enough for a radiologist to make a recommendation. Knowing this, we just encoded this information in a 6-element feature vector as follow:

- The first 3 elements encode the number of nodules of each of the 3 sizes (less than 100 mm^3, between 100 and 250 mm^3 and more than 250 mm^3),
- The last 3 elements encode the number of nodules of each texture type (ground glass opacities, part-solid and solid).

From the predictions of the joint nodule segmentation and texture classification model, we directly know the texture type of the nodule and from the segmentation mask we can compute the nodule size (every nodule has a common spatial resolution).

We first evaluated the prediction capability of such an approach on the ground truth data, using the Random Forest model as a predictor and it showed a remarkable performance – over 90% balanced accuracy. For the leaderboard submissions we just replaced the ground truth segmentation and texture with our own predictions. To overcome the effect of cascade error, we also tried to predict Fleischner target based only on the nodules size (without information of its texture) and surprisingly it had a quite similar prediction capability. Based on these observations, it becomes quite clear that *the crucial factor of the follow-up recommendation is the number of nodules in the lung, which is achieved by the accurate nodule detection or segmentation algorithm.*

The test set of the challenge was extremely noisy due to the false positive nodules in order not to invalidate other tracks targets. Since our team didn't participate in the detection track and the non-nodule filtration is crucial for the main target prediction (because it heavily relies on the information about the number of found nodules in the patient), a strong need arose for some non-nodule filtration mechanism. While submitting the train/val results, this task was assigned to the nodule segmentation network, i.e. a candidate was considered as a non-nodule (false positive) if his volume, based on the predicted nodule segmentation, was nearly zero. We measured the precision of such an approach to non-nodule recognition and it was around 0.64, which, as it turned out, was enough for the slightly noisy train/val data. Looking at the test data, we correctly decided that it would not be sufficient for the highly noisy test data. To solve this problem, we forced to train another auxiliary model, i.e. a separate non-nodule recognizer. For this purpose, we took our joint network without its decoder part, initialized with the best checkpoint, and trained it for the 2-class (nodule/ non-nodule) classification problem. Precision of such a model was much higher – 0.78. Incidentally, it was even higher than for a joint model trained for 4 (3 actual classes and 1 non-nodule class) instead of 3 classes. Unfortunately, it turned out that this is still not enough for accurate non-nodule filtration in the test set data which led to the great metrics decrease in the test submission compared to the train/val one (see Sect. 3 for the details).

2.5 Model Optimization

Dice loss is default choice nowadays for the training of segmentation models. It worked well in our case too. We experimented with the Generalized Dice overlap loss [15] but it did not give us an improvement. For the classification head we used plain Cross Entropy. We used inversely proportional class weights for Cross Entropy, and it boosted the accuracy, while weighting of the Dice loss didn't provide us with any improvements. Our final loss was an average of the Dice and Cross Entropy, where Cross Entropy was multiplied by 0.2.

As for optimizers, we used very popular Adam optimizer. We also tried recently introduced diffGrad [16] and Adamod [17] but they didn't provide us with any improvements (we didn't perform a hyper-parameter tuning). Comparison of optimizers depicted on Fig. 4. We didn't start optimization of texture classification head (by multiplying Cross Entropy loss by 0) until nodule segmentation achieves 0.45 IoU (intersection over union).

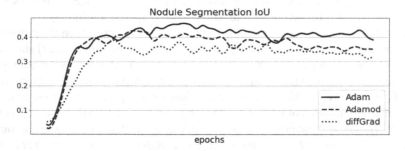

Fig. 4. Training curves of the nodule segmentation model for different optimizers.

2.6 Data Augmentations

In our work, we used some quite standard augmentations set: random flipping, random rotations by 90°, elastic deformation and noise. We couldn't use rotation for arbitrary angle because it could break the structure of the scan and has padding uncertainty. Our experiments show that augmentations can boost nodule segmentation IoU by 0.05 in average (see Fig. 5).

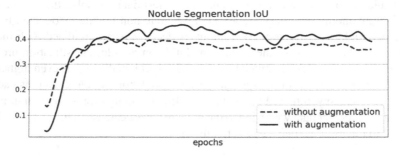

Fig. 5. Training curves for two identical nodule segmentation networks: with and without data augmentation.

3 Results

3.1 Train/Val Leaderboard

The organizers provided us with train/val set with 4 predefined folds. Results for the public train/val leaderboard must be submitted using a 4-fold procedure, so we trained 4 joint nodule segmentation and texture classification models. Its predictions were used for the segmentation and texture classification tracks in a straightforward way while for the main target prediction, we first collected the features (volumes and textures) for every nodule, then trained the corresponding predictor for the Fleischner classification (see Sect. 2.4).

Table 1 summarizes our nodule segmentation result on the train/val leaderboard. Here, J stands for Jaccard index, MAD stands for mean average distance, HD stands for Hausdorff distance, C stands for the Pearson correlation coefficient, $Bias$ stands for the mean absolute difference, Std stands for the standard deviation, symbol * stands for the inversion of the metric, e.g. J * means $1 - J$. The final score in the leaderboard was calculated as an average of all six metrics, which were preliminarily normalized by the maximum value over all the submissions in the leaderboard, for each metric separately. See the LNDb challenge evaluation page [18] for a detailed description of the evaluation metrics.

Table 1. Top-5 segmentation results in the train/val leaderboard. Our result highlighted in bold.

J*	MAD	HD	C*	Bias	Std
0.0865	0.0827	0.4123	0.0017	16.2162	86.1182
0.4178	0.4122	2.136	0.0671	92.7103	472.6242
0.433	0.3888	2.0493	0.0791	75.536	507.3688
0.4892	**0.5668**	**2.4819**	**0.0781**	**103.3177**	**486.8666**
0.3694	0.3484	1.8991	0.1116	111.4641	719.5786

Our Fleischner classification score is 0.5281 Fleiss-Cohen weighted kappa [19], which is the third best result in the leaderboard (after 0.603 and 0.5323 kappa).

3.2 Test Leaderboard

We used 70% of the train/val data for training our joint nodule segmentation and texture classification model while the remaining 30% were used for validation and also for the training of our main target predictor.

Nodule segmentation and its texture classification procedures were the same as for the train/val submission – results were obtained in a straightforward way from the joint model.

Additionally, for every sample, we predicted whether or not it is a nodule using our non-nodule recognition model (see Sect. 2.4) and saved this information to make a later prediction of the main target. Then we took the remaining 30% of the train/val set, which was not used for training the joint model, and collected the segmentation, texture classification and non-nodule recognition results for this data. From this prediction we formed a sampling for training a Random Forest predictor of the main target (see Sect. 2.4). Finally, this model was used for the prediction of the main target of the test set.

Table 2 summarizes our nodule segmentation results on the test leaderboard. See the LNDb challenge evaluation page [18] for a detailed description of the evaluation metrics.

Table 2. Top-3 segmentation results in the test leaderboard. Our result highlighted in bold.

J*	MAD	HD	C*	Bias	Std
0.4779	**0.4203**	**2.0275**	**0.055**	**44.2826**	**86.3227**
0.468	0.4686	2.1371	0.081	40.701	98.741
0.4447	0.4115	2.0618	0.1452	41.4341	129.47

Our Fleischner classification score is −0.0229 Fleiss-Cohen weighted kappa [19]. We explained the reasons of such a poor result and the significant difference with the train/val submission in Sect. 2.4 in detail.

4   Conclusions

In this paper we described a solution for lung nodules segmentation, their texture classification and a consequent follow-up recommendation for the patient. Our approach consists of a joint nodule segmentation and texture classification neural network, which is essentially a deep residual U-Net [4] with batch normalization [8] replaced by a group normalization [9] and ReLU replaced by ELU [7]. For the patient's follow-up recommendation [2], we used an ensemble-based model. We evaluated our approach by participating in the LNDb challenge [1] and took the first place in the segmentation track with a result of 0.5221 IoU. Our approach is simple yet effective and can potentially be used in real diagnostic systems reducing the routine workload on medical personnel, which clearly defines the direction of our future work.

Acknowledgments. We are grateful to xperience.ai for support of the research and Andrey Savchenko for his assistance in preparation of this paper.

References

1. Pedrosa, J., et al.: LNDb: a lung nodule database on computed tomography. arXiv preprint arXiv:1911.08434 (2019)
2. MacMahon, H., et al.: Guidelines for management of incidental pulmonary nodules detected on CT images: from the fleischner society 2017. Radiology **284**(1), 228–243 (2017)
3. Graham, B., Maaten, L.: Submanifold sparse convolutional networks. arXiv preprint arXiv: 1706.01307 (2017)

4. Ronneberger, O., Fischer, P., Brox, T.: U-Net: convolutional networks for biomedical image segmentation. arXiv preprint arXiv:1505.04597 (2015)
5. He, K., Zhang, X., Ren, S., Sun, J.: Deep residual learning for image recognition. arXiv preprint arXiv:1512.03385 (2015)
6. Lee, K., Zung, J., Li, P., Jain, V., Seung, H. S.: Superhuman accuracy on the SNEMI3D connectomics challenge. arXiv preprint arXiv:1706.00120 (2017)
7. Clevert, D. A., Unterthiner, T., Hochreiter, S.: Fast and accurate deep network learning by Exponential Linear Units (ELUs). arXiv preprint arXiv:1511.07289 (2015)
8. Ioffe, S., Szegedy, C.: Batch normalization: accelerating deep network training by reducing internal covariate shift. arXiv preprint arXiv:1502.03167 (2015)
9. Wu, Y., He, K.: Group normalization. arXiv preprint arXiv:1803.08494 (2018)
10. Wolny, A., Cerrone, L., Kreshuk, A.: Accurate and versatile 3D segmentation of plant tissues at cellular resolution. bioRxiv preprint https://doi.org/10.1101/2020.01.17.910562 (2020)
11. Woo, S., Park, J., Lee, J.Y., Kweon, I.S.: CBAM: convolutional block attention module. arXiv preprint arXiv:1807.06521 (2018)
12. Simonyan, K., Zisserman, A.: Very deep convolutional networks for large-scale image recognition. arXiv preprint arXiv:1409.1556 (2014)
13. Luo, P., Ren, J., Peng, Z., Zhang, R., Li, J.: differentiable learning-to-normalize via switchable normalization. arXiv preprint arXiv:1806.10779 (2018)
14. Lin, T.Y., RoyChowdhury, A., Maji, S.: Bilinear CNNs for fine-grained visual recognition. arXiv preprint arXiv:1504.07889 (2015)
15. Sudre, C., Li, W., Vercauteren, T., Ourselin, S., Cardoso, M.: Generalised dice overlap as a deep learning loss function for highly unbalanced segmentations. arXiv preprint arXiv: 1707.03237 (2017)
16. Dubey, S., Chakraborty, S., Roy, S., Chaudhuri, B.: diffGrad: an optimization method for convolutional neural networks. arXiv preprint arXiv:1909.11015 (2019)
17. Ding, J., Ren, X., Luo, R., Sun, X.: An adaptive and momental bound method for stochastic learning. arXiv preprint arXiv:1910.12249 (2019)
18. LNDb challenge evaluation page. https://lndb.grand-challenge.org/Evaluation/. Accessed 08 Feb 2020
19. Spitzer, R., Cohen, J., Fleiss, J., Endicott, J.: Quantification of agreement in psychiatric diagnosis: a new approach. Arch. Gen. Psychiatry 17(1), 83–87 (1967)

Author Index

Printed in the United States
By Bookmasters